HORIZONS IN WORLD PHYSICS SERIES

THE PHYSICS OF QUARKS: NEW RESEARCH (HORIZONS IN WORLD PHYSICS, VOLUME 265)

HORIZONS IN WORLD PHYSICS SERIES

Volume 268 **Horizons in World Physics**, Edited by Michael Everett and Louis Pedroza
Volume 267 **Superconductivity and Superconducting Wires,** Edited by Dominic Matteri and Leone Futino
Volume 266 **Applied Physics in the 21st Century,** Edited by Raymond P. Valencia
Volume 265 **The Physics of Quarks: New Research,** Edited by Nicolas L. Watson and Theo M. Grant
Volume 264 **Lev Davidovich Landau and his Impact on Contemporary Theoretical Physics** Edited by Ammar Sakaji and Ignazio Licata
Volume 263 **New Trends in Quantum Coherence and Nonlinear Optics** Edited by Rafael Drampyan
Volume 262 **Astrophysics and Condensed Matter** Edited Thomas G. Hardwell
Volume 261 **Superconductivity Research Advances** Edited by James E. Nolan
Volume 260 **Instabilities of Relativistic Electron Beam in Plasma,** Edited by Valery B. Krasovitskii
Volume 259 **Self-Focusing of Relativistic Electron Bunches in Plasma,** Edited by Valery B. Krasovitskii
Volume 258 **New Topics in Theoretical Physics,** Edited by Henk F. Arnoldus and Thomas George
Volume 257 **Boson's Ferromagnetism and Crystal Growth Research,** Edited by Emerson D. Seifer
Volume 256 **Horizons in World Physics,** Edited by Victor H. Marselle
Volume 255 **Frontiers in General Relativity and Quantum Cosmology Research,** Edited by Victor H. Marselle
Volume 254 **Optics and Electro-Optics Research,** Edited by Albert V. Berzilla
Volume 253 **Trends in General Relativity and Quantum Cosmology,** Edited by Albert Reimer
Volume 252 **Chemical Physics Research Trends,** Edited by Benjamin V. Arnold
Volume 251 **Quantum Dots: Research Developments,** Edited by Peter A. Ling
Volume 250 **Quantum Gravity Research Trends,** Edited by Albert Reimer
Volume 249 **General Relativity Research Trends,** Edited by Albert Reimer
Volume 248 **Spacetime Physics Research Trends,** Edited by Albert Reimer
Volume 247 **New Developments in Quantum Cosmology Research,** Edited by Albert Reimer
Volume 246 **Quantum Cosmology Research Trends,** Edited by Albert Reimer
Volume 245 **Horizons in World Physics,** Edited by Tori V. Lynch
Volume 244 **Horizons in World Physics,** Edited by Albert Reimer
Volume 243 **Theoretical Physics 2002, Part 2,** Edited by Henk F. Arnoldus and Thomas F. George
Volume 242 **Models and Methods of High-T_c Superconductivity: Some Frontal Aspects, Volume 2,** Edited by J. K. Srivastava and S. M. Rao
Volume 241 **Models and Methods of High-T_c Superconductivity: Some Frontal Aspects, Volume 1,** Edited by J. K. Srivastava and S. M. Rao
Volume 240 **Horizons in World Physics,** Edited by Albert Reimer
Volume 239 **Theoretical Physics 2002, Part 1,** Edited by Henk F. Arnoldus and Thomas F. George
Volume 238 **Theoretical Physics 2001,** Edited by Henk F. Arnoldus and Thomas F. George
Volume 237 **Generalized functions in mathematical physics: Main ideas and concepts,** Edited by A.S. Demidov
Volume 236 **Edge Excitations of Low-Dimensional Charged Systems,** Edited by Oleg Kirichek
Volume 235 **On the Structure of the Physical Vacuum and a New Interaction in Nature (Theory, Experiment, Applications),** Edited by Yu. A. Baurov
Volume 234 **Electron Scattering on Complex Atoms (Ions),** Edited by V. Lengyel, O. Zatsarinny and E. Remeta
Volume 233 **An Introduction to Hot Laser Plasma Physics,** Edited by A.A. Andreev, A.A. Mak and N.A. Solovyev
Volume 232 **Electron-Atom Scattering: An Introduction**, Maurizio Dapor
Volume 231 **Theory of Structure Transformations in Non-equilibrium Condensed Matter**, Edited by Alexander L. Olemskoi
Volume 230 **Topics in Cosmic-Ray Astrophysics**, Edited by M. A. DuVernois
Volume 229 **Interactions of Ions in Condensed Matter**, Edited by A. Galdikas and L. Pranevièius
Volume 228 **Optical Vortices**, Edited by M. Vasnetsov and K. Staliunas
Volume 227 **Frontiers in Field Theory**, Edited by Quantum Gravity and Strings, R.K. Kaul, J. Maharana, S. Mukhi and S. K. Rama
Volume 226 **Quantum Groups, Noncommutative Geometry and Fundamental Physical Interactions**, Edited by D. Kastler, M. Russo and T. Schücker
Volume 225 **Aspects of Gravitational Interactions**, Edited by S.K. Srivastava and K.P. Sinha
Volume 224 **Dynamics of Transition Metals and Alloys**, Edited by S. Prakash
Volume 223 **Fields and Transients in Superhigh Pulse Current Devices** Edited by G.A. Shneerson

Volume 222	**Gaussian Beams and Optical Resonators,** Edited by A. N. Oraevsky
Volume 221	**Laser Cathode-Ray Tubes,** Edited by Yu. M. Popov
Volume 220	**Quantum Electrodynamics with Unstable Vacuum,** Edited by V.L. Ginzburg
Volume 219	**The Muon Method in Science,** Edited by V.P. Smilga and Yu.M. Belousov
Volume 218	**Radiative and Collisional Characteristics of Ions in Hot Plasmas,** Edited by I.I. Sobel'man
Volume 217	**Nonlinear and Quantum Optical Phenomena in Nonequilibrium Media** , Edited by V.A. Shcheglov
Volume 216	**Semiconductors and Insulators: Optical and Spectroscopic ...,** Edited by Yu.I. Koptev
Volume 215	**Relativistic Theory of Gravity,** Edited by A.A. Logunov
Volume 214	**Research in Quantum Field Theory**, Edited by V.I. Man'ko
Volume 213	**Nonlinear Theory of Strong Electromagnetic Wave-plasma...,** Edited by O.N. Krokhin
Volume 212	**Dynamics of Elementary Atomic-Molecular...**, Edited by O. S. Scheglov
Volume 211	**Optics and Lasers,** Edited by G.G. Petrash
Volume 210	**Theory and Technology of High Temperature Superconductors,** Edited by Yu.I. Koptev
Volume 209	**Theory of Interaction of Multilevel Systems...**, Edited by V.I. Man'ko
Volume 208	**Physical Properties of High Temperature Superconductors,** Edited by Yu.I. Koptev
Volume 207	**Generalized Mayer Series in Classical Statistical Mechanics**, Edited by Igor J. Ivanchik
Volume 206	**Mathematical Physics, Applied Mathematics and Informatics,** Edited by Yu.I. Koptev
Volume 205	**Squeezed and Correlated States of Quantum Systems**, Edited by M.A. Markov
Volume 204	**Sakharov Memorial Lectures in Physics**, Edited by L.V. Keldysh and V.Ya. Fainberg
Volume 203	**Fiber Optics: Research and Development,** Edited by Ye.M. Dianov
Volume 202	**Gas Dynamics**, Edited by Yu.I. Koptev
Volume 201	**Generation of Nonlinear Waves and Quasistationary ...**, Edited by L.M. Kovrizhnykh
Volume 200	**Laser Thermonuclear Targets and Superdurable Microballoons**, Edited by A.I. Isakov
Volume 199	**Spatiotemporal Characteristics of Laser Emission,** Edited by M.V. Pyatakhin and A.F. Suchkov
Volume 198	**Theory of Nonstationary Quantum Oscillators**, Edited by M.A. Markov
Volume 197	**Reliability Problems of Semiconductor Lasers**, P.G. Eliseev
Volume 196	**Pulsars,** Edited by A.D. Kuzmin
Volume 195	**Physical Characteristics and Critical Temperature of High Temperature Superconductors**, Edited by M.M. Sushchinskiy
Volume 194	**Biophysical Approach to Complex Biological Phenomena,** Edited by E. Volkov
Volume 193	**Research on Chemical Lasers** Edited by A.N. Orayevskiy
Volume 192	**Atomic and Ionic Spectra and Elementary Processes in Plasma** , Edited by I.I. Sobelman
Volume 191	**Polarization of the Vacuum and a Quantum Relativistic Gas ...**, A.Ye. Shabad
Volume 190, (Supp. Vol.)	**Superconductors with A15 Lattice and Bridge Contacts Based ...**, Edited by M. M. Sushchinskiy
Volume 190	**Metal-Optics and Superconductivity**, Edited by A.I. Golovashkin
Volume 189	**Clusters of Galaxies and Extragalactic Radio Sources**, Edited by A.D. Kuz'min
Volume 188	**Integral Systems, Solid State Physics and Theory of Phase Transitions. Pt. 2. Symm. and Alg. Structures in Physics**, Edited by V.V. Dodonov and V.I. Man'ko
Volume 187	**Quantum Field Theory, Quantum Mechanics and Quantum Optics: Pt. 1. Symm. and Alg. Structures in Physics,** Edited by V.V. Dodonov and V.I. Man'ko
Volume 186	**Photoproduction of Pions on Nucleons and Nuclei**, Edited by A.A. Komar
Volume 185	**Injection Lasers in Optical Communication and Information Processing Systems**, Edited by Yu. M. Popov
Volume 184	**Electron Processes in MIS-Structure Memories**, Edited by A.F. Plotnikov
Volume 183	**Invariants and the Evolution of Nonstationary Quantum Systems**, Edited by M.A. Markov
Volume 182	**Luminescence of Wideband Semiconductors**, Edited by M.D. Galanin
Volume 181	**Pulsed Metal and Metal Halogenide Vapor Lasers**, Edited by G.G. Petrash
Volume 180	**Inelastic Light Scattering in Crystals**, Edited by M.M. Sushchinskiy
Volume 179, (Supp. Vol. 2)	**Electron-Excited Molecules in Nonequilibrium Plasma**, Edited by N.N. Sobolev
Volume 179, (Supp. Vol. 1)	**Interaction of Ultrashort Pulses with Matter** , Edited by M.D. Galanin
Volume 179	**X-Ray Plasma Spectroscopy and the Properties of Multiply-Charged Ions**, Edited by I.I. Sobel'man
Volume 178	**The Delfin Laser Thermonuclear Installation: Operational Complex ...**, Edited by G.V. Sklizkov
Volume 177	**Stoichiometry in Crystal Compounds and Its Influence on Their Physical Properties**, Edited by Yu. M. Popov
Volume 176,	**Quantization, Gravitation and Group Methods in Physics**

(Supp. Vol)	Edited by A.A. Komar
Volume 176	**Classical and Quantum Effects in Electrodynamics** Edited by A.A. Komar
Volume 175,	**Luminescence and Anisotropy of Zinc Sulfide Crystals**
(Supp. Vol.)	Edited by M.D. Galanin
Volume 175	**Luminescence Centers of Rare Earth Ions in Crystal Phosphors**, Edited by M.D. Galanin
Volume 174,	**Electron Liquid Theory of Normal Metals**
(Supp. Vol. 2)	Edited by V.P. Silin
Volume 174,	**Nonequilibrium Superconductivity**,
(Supp. Vol. 1)	Edited by V.L. Ginzburg
Volume 174	**Thermodynamics and Electrodynamics of Superconductors,** Edited by V.L. Ginzburg
Volume 173	**Quantum Mechanics and Statistical Methods** , Edited by M.M. Sushchinskiy
Volume 172	**Phase Conjugation of Laser Emission** , Edited by N.G. Basov
Volume 171	**Research on Laser Theory** , Edited by A.N. Orayevskiy
Volume 170	**The Theory of Target Compression by Longwave Laser Emission** , Edited by G.V. Sklizkov
Volume 169	**The Physical Effects in the Gravitational Field Black Holes**, Edited by M.A. Markov
Volume 168	**Issues in Intense-Field Quantum Electrodynamics**, Edited by V.L. Ginzburg
Volume 167	**Group Theory, Gravitation and Elementary Particle Physics**, Edited by A.A. Komar
Volume 166	**The Nonlinear Optics of Semiconductor Lasers**, Edited by N.G. Basov
Volume 165	**Solitons and Instantons, Operator Quantization**, Edited by V.L. Ginzburg

HORIZONS IN WORLD PHYSICS SERIES

THE PHYSICS OF QUARKS: NEW RESEARCH (HORIZONS IN WORLD PHYSICS, VOLUME 265)

NICOLAS L. WATSON
AND
THEO M. GRANT
EDITORS

Nova Science Publishers, Inc.
New York

For permission to use material from this book please contact us:
Telephone 631-231-7269; Fax 631-231-8175
Web Site: http://www.novapublishers.com

LIBRARY OF CONGRESS CATALOGING-IN-PUBLICATION DATA

Watson, Nicola J., 1958-
The physics of quarks : new research / Nicolas L. Watson and Theo M. Grant.
p. cm. -- (Horizons in world physics ; v. 265)
Includes index.
ISBN 978-1-60456-802-8 (hardcover)
1. Quarks. 2. Particles (Nuclear physics) I. Grant, Theo M. II. Title.
QC793.5.Q252W16 2009
539.7'2167--dc22
2009015090

ISBN 978-1-60456-802-8

Published by Nova Science Publishers, Inc. ✢ New York

CONTENTS

PREFACE

A quark is a type of elementary particle found in protons and neutrons and other subatomic particles. They are a major constituent of matter, along with leptons. In nature, quarks are never found on their own, as isolated, single particles; rather, they are bound together in composite particles named hadrons. For this reason, much of what is known about quarks has been inferred from observations on the hadrons themselves. This book provides leading edge research on this field from around the globe.

In Chapter 1, substantial theoretical research has been carried out to study the phase transition between hadronic matter and the quark-gluon phase (QGP). When calculating the QGP signatures in relativistic nuclear collisions, the distribution functions of quarks and gluons are traditionally described by Boltzmann-Gibbs (BG) statistics. In the past few years the non-extensive form of statistical mechanics proposed by Tsallis has found applications in astrophysical self-gravitating systems, solar neutrinos, high energy nuclear collisions, cosmic microwave back ground radiation, high temperature superconductivity and many others. In these cases a small deviation of the Tsallis parameter, q, from 1 (BG statistics) reduces the discrepancies between experimental data and theoretical models. Recently Hagedorn's statistical theory of the momentum spectra produced in heavy ion collisions has been generalized using Tsallis statistics to provide a good description of e^+e^- annihilation experiments. Furthermore, Walton and Rafelski studied a Fokker- Planck equation describing charmed quarks in a thermal quark-gluon plasma and showed that Tsallis statistics were relevant. These results suggest that perhaps BG statistics may not be adequate in the quark-gluon phase.

Two key experiments, in nuclear/hadron physics and in astrophysics have been started at the beginning of this century. They are RHIC (Relativistic Heavy Ion Collider) andWMAP (Wilkinson Microwave Anisotropy Probe). Although what they actually measure are quite different, physics goals have some overlaps with each other. In fact, "the origin of masses", which is the most fundamental problem in modern physics, is s key question to be studied in RHIC, WMAP and in future facilities. Chapter 2 discusses some of the recent topics in hot and dense QCD and their relations to the physics in cosmology and atomic physics.

Existence of metastable quarks of new generation can be embedded into phenomenology of heterotic string together with new long range interaction, which only this new generation possesses. Chapter 3 discusses primordial quark production in the early Universe, their successive cosmological evolution and astrophysical effects, as well as possible production in present or future accelerators. In case of a charge symmetry of 4th generation quarks in

Universe, they can be stored in neutral mesons, doubly positively charged baryons, while all the doubly negatively charged "baryons" are combined with He-4 into neutral nucleus-size atom-like states. The existence of all these anomalous stable particles may escape present experimental limits, being close to present and future experimental test. Due to the nuclear binding with He-4 primordial lightest baryons of the 4th generation with charge +1 can also escape the experimental upper limits on anomalous isotopes of hydrogen, being compatible with upper limits on anomalous lithium. While 4th quark hadrons are rare, their presence may be nearly detectable in cosmic rays, muon and neutrino fluxes and cosmic electromagnetic spectra. In case of charge asymmetry, a nontrivial solution for the problem of dark matter (DM) can be provided by excessive (meta)stable anti-up quarks of 4th generation, bound with He-4 in specific nuclear-interacting form of dark matter. Such candidate to DM is surprisingly close to Warm Dark Matter by its role in large scale structure formation. It catalyzes primordial heavy element production in Big Bang Nucleosynthesis and new types of nuclear transformations around us.

Mesons are the simplest bound states in Quantum Chromodynamics (QCD). Their decays provide an important tool for understanding non-perturbative (long range) behavior of strong interactions which till date is not completely understood. Towards this end, we employ a Bethe-Salpeter framework under Covariant Instantaneous Ansatz for carrying out extensive studies on various processes in hadronic physics. We first derive the non-perturbative Hadron-quark vertex function which incorporates various Dirac covariants in accordance with our power counting scheme order-by-order in powers of inverse of meson mass since various studies have shown that the incorporation of various Dirac covariants is necessary to obtain quantitatively accurate observables. The power counting scheme we proposed in Chapter 4 gives us a lot of insight as to which of the covariants from their complete set are expected to contribute maximum to the calculation of various meson observables since all Dirac covariants do not contribute equally. Calculations employing this vertex function that have been done on leptonic decays of vector mesons and unequal mass pseudoscalar mesons along with the two photon decays of pions have yielded excellent agreements with experimental results and thus validating the power counting rule we have proposed.

In Chapter 5, after a brief summary for experiments, we discuss mostly theoretical aspects of the recent research on the pentaquark baryon. For the discussion of the structure, we use quark models. We discuss the parity and decay properties in a simple framework. We then show the results of the recent serious calculation for the five-quark $uudd\bar{s}$ system for Θ^+. Finally, we discuss production reactions with some remarks on the recent experimental status.

In Chapter 6 we report on recent research on the properties of elementary particle matter governed by the strong nuclear force, at extremes of high temperature and energy density. At about 10^{12} Kelvin, the theory of the strong interaction, Quantum Chromodynamics (QCD), predicts the existence of a new state of matter in which the building blocks of atomic nuclei (protons and neutrons) dissolve into a plasma of quarks and gluons. The Quark-Gluon Plasma (QGP) is believed to have prevailed in the Early Universe during the first few microseconds after the Big Bang. Highly energetic collisions of heavy atomic nuclei provide the unique opportunity to recreate, for a short moment, the QGP in laboratory experiments and study its properties. After a brief introduction to the basic elements of QCD in the vacuum, most notably quark confinement and mass generation, we discuss how these phenomena relate to the occurrence of phase changes in strongly interacting matter at high temperature, as inferred from first-principle numerical simulations of QCD (lattice QCD). This will be followed by a short review of the main experimental findings at the Relativistic Heavy Ion Collider (RHIC) at Brookhaven National Laboratory. The data taken in collisions of gold nuclei thus far provide strong evidence that a QGP has indeed been produced, but with rather remarkable properties indicative for an almost perfect liquid with unprecedentedly small viscosity and high opacity. We then discuss how heavy quarks (charm and bottom) can be utilized to quantitatively probe the transport properties of the strongly-coupled QGP (sQGP). The large heavy-quark mass allows to set up a Brownian motion approach, which can serve to evaluate different approaches for heavy-quark interactions in the sQGP. In particular, we discuss an implementation of lattice QCD computations of the heavy-quark potential in the QGP. This approach generates "pre-hadronic" resonance structures in heavy-quark scattering off light quarks from the medium, leading to large scattering rates and small diffusion coefficients. The resonance correlations are strongest close to the critical temperature (T_c), suggesting an intimate connection to the hadronization of the QGP. The implementation of heavy-quark transport into Langevin simulations of an expanding QGP fireball at RHIC enables quantitative comparisons with experimental data. The extracted heavyquark diffusion coefficients are employed for a schematic estimate of the shear viscosity, corroborating the notion of a strongly-coupled QGP in the vicinity of T_c.

Since quarks are regarded as the most fundamental particles which constitute hadrons that we observe in the real world, there are many theories about how many of them are needed and what quantum numbers they carry. Another important question is what keeps them inside the hadron, which is known to have space-time extension. Since they are relativistic objects, how would the hadron appear to observers in different Lorentz frames? The hadron moving with speed close to that of light appears as a collection of Feynman's partons. In other words, the same object looks differently to observers in two different frames, as Einstein's energy-momentum relation takes different forms for those observers. In order to explain this, it is necessary to construct a quantum bound-state picture valid in all Lorentz frames. It is noted that Paul A. M. Dirac studied this problem of constructing relativistic quantum mechanics beginning in 1927. It is noted further that he published major papers in this field in 1945, 1949, 1953, and in 1963. By combining these works by Dirac, it is possible to construct a Lorentz-covariant theory which can explain hadronic phenomena in the static and high-speed limits, as well as in between. It is shown also in Chapter 7 that this Lorentz-covariant bound-state picture can explain what we observe in high-energy laboratories, including the parton distribution function and the behavior of the proton form factor.

Chapter 8 discusses different resummations of large logarithms that arise in hard-scattering cross sections of quarks and gluons in regions of large and small x. The large-x logarithms are typically dominant near threshold for the production of a specified final state. These soft and collinear gluon corrections produce large enhancements of the cross section for many processes, notably top quark and Higgs production, and typically the higher-order corrections reduce the factorization and renormalization scale dependence of the cross section. The small-$x$logarithms are dominant in the regime where the momentum transfer of the hard sub-process is much smaller than the total collision energy. These logarithms are important to describe multijet final states in deep inelastic scattering and hadron colliders, and in the study of parton distribution functions. The resummations at small and large $x$are linked by the eikonal approximation and are dominated by soft gluon anomalous dimensions. We will review their role in both contexts and provide some explicit calculations at one and two loops.

Chapter 9 studies the soliton type solutions arising in two-dimensional quantum chromodynamics (QCD_2). In bosonized QCD_2 these type of solutions emerge as describing baryons and quark solitons (excitations with "colored" states), respectively. The socalled generalized sine-Gordon model (GSG) arises as the low-energy effective action of bosonized QCD_2 for unequal quark mass parameters, and it has been shown that the relevant solitons describe the normal and exotic baryonic spectrum of QCD_2 [JHEP(03)(2007)(055)]. In the first part of this chapter we classify the soliton and kink type solutions of the sl(3) GSG model with three real fields, which corresponds to QCD_2 with three flavors. Related to the GSG model we consider the sl(3) affine Toda model coupled to matter fields (Dirac spinors) (ATM). The strong coupling sector is described by the sl(3) GSG model which completely decouples from the Dirac spinors. In the spinor sector we are left with Dirac fields coupled to GSG fields. Based on the equivalence between the U(1) vector and topological currents, which holds in the ATM model, it has been shown the confinement of the spinors inside the solitons and kinks of the GSG model providing an extended hadron model for "quark" confinement [JHEP(01)(2007)(027)]. Moreover, it has been proposed that the constituent quark in QCD is a topological soliton. These qualitons (quark solitons), topological excitations with the quantum numbers of quarks, may provide an accurate description of what is meant by constituent quarks in QCD. In the second part of this chapter we discuss the appearance of these type of quark solitons in the context of bosonized QCD_2 (with N_f= 1 and $N_c$colors) and the relevance of the sl(2) ATM model in order to describe the confinement of the color degrees of freedom. We have shown that QCD_2 has quark soliton solutions if the quark mass is sufficiently large.

The Physics of Quarks: New Research
Editors: N.L. Watson and T.M. Grant, pp. 1–8
ISBN 978-1-60456-802-8

Chapter 1

Generalized Statistics and the Formation of a Quark-gluon Plasma

H.G. Miller, A.M. Teweldeberhan[1] *and R. Tegen*[2]

Department of Physics, University of Pretoria
Pretoria, South Africa

1. Introduction

Substantial theoretical research has been carried out to study the phase transition between hadronic matter and the quark-gluon phase (QGP). When calculating the QGP signatures in relativistic nuclear collisions, the distribution functions of quarks and gluons are traditionally described by Boltzmann-Gibbs (BG) statistics. In the past few years the non-extensive form of statistical mechanics proposed by Tsallis [1] has found applications in astrophysical self-gravitating systems [2], solar neutrinos [3], high energy nuclear collisions [4], cosmic microwave back ground radiation [5], high temperature superconductivity [6, 7] and many others. In these cases a small deviation of the Tsallis parameter, q, from 1 (BG statistics) reduces the discrepancies between experimental data and theoretical models. Recently Hagedorn's [8] statistical theory of the momentum spectra produced in heavy ion collisions has been generalized using Tsallis statistics to provide a good description of e^+e^- annihilation experiments [9, 10]. Furthermore, Walton and Rafelski [11] studied a Fokker-Planck equation describing charmed quarks in a thermal quark-gluon plasma and showed that Tsallis statistics were relevant. These results suggest that perhaps BG statistics may not be adequate in the quark-gluon phase.

---

[1] Present address: Tyndall National Institute, Lee Maltings, Prospect Row, Cork, Ireland.

[2] Present address: Stiftung Louisenlund, Buchenhaus 1, D-24357 Guby Germany.

It has been demonstrated [12, 13] that the non-extensive statistics can be considered as the natural generalization of the extensive BG statistics in the presence of *long-range* interactions, *long-range* microscopic memory, or *fractal space-time* constraints. It was suggested in [4] that the extreme conditions of high density and temperature in ultra relativistic heavy ion collisions can lead to memory effects and *long-range* color interactions. For this reason, the effect of the non-extensive form of statistical mechanics proposed by Tsallis (and more recently an extensive form proposed by Kanniadakis [14]) on the formation of a QGP has been recently investigated in [15,16].

2. Generalized Statististics

The generalized entropy proposed by Tsallis [1] takes the form:

$$S_q = \kappa \frac{(1 - \sum_{i=1}^{w} p_i^q)}{q-1} \quad (q \in \Re), \tag{1}$$

where κ is a positive constant (from now on set equal to 1), w is the total number of microstates in the system, p_i are the associated probabilities with $\sum_{i=1}^{w} p_i = 1$, and the Tsallis parameter (q) is a real number. It is straightforward to verify that the usual BG logarithmic entropy, $S = -\sum_{i=1}^{w} p_i \ln p_i$, is recovered in the limit $q \to 1$. Only in this limit is the ensuing statistical mechanics extensive [1, 13, 17]. For general values of q, the measure S_q is non-extensive. That is, the entropy of a composite system $A \oplus B$ consisting of two subsystems A and B, which are statistically independent in the sense that $p_{i,j}^{(A \oplus B)} = p_i^{(A)} p_j^{(B)}$, is not equal to the sum of the individual entropies associated with each subsystem. Instead, the entropy of the composite system is given by Tsallis' q-additive relation [1],

$$S_q(A \oplus B) = S_q(A) + S_q(B) + (1-q) S_q(A) S_q(B) \tag{2}$$

The quantity $|1 - q|$ can be regarded as a measure of the degree of non-extensivity exhibited by S_q.

The standard quantum mechanical distributions can be obtained from a maximum entropy principle based on the entropic measure [18, 19],

$$S = -\sum_i [\bar{n}_i \ln \bar{n}_i \mp (1 \pm \bar{n}_i) \ln(1 \pm \bar{n}_i)], \tag{3}$$

where the upper and lower signs correspond to bosons and fermions, respectively, and $\bar{n}_i$ denotes the number of particles in the ith energy level with energy ϵ_i. The extremization of the above measure under the constraints imposed by the total number of particles,

$$\sum_i \bar{n}_i = N, \tag{4}$$

and the total energy of the system,

$$\sum_i \bar{n}_i \epsilon_i = E, \tag{5}$$

leads to the standard quantum distributions,

$$\bar{n}_i = \frac{1}{\exp \beta(\varepsilon_i - \mu) \mp 1}, \tag{6}$$

where $\beta = \frac{1}{T}$ and the upper and lower signs correspond to the Bose-Einstein and Fermi-Dirac distributions, respectively.

To deal with non-extensive scenarios (characterized by $q \neq 1$), the extended measure of entropy for fermions proposed in [6, 20] is:

$$S_q^{(F)}[\bar{n}_i] = \sum_i \{(\frac{\bar{n}_i - \bar{n}_i^q}{q-1}) + [\frac{(1-\bar{n}_i) - (1-\bar{n}_i)^q)}{q-1}]\}, \tag{7}$$

which for $q \to 1$ reduces to the entropic functional (3) (with lower signs).

The constraints

$$\sum_i \bar{n}_i^q = N \tag{8}$$

and

$$\sum_i \bar{n}_i^q \epsilon_i = E \tag{9}$$

lead to

$$\bar{n}_i = \frac{1}{[1 + (q-1)\beta(\epsilon_i - \mu)]^{\frac{1}{q-1}} + 1} \tag{10}$$

In the limit $q \to 1$ one recovers the usual Fermi-Dirac distribution (6) (with lower sign). Similarly,

$$\bar{n}_i = \frac{1}{[1 + (q-1)\beta(\epsilon_i - \mu)]^{\frac{1}{q-1}} - 1} \tag{11}$$

for bosons. We now turn to the description of the system in the QGP phase.

If we use the generalized statistics to describe the entropic measure of the whole system, the distribution function can not , in general, be reduced to a finite, closed, analytical expression [20-25]. For this reason we use generalized statistics to describe the entropies of the individual particles, rather than of the system as a whole. For a more detailed account of this important point see ref. [20].

The single particle distribution function of quarks, antiquarks and gluons is given by

$$\bar{n}_{Q(\bar{Q})} = \frac{1}{[1 + \frac{1}{T}(q-1)(k \mp \mu_Q)]^{\frac{1}{q-1}} + 1} \tag{12}$$

and

$$\bar{n}_G = \frac{1}{[1 + \frac{1}{T}(q-1)k]^{\frac{1}{q-1}} - 1} \tag{13}$$

respectively. In the limit $q \to 1$ one recovers the usual BG results [15].

The expression for the pressure is given by

$$p_{QGP} = \frac{d_Q T}{2\pi^2}\int_0^\infty dk k^2 (\frac{f_Q^{q-1} - 1}{q-1} + \frac{f_{\bar{Q}}^{q-1} - 1}{q-1}) - \frac{d_G T}{2\pi^2}\int_0^\infty dk k^2 (\frac{f_G^{q-1} - 1}{q-1}) - B \tag{14}$$

where

$$f_{Q(\bar{Q})} = 1 + [1 + \frac{1}{T}(q-1)(k \mp \mu_Q)]^{\frac{1}{1-q}} \tag{15}$$

and

$$f_G = 1 - [1 + \frac{1}{T}(q-1)k]^{\frac{1}{1-q}} \tag{16}$$

which in the limit $q \to 1$ reduces the BG results [15]. Here d_i is the degeneracy factor and B is the bag pressure.

Since the integrals in (12)-(14) in ref. 15 are not integrable analytically one has to calculate these integrals numerically. For $q > 1$, the quantity $[1 + \frac{1}{T}(q-1)(k - \mu_Q)]$ becomes negative if $\mu_Q > k$. To avoid this problem we use [26],

$$f_Q = 1 + [1 + \frac{1}{T}(q-1)(k - \mu_Q)]^{\frac{1}{1-q}}, \ k \geq \mu_Q \tag{17}$$

and

$$f_Q = 1 + [1 + \frac{1}{T}(1-q)(k - \mu_Q)]^{\frac{1}{q-1}}, \ k < \mu_Q \tag{18}$$

In the limit $q \to 1$ one recovers, of course, the appropriate Fermi-Dirac distribution in both cases.

Starting from a one parameter deformation of the exponential function $\exp_{\{\kappa\}}(x) = (\sqrt{1+\kappa^2 x^2} + \kappa x)^{\frac{1}{\kappa}}$, a generalized statistical mechanics has been recently constructed by Kaniadakis [14], which reduces to the ordinary BG statistical mechanics as the deformation parameter, κ, approaches to zero. The difference between Tsallis and Kaniadakis statistics is the following: Tsallis statistics is non-extensive and reduces to BG statistics (extensive) as the Tsallis parameter, q, tends to one. On the other hand, Kaniadakis statistics is extensive and tends to BG statistics as the deformation parameter, κ, tends to zero. In the present effort we use the extensive κ-deformed statistical mechanics constructed by Kaniadakis to represent the constituents of the QGP and compare the results with [15].

For a particle system in the velocity space, the entropic density in κ-deformed statistics is given by [14]

$$\sigma_\kappa(\bar{n}) = -\int d\bar{n} \ln_{\{\kappa\}}(\alpha\,\bar{n}), \tag{19}$$

where α is a real positive constant and κ is the deformation parameter ($-1 < \kappa < 1$). As $\kappa \to 0$, the above entropic density reduces to the standard Boltzmann-Gibbs-Shannon (BGS) entropic density if α is set to be one. The entropy of the system, which is given by $S_\kappa = \int_{\Re} d^n v\, \sigma_\kappa(\bar{n})$, assumes the form

$$S_\kappa = -\frac{1}{2\kappa}\int_{\Re} d^n v \left(\frac{\alpha^\kappa}{1+\kappa}\,\bar{n}^{1+\kappa} - \frac{\alpha^{-\kappa}}{1-\kappa}\,\bar{n}^{1-\kappa}\right) \tag{20}$$

and reduces to the standard BGS entropy $S_0 = -\int_{\Re} d^n v[\ln(\alpha\,\bar{n}) - 1]\bar{n}$ as the deformation parameter approaches to zero. This κ-entropy is linked to the Tsallis entropy $S_q^{(T)}$ through the following relationship [14]:

$$S_\kappa = \frac{1}{2}\frac{\alpha^\kappa}{1+\kappa}\, S_{1+\kappa}^{(T)} + \frac{1}{2}\frac{\alpha^{-\kappa}}{1-\kappa}\, S_{1-\kappa}^{(T)} + const. \tag{21}$$

For α=1, the stationary statistical distribution corresponding to the entropy S_κ can be obtained by maximizing the functional

$$\delta[S_\kappa + \int_{\Re} d^n v(\beta\,\mu\,\bar{n} - \beta\,\epsilon\,\bar{n})] = 0. \tag{22}$$

In doing so, one obtains

$$\bar{n} = \exp_{\{\kappa\}} \beta(\mu - \epsilon), \tag{23}$$

which reduces to the standard classical distribution as $\kappa \to 0$.

The entropic density for quantum statistics is given by [14]

$$\sigma_\kappa(\bar{n}) = -\int d\bar{n}\ln_{\{\kappa\}}\left(\frac{\bar{n}}{1+\eta\,\bar{n}}\right) \tag{24}$$

where η is a real number. After maximization of the constrained entropy or, equivalently, after obtaining the stationary solution of the proper evolution equation, one arrives to the following distribution [14]:

$$\bar{n} = \frac{1}{\exp_{\{\kappa\}} \beta(\epsilon - \mu) - \eta}, \tag{25}$$

where $\eta = 1$ for κ-deformed Bose-Einstein distribution and $\eta = -1$ for κ-deformed Fermi-Dirac distribution.

If we use κ-deformed statistics to describe the entropic measure of the whole system, the distribution function can not , in general, be reduced to a finite, closed, analytical expression. For this reason, we use the κ-deformed statistics to describe the entropies of the individual particles, rather than of the system as a whole. In this case

$$\bar{n}_{Q(\bar{Q})} = \left\{\sqrt{1+\kappa^2\,T^{-2}\,(k\mp\mu_Q)^2} + \kappa\,T^{-1}\,(k\mp\mu_Q) + 1\right\}^{-1} \tag{26}$$

and

$$\bar{n}_G = \left\{\sqrt{1+\kappa^2\,T^{-2}\,k^2} + \kappa\,T^{-1}\,k - 1\right\}^{-1}. \tag{27}$$

In the limit $\kappa \to 0$ one recovers the corresponding BG quantum distributions for quarks, antiquarks and gluons (see (15) and (16) in [15]).

The expression for the pressure is given by

$$P_{QGP} = \frac{d_Q\,T}{2\,\pi^2}\int_0^\infty dk\,k^2\left(\frac{f_Q^\kappa - f_Q^{-\kappa}}{2\,\kappa} + \frac{f_{\bar{Q}}^\kappa - f_{\bar{Q}}^{-\kappa}}{2\,\kappa}\right) -$$

$$\frac{d_G\,T}{2\,\pi^2}\int_0^\infty dk\,k^2\left(\frac{f_G^\kappa - f_G^{-\kappa}}{2\,\kappa}\right) - B\,, \tag{28}$$

where

$$f_{Q(\bar{Q})} = 1 + \left[\sqrt{1+\kappa^2\,T^{-2}(k\mp\mu_Q)^2} + \kappa\;T^{-1}(k\mp\mu_Q)\right]^{-\frac{1}{\kappa}}, \tag{29}$$

$$f_G = 1 - \left[\sqrt{1+\kappa^2\,T^{-2}\,k^2} + \kappa\,T^{-1}\,k\right]^{-\frac{1}{\kappa}} \tag{30}$$

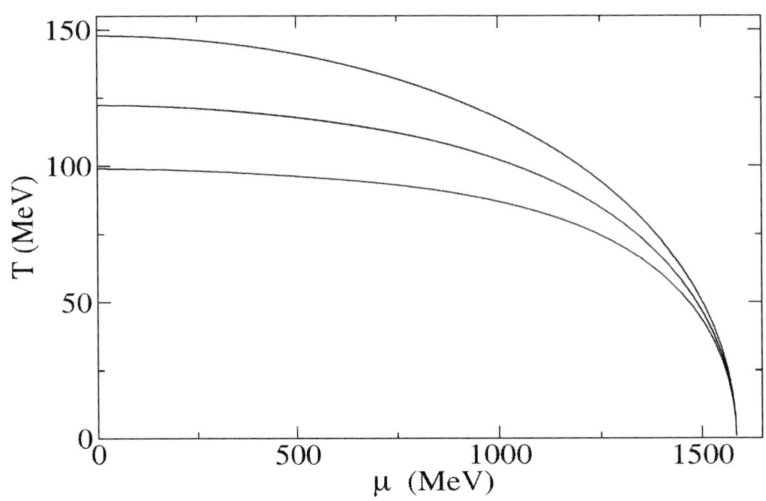

Figure 1. Phase transition curves between the hadronic matter and QGP for κ=0 (solid line), κ=0.23 (dotted line) and κ=0.29 (dashed line).

and B is the bag constant which is taken to be (210 MeV)^4.

The hadron phase is taken to contain only interacting nucleons and antinucleons and an ideal gas of massless pions. Since hadron-hadron interactions are of *short-range*, the BG statistics is successful in describing particle production ratios seen in relativistic heavy ion collisions below the phase transition. The interactions between nucleons is treated by means of a mean field approximation as in [15].

3. Results

Assuming a first order phase transition between hadronic matter and QGP, one matches an equation of state (EOS) for the hadronic system and the QGP via Gibbs conditions equilibrium:

$$P_H = P_{QGP},\ T_H = T_{QGP},\ \mu_H = \mu_{QGP}$$

With these conditions the pertinent regions of temperature, T, and baryon chemical potential, μ, are shown in figure 1 for κ= 0, 0.23 and 0.29. For κ=0.23 (see figure 2), we obtain essentially the same phase diagram as in the case of Tsallis statistics with q=1.1.

Since both Tsallis and κ-deformed statistics are fractal in nature, we observe a similar flattening of the $T(\mu)$ curves. This can be interpreted as follows: the formation of a QGP occurs at a critical temperature which is almost independent of the total number of baryons participating in heavy ion collision. The resulting insensitivity of the critical temperature to the total number of baryons presents a clear experimental signature for the existence of fractal statistics for the constituents of the QGP.

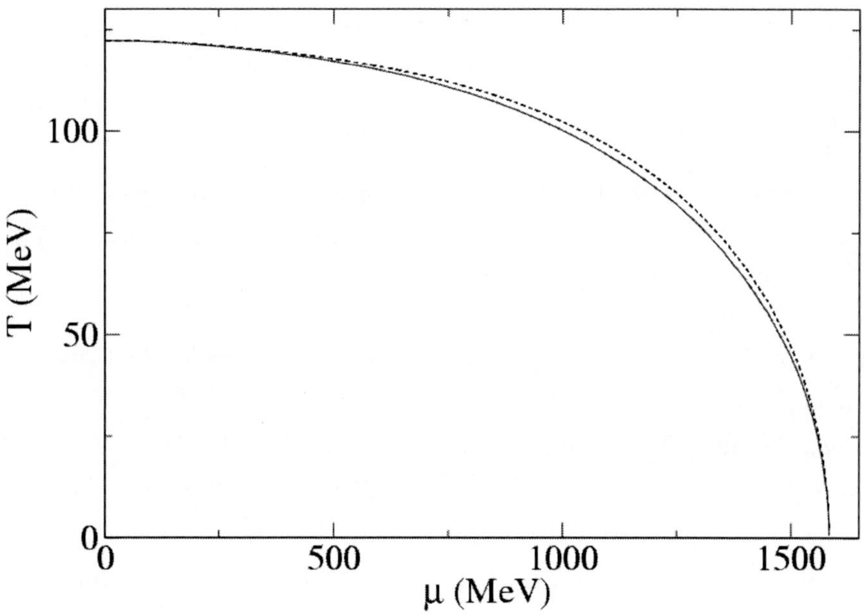

Figure 2. Phase transition curves between the hadronic matter and QGP for κ=0.23 (dashed line) and q=1.1 (solid line).

References

[1] C. Tsallis, *J. Stat. Phys.* **52** (1988) 479.

[2] A.R. Plastino and A. Plastino, *Phys. Lett.* **A 193** (1994) 251.

[3] G. Kaniadakis, A. Lavagno and P. Quarati, *Phys. Lett.* **B 369** (1996) 308, G. Kaniadakis, A. Lavagno, M. Lissia and P. Quarati, *Physica* **A 261** (1998) 359.

[4] W.M. Alberico, A. Lavagno and P. Quarati, *Eur. Phys. J.* **C 12** (2000) 499 and *Nucl. Phys.* A 680 (2000) 94.

[5] C. Tsallis, F.C.S. Barreto and E.D. Loh, *Phys. Rev.* **E 52** (1995) 1447.

[6] H. Uys, H.G. Miller and F.C. Khanna,*Phys. Lett.* **A 289** (2001) 264.

[7] Lizardo H.C.M. Nunes and E.V.L. de Mello, *Physica* **A 305** (2002) 340.

[8] R. Hagedorn, *Nuovo Cimento,* Suppl. 3 (1965) 147.

[9] I. Bediaga, E. M. F. Curado and J. Miranda hep-th/9905 255.

[10] C. Beck, *Physica A* **286** (2000) 164.

[11] D. B. Walton and J. Rafelski *Phys. Rev Lett.* **84** (2000) 31.

[12] C. Tsallis, *Phys. World* **10** (1997) 42.

[13] E.M.F Curado and C. Tsallis, *J. Phys.* **A 24** (1991) L69.

[14] G. Kaniadakis, *Physica* **A 296** (2001) 405 and *Phys. Rev.* **E 66** (2002) 056125.

[15] A.M. Teweldeberhan, H.G. Miller and R. Tegen, *Int. J. Mod. Phys.* **E12** (2003) 395.

[16] A.M. Teweldeberhan, H.G. Miller and R. Tegen, *Int. J. Mod. Phys.* **E12** (2003) 669.

[17] C. Tsallis, *Non-extensive Statistical Mechanics and Thermodynamics: Historical Back-*

ground and Present Status , Pag. 3, in S. Abe and Y. Okamoto (Eds.), *Non-extensive Statistical Mechanics and Its Applications*, Springer, Berlin, 2001.

[18] J.N. Kapur and H.K. Kesavan, *Entropy Optimization Principles with Applications*, Academic Press, 1992.

[19] A. Sommerfeld, *Thermodynamics and Statistical Mechanics*, Academic Press, 1993.

[20] A. R. Plastino, A. Plastino, H. G. Miller and H. Uys, *Foundations of Nonextensive Statistical Mechanics and its Cosmological Apllications*, *Astrophysics and Space Science* **290** (2004) 275.

[21] F. Buyukkilic, D. Demirhan and A. Gulec, *Phys. Lett.* **A 197** (1995) 209.

[22] U. Tirnakli, F. Buyukkilic and D. Demirhan, *Phys. Lett.* **A 245** (1998) 62.

[23] F. Buyukkilic and D. Demirhan, *Eur. Phys. Jr.* **B 14** (2000) 705.

[24] Q.A. Wang and A. Le Mehaute, *Phys. Lett.* **A 235** (1997) 222.

[25] F. Pennini, A. Plastino and A.R. Plastino, *Phys. Lett.* **A 208** (1995) 309.

[26] A. M. Teweldeberhan, A. R. Plastino and H. G. Miller, *Phys. Lett* **A343** (2005) 71.

In: The Physics of Quarks: New Research
Editors: N.L. Watson and T.M. Grant, pp. 9-18
ISBN 978-1-60456-802-8

Chapter 2

QUARK-GLUON PLASMA AND QCD

T. Hatsuda*
Physics Department, University of Tokyo,
Hongo, Tokyo 113-0033, Japan

1. Introduction

Two key experiments, in nuclear/hadron physics and in astrophysics have been started at the beginning of this century. They are RHIC (Relativistic Heavy Ion Collider) [1] and WMAP (Wilkinson Microwave Anisotropy Probe) [2]. Although what they actually measure are quite different, physics goals have some overlaps with each other. In fact, "the origin of masses", which is the most fundamental problem in modern physics, is s key question to be studied in RHIC, WMAP and in future facilities. In this talk, I will discuss some of the recent topics in hot and dense QCD and their relations to the physics in cosmology and atomic physics [3].

2. Origin of Masses

What is the main ingredient of the energy density of the Universe? It has been a long standing problem in cosmology. WMAP has recently revealed that 73% and 23% of the total energy density is due to the dark energy and non-baryonic dark matter, respectively. Baryons which make stars and our bodies contribute only by 4%. The main component of the Universe, the dark energy, is related to the unusual form of the equation of state ($\varepsilon = -P$ with ε and P being the energy density and pressure, respectively) and may be related to the vacuum condensate:

$$\langle T^{\mu\nu} \rangle = \varepsilon_{\rm vac} g^{\mu\nu}, \ \ \varepsilon_{\rm vac} = -P_{\rm vac}, \tag{1}$$

where $T^{\mu\nu}$ is the energy-momentum tensor. $\varepsilon_{\rm vac}$ is nothing but the cosmological constant originally introduced by A. Einstein in 1917. Obvious problem here is that, although the

*E-mail address: xxxx

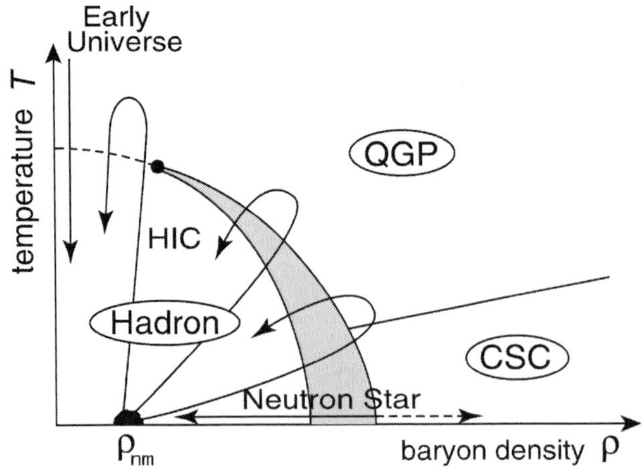

Figure 1. A schematic phase diagram of QCD. "Hadron", "QGP" and "CSC" denote the hadronic phase, the quark-gluon plasma and the color superconducting phase, respectively. $\rho_{\rm nm}$ denotes the baryon density of the normal nuclear matter. Hot plasma in early Universe, the interior of the neutron stars and the matter created in heavy ion collisions (HIC) are relevant places to find various phases in QCD.

dark energy is a dominant contribution in the Universe, its absolute value is many orders of magnitude smaller than expected from the known condensates in strong and electroweak vacuum. More data from WMAP and also accurate data from planned satellite Planck will shed lights on this fundamental problem.

Let us turn to the origin of the baryon mass. The proton mass is about 1 GeV while the light u, d quark masses are about 10 MeV. Namely, the proton mass is not from the quark masses but from the quark-gluon interactions. The dynamical breaking of chiral symmetry and associated fermion-anti-fermion condensate $\langle\bar{\psi}\psi\rangle$ as an origin of the baryon masses were originally proposed by Y. Nambu around 1960 in analogy with the BCS theory of superconductivity. One of the main aim of the RHIC physics is to explore the origin of the baryon and meson masses by creating a hot matter where the chiral symmetry is restored. Heavy-ion experiments at future LHC (Large Hadron Collider) will also help to understand the physics of the QCD vacuum in quantitative manner.

The quark masses (about 10 MeV for u and d quarks) are supposed to be generated by the Higgs condensate $\langle\phi\rangle$ in the standard model. This idea, now called the Higgs mechanism, as an origin of the masses of quarks, leptons and gauge bosons was introduced by Englert and Brout and by Higgs in 1964. One of the aims of the pp collision experiments at future LHC is to look for the Higgs boson and its coupling to other particles to unravel the origin of the particle masses.

As shown above, the origin of the masses are intimately related to the complex vacuum structures with various condensates.

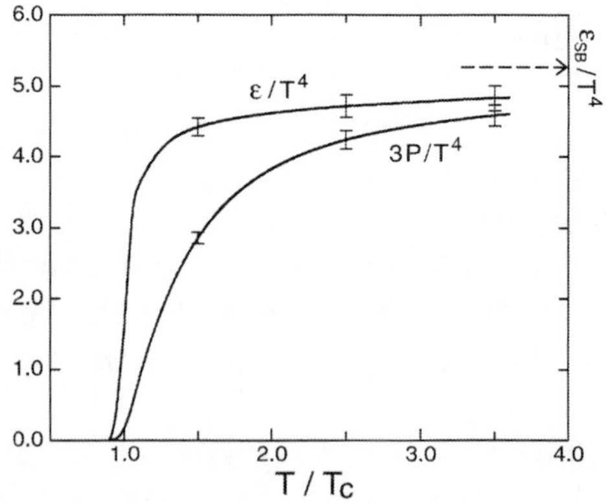

Figure 2. Equations of state for pure Yang-Mills theory in Monte Carlo simulations. The dashed arrow shows the Stefan-Boltzmann limit of the energy density. The figure is adapted from [4].

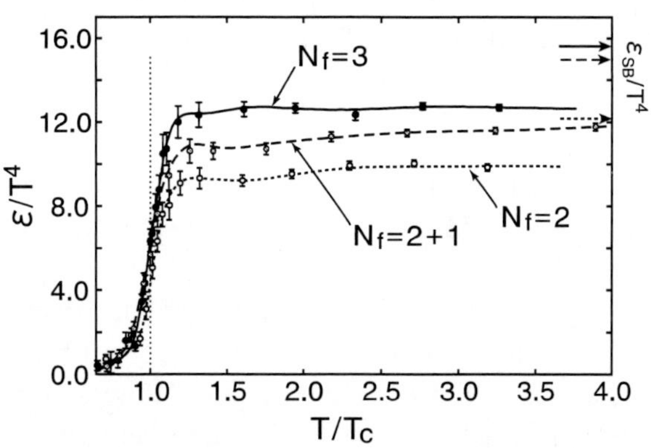

Figure 3. The energy density of QCD with dynamical quarks in lattice Monte Carlo simulations. The figure is adapted from [7].

3. Recent Progress in Hot QCD

Shown in Fig.1 is a schematic phase diagram of QCD at finite temperature and baryon density. There are basically three phases: the hadronic phase, the quark-gluon plasma and the color superconducting phase. The exact locations of the phase boundaries and the critical points as well as the order of the phase transitions are not clearly understood yet except for the case with small baryon density where lattice QCD simulations are possible.

Shown in Fig.2 is ε/T^4 and $3P/T^4$ for SU$_c$(3) pure Yang-Mills theory (gauge theory without fermions or equivalently $N_f = 0$) [4]. ε/T^4 is suppressed below T_c, and has a big jump in a narrow interval in temperature, while $3P/T^4$ has a smooth change across T_c. In the pure Yang-Mills theory, glueballs having masses more than 1 GeV are the only excitations below T_c. Therefore, ε and P are highly suppressed. Above T_c, the system is supposed to be in a deconfined gluon plasma. The arrow in the figure shows the Stefan-Boltzmann limit $\varepsilon_{\rm SB}/T^4$ corresponding to the non-interacting gluon gas. Deviation of ε/T^4 from the arrow indicates that the gluons are still interacting above T_c.

Although the order of the phase transition is not clear from the figure, finite size scaling analysis, in which observables as a function of the lattice volume are studied, shows that the phase transition is of first order and thus there is a discontinuous jump in ε [5]. This has been indeed anticipated from a theoretical argument on the basis of the center symmetry in pure Yang-Mills theories [6]. Critical temperature of the phase transition turns out to be $T_c(N_f = 0) \simeq 273$ MeV with an statistical and systematic error not more than a few %.

Shown in Fig.3 is a result of lattice QCD simulations with dynamical quarks ($N_f \neq 0$) [7]. Quark masses employed in the figure are $m_{\rm u,d}/T = 0.4$ for $N_f = 2$, $m_{\rm u,d,s}/T = 0.4$ for $N_f = 3$, and $m_{\rm u,d}/T = 0.4$, $m_{\rm s}/T = 1.0$ for $N_f =$"2+1". Sudden jumps of the energy density at T_c are seen in all three cases. The deviations from the Stefan-Boltzmann limit are also seen at high T. The critical temperatures extrapolated into the chiral limit ($m_q = 0$) are found to be

$$T_c(N_f = 2) \simeq 175 \text{ MeV}, \quad T_c(N_f = 3) \simeq 155 \text{ MeV}, \tag{2}$$

with at least ± 10 MeV statistical and systematic errors at the moment. Corresponding critical energy density can be read off from Fig.3 as

$$\varepsilon_{\rm crit} \sim 1 \text{ GeV} \cdot \text{fm}^{-3}. \tag{3}$$

Also the behaviors of the chiral condensate $\langle \bar{q}q \rangle$ on the lattice show evidence of second (first) order transition for $N_f = 2$ ($N_f = 3$). This is consistent with the results of renormalization group analysis [8].

As far as the baryon chemical potential is small enough, direct numerical simulations are shown to be feasible to explore the phase boundary separating the hadronic phase and the quark-gluon plasma. Shown in Fig.4 is such a calculation [9]: one can see the existence of a critical end point (CEP) at which the first-order phase transition at high chemical potential changes into crossover at low chemical potential. This is qualitatively consistent with a prediction of the Nambu-Jona-Lasinio model originally done by Asakawa and Yazaki [10]. The exact location of CEP is, however, still an open question: careful studies with an extrapolation to continuum and thermodynamic limits are necessary on the lattice to draw a definite conclusion.

Recently, there are growing evidences that the QGP above T_c is still a strongly interacting matter. Indeed it has been known for many years that simple perturbation theories at high T do not work even at very high T [11]. Furthermore, it was claimed that hadronic bound states can survive even above T_c due to the strong residual interactions among quarks [12]. The latter conjecture has been confirmed recently in quenched lattice QCD simulations [13] with the use of the maximal entropy method (MEM) [14].

The spectral function of a hadron $\rho(\omega, p)$ is related to the imaginary time (τ) correlation function with a fixed three-momentum (p) as

$$G(\tau, p) \quad = \quad \int_0^{+\infty} K(\tau, \omega)\, \rho(\omega, p)\, d\omega \quad (0 \leq \tau < \beta), \tag{4}$$

where $K = ({\rm e}^{-\tau\omega} + {\rm e}^{-(\beta-\tau)\omega})/(1 - e^{-\beta\omega})$ is an integral kernel which reduces to the Laplace kernel at $T = 0$. $\rho(\omega, p)$ gives a spectral distribution as a function of the energy ω and the three-momentum (p).

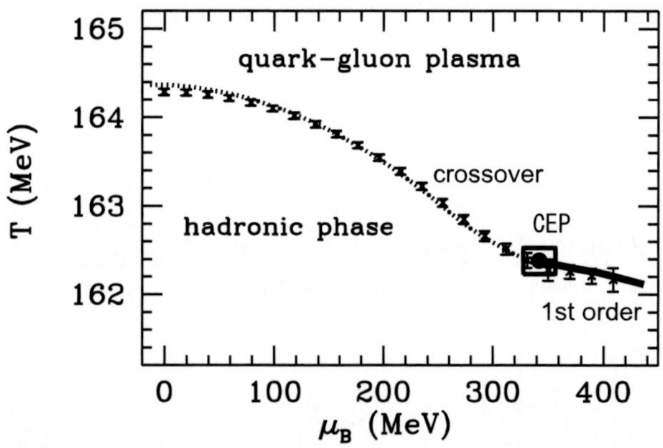

Figure 4. The phase diagram in the space of temperature (T) and baryon chemical potential $\mu_{\rm B} = 3\mu$ obtained from lattice QCD with the reweighing method. The dotted (solid) line shows the crossover (the first order transition). The small box shows the uncertainties of the location of the endpoint. $m_{\rm s}/m_{\rm ud} \simeq 27$ with $m_{\rm ud}$ roughly correspond to the physical quark mass. The figure is adapted from [9].

We show in Fig.5 the spectral function of the J/ψ channel below and above the critical temperature of the deconfinement transition in the quenched approximation. Unlike the naive expectation, the J/ψ peak around 3 GeV still exists at least up to $T \sim 1.6T_c$ (Fig.5(a)) and then it disappears above $T \sim 1.8T_c$ (Fig.5(b)). This may suggest that the plasma is rather strongly interacting so that it can hold bound states although it is deconfined. Further studies with dynamical quarks are necessary to reveal the true nature of the plasma just above T_c.

4. Relativistic Heavy Ion Collisions

The study of the "big bang" by satellite observations and that of the "little bang" by relativistic heavy-ion collision experiments are pretty much analogous not only in their ultimate physics goal, but also in their ways to analyze data. Such analogy is shown in Fig.6 and is summarized as follows:

(i) The initial condition of the Universe is not precisely known and it is one of the most challenging problems in the current cosmology. One promising scenario is the exponential growth of the Universe (inflation) at around the time 10^{-35} sec [15]. Due to the conversion of the energy of the scalar field driving the inflation (inflaton) to the thermal energy, thermal era of the Universe starts after the inflation. In relativistic heavy-ion collisions, the initial condition right after the impact is also not precisely known. The color glass condensate (CGC) [16], which is a coherent but highly excited gluonic configuration, could be a possible initial state at around the time 10^{-24} sec. Then the decoherence of CGC due to particle production initiates the thermal era, the quark-gluon plasma.

(ii) Once the inflation era of the Universe comes to an end and the system becomes thermalized, the subsequent slow expansion of the Universe can be described by the

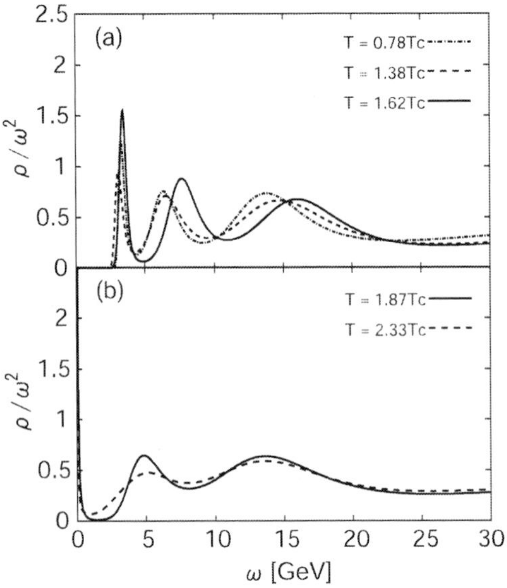

Figure 5. The dimensionless spectral function $\rho(\omega, p = 0)/\omega^2$ in the J/ψ channel as a function of ω for several different temperatures. Since $p = 0$ is taken, we have $\rho = \rho_{\rm T} = \rho_{\rm L}$ where T and L stand for the transverse and longitudinal parts. The figure is adapted from [13].

Friedmann equation with an appropriate equation of state of matter and radiation. In the case of the little bang, the expansion of the locally thermalized plasma is governed by the laws of relativistic hydrodynamics originally introduced by Landau [17]. If the constituent particles of the plasma interact strong enough, one may assume a perfect fluid which simplifies the hydrodynamic equations.

(iii) The Universe expands, cools down and undergoes several phase transitions such as the electro-weak and QCD phase transitions. Eventually, the neutrinos and photons decouple (freeze-out) from the matter and become sources of the cosmic neutrino background (CνB) and cosmic microwave background (CMB). Even the cosmic gravitational background (CGB) could be produced. They carry not only the information of the thermal era of the Universe but also the information of the initial conditions before the thermal era. In the case of the little bang, the system also expands, cools down and experiences the QCD phase transition. The plasma eventually undergoes a chemical freeze-out, and a thermal freeze-out and then falls apart into many hadrons. Not only the hadrons, but also photons, dileptons and jets come out from the various stages of the expansion. They carry information of the thermal era and also the initial conditions.

(iv) What we want to know is the state of matter in the early epoch of the big bang and the little bang. The CMB data and its anisotropy from the big bang is analyzed in the following way: First we define certain key cosmological parameters (usually 8 to 10 parameters) such as the initial density fluctuations, cosmological constant, Hubble constant, etc. Then we make a detailed comparison of the data with the

	Big Bang	Little Bang
Initial state	Inflation ? (10^{-35} sec)	Color glass ? (10^{-24} sec)
Thermalization	Inflaton decay ?	decoherence ?
Expansion	$R^{\mu\nu} - \frac{1}{2}Rg^{\mu\nu} = 8\pi G T^{\mu\nu}$ 	$\partial_\mu T^{\mu\nu} = 0$ $\partial_\mu j^\mu_B = 0$ 
Freezeout	(T = 1.95 K neutrino) T = 2.73 K photon	T_{chem} = 170 MeV T_{therm} = 120 MeV
Observables	CMB & anisotropy ($C\nu B$, CGB & anisotropy)	Flow & its anisotropy, Hadrons, Jets, leptons, photons
Key Parameters	**8~10** cosmological parameters · Initial density fluctuation · Cosmological const. Λ etc	Plasma parameters · Initial energy density · thermalization time etc
Evolution Code	CMBFAST	3D-hydrodynamics

Figure 6. Comparison of the physics and analysis of the big bang and the little bang.

theoretical CMB obtained by solving the Boltzmann equation for the photons. By doing this, we can bridge what happened in the past to what is observed now. WMAP data provide an impressive determination of many of the cosmological parameters in great precision from this method as shown in Table 1. The strategy in the little bang is similar: We first define a few key plasma parameters such as the initial energy density and its profile, the initial thermalization time, the freeze-out temperatures, etc. Then a full three dimensional relativistic hydrodynamics code is solved to relate these parameters to the plenty of data from laboratory experiments. Such precision study has now begun to be possible under the assumption of the perfect fluid [18].

COBE launched by NASA in 1989 (SPS at CERN started in 1987) exposed a tantalizing evidence of the initial state of the Universe (heavy ion collisions). WMAP launched by NASA in 2001 (RHIC at BNL started in 2000) gives better images of the newly born state and has initiated the precision cosmology (precision QGP physics). Future Planck satellite by ESA to be launched in 2007 (LHC at CERN to be started in 2007) will shed further lights on the detailed information on the initial conditions and the origin of dark energy (initial conditions and the dynamics of the quark-gluon system in hot environment).

Table 1. Some of the cosmological parameters determined by WMAP [2].

description	symbol	value	± uncertainty
total density	Ω_{tot}	1.02	± 0.02
dark energy density	Ω_Λ	0.73	± 0.04
baryon density	Ω_{B}	0.044	± 0.004
matter density	$\Omega_{\text{B}} + \Omega_{\text{DM}}$	0.27	± 0.04
baryon-to-photon ratio	η	6.1×10^{-10}	${}^{+0.3}_{-0.2} \times 10^{-10}$
red shift of decoupling	z_{dec}	1089	± 1
age of decoupling (year)	t_{dec}	379×10^3	${}^{+8}_{-7} \times 10^3$
Hubble constant	h	0.71	${}^{+0.04}_{-0.03}$
age of the Universe (year)	t_0	13.7×10^9	$\pm 0.2 \times 10^9$

5. Recent Progress in Dense QCD

The color superconductivity is one of the most exciting development in recent years in the study of high density matter [19]. The color superconductor is unique in the sense that it is a high temperature superconductor because of the long-range magnetic interaction between quarks and also there are color-flavor entanglement. The former leads to the small size Cooper pairs and a large gap, while the latter leads to various phase structures. A most attractive channel between the quark pairs is the color anti-symmetric and flavor-antisymmetric with $J^P = 0^+$:

$$\langle q_a^i C\gamma_5 q_b^j \rangle = \epsilon^{ijk}\epsilon_{abc}\Delta_c^i, \tag{5}$$

where i, j, k (a, b, c) denote flavor (color) indices, and C denotes the charge conjugation. The gap matrix Δ_c^i is a 3x3 matrix in color and flavor space and belongs to the $(3^*, 3^*)$ representation of SU$_c$(3)$\times$ SU$_f$(3). If we take a diagonal ansatz for this matrix, $\Delta_c^i = \text{diag}(\Delta_1, \Delta_2, \Delta_3)$, we can define various phases as, mCFL ($\Delta_{1,2,3} \neq 0$), uSC ($\Delta_1 = 0, \Delta_{2,3} \neq 0$), dSC ($\Delta_2 = 0, \Delta_{1,3} \neq 0$), sSC ($\Delta_3 = 0, \Delta_{1,2} \neq 0$) and 2SC ($\Delta_{1,2} = 0, \Delta_3 \neq 0$), where dSC (uSC, sSC) stands for superconductivity in which for d (u, s) quarks all three colors are involved in the pairing.

Due to the strange quark mass and the effect of charge neutrality, it has been shown that successive phase transitions may take place around the CSC-QGP phase transition at finite T [20], namely

$$\text{mCFL} \to \text{dSC} \to \text{2SC} \to \text{QGP}. \tag{6}$$

The existence of such successive phase transitions can be proven at asymptotically high baryon densities with very few assumptions on the basis of the Ginzburg-Landau approach. However, the situation could be different in low baryon density (strong coupling) region.

There have been suggested an interesting possibility that the large-size Cooper pair at high density turns into tightly-bound bosonic particle at intermediate density and the system may undergo BCS-BEC crossover before the confinement takes place at low density [21].

This phenomena may have close connection to the recently found BCS-BEC crossover of the fermionic condensates in ultracold atomic systems such as ^{40}K atoms and ^{6}Li atoms [22]. The question of how the low-temperature baryonic matter undergoes a phase transition to quark matter at high density is still unanswered clearly although there are long history of research in the past on the basis of various models.

What is called for at present is a new idea to attach this problem from first principle lattice QCD simulations. Also new dedicated experiments for this purpose are necessary: The 50 GeV PS (J-PARC) under construction at Tokai, Japan, and a planned 90 GeV PS (SIS 100/300) at GSI may shed light toward the resolution of this problem in the future.

6. Summary

Hot QCD studies started to become a mature field. Within a few year, one may hope to combine lattice QCD results (for soft physics) and perturbative QCD results (for hard physics) together as basic inputs to the hydrodynamics simulation code. Then one may make quantitative comparison of the prediction with the RHIC data and extract the information on QGP created in the early stage of the relativisitic heavy-ion collisions. There are several key questions we need to answer through such studies:

1. How the thermalization takes place in the relativistic heavy ion collisions? This is a long standing issue after the seminal paper by Landau [17] where the rapid thermalization is tacitly assumed.

2. How complex is the QCD plasma at high T? Is it a strongly interacting matter even above T_c. Are there hadronic bound states above T_c?

3. What is the direct experimental evidence of the evaporation of the QCD condensates at high T?

Dense QCD is still an open field with lots of unknown physics. Nevertheless, there are interesting theoretical developments in recent years, such as the variety of phases in color superconductivity and the possibility of BCS-BEC crossover. Exploring similarities with other condensed matter systems such as the liquid ^{3}He and ^{4}He, high temperature superconductors and ultracold atomic gases may give us hints to the physics of high density matter. Also new ideas to attach dense QCD in lattice QCD simulations are called for.

References

[1] http://www.bnl.gov/RHIC/.

[2] C. L. Bennett, et al., *Astrophys. J. Suppl.*, **148** (2003) 1.

[3] For more details, see, K. Yagi, T. Hatsuda and Y. Miake, *Quark-Gluon Plasma*, (Cambridge Monographs on Particle Physics, Nuclear Physics and Cosmology, Vol. 23, Combridge Univ. Press, 2005).

[4] M. Okamoto et al., *Phys. Rev.* **D60** (1999) 094510.

[5] M. Fukugita, M. Okawa and A. Ukawa, *Phys. Rev. Lett.* **63** (1989) 1768.

[6] L. G. Yaffe and B. Svetitsky, *Phys. Rev.* **D26** (1982) 963.

[7] F. Karsch, *Lecture Notes in Physics,* **583** (2002) 209.

[8] R. D. Pisarski and F. Wilczek, *Phys. Rev.* **D29** (1984) 338.

[9] Z. Fodor and S. D. Katz, *JHEP* **0404** (2004) 050.

[10] M. Asakawa and K. Yazaki, *Nucl. Phys.* **A504** (1989) 668.

[11] A. D. Linde, *Phys. Lett.* **B96** (1980) 289.

[12] T. Hatsuda and T. Kunihiro, Phys. Rev. Lett. **55** (1985) 158.
C. DeTar, *Phys. Rev.* **D32** (1985) 276.

[13] M. Asakawa and T. Hatsuda, *Phys. Rev. Lett.* **92** (2004) 012001.

[14] M. Asakawa, T. Hatsuda and Nakahara, *Prog. Part. Nucl. Phys.* **46** (2001) 459.

[15] A.D. Linde, *Particle Physics and Inflationary Cosmology* (Harwood, New York, 1990).

[16] E. Iancu and R. Venugopalan, hep-ph/0303204.

[17] S. Z. Belensky and L. D. Landau, *Ups.Fiz.Nauk.* **56** (1955) 309.

[18] T. Hirano and Y. Nara, *J. Phys.* **G30** (2004) S1139.

[19] K. Rajagopal and F. Wilczek, hep-ph/0011333.

[20] K. Iida, T. Matsuura, M. Tachibana and T. Hatsuda, *Phys. Rev. Lett* **93** (2004) 132001.
K. Fukushima, C. Kouvaris and K. Rajagopal, *Phys.Rev*. **D71** (2005) 034002.

[21] H. Abuki, K. Itakura and T. Hatsuda, *Phys. Rev.* **D65** (2002) 074014.
Y. Nishida and H. Abuki, hep-ph/0504083.

[22] Q. Chen, J. Stajic, S. Tan and K. Levin, *Phys. Rept.* **412** (2005) 1.

In: The Physics of Quarks: New Research ISBN 978-1-60456-802-8
Editors: N.L. Watson and T.M. Grant , pp. 19-47

Chapter 3

STABLE QUARKS OF THE 4TH FAMILY?

***K. Belotsky*[1], *M. Khlopov*[1,2] *and K. Shibaev*[1]**
[1]Moscow Engineering Physics Institute, Moscow, Russia
Center for Cosmoparticle physics "Cosmion", Moscow, Russia
[2]APC laboratory, Paris, France

Abstract

Existence of metastable quarks of new generation can be embedded into phenomenology of heterotic string together with new long range interaction, which only this new generation possesses. We discuss primordial quark production in the early Universe, their successive cosmological evolution and astrophysical effects, as well as possible production in present or future accelerators. In case of a charge symmetry of 4th generation quarks in Universe, they can be stored in neutral mesons, doubly positively charged baryons, while all the doubly negatively charged "baryons" are combined with He-4 into neutral nucleus-size atom-like states. The existence of all these anomalous stable particles may escape present experimental limits, being close to present and future experimental test. Due to the nuclear binding with He-4 primordial lightest baryons of the 4th generation with charge $+1$ can also escape the experimental upper limits on anomalous isotopes of hydrogen, being compatible with upper limits on anomalous lithium. While 4th quark hadrons are rare, their presence may be nearly detectable in cosmic rays, muon and neutrino fluxes and cosmic electromagnetic spectra. In case of charge asymmetry, a nontrivial solution for the problem of dark matter (DM) can be provided by excessive (meta)stable anti-up quarks of 4th generation, bound with He-4 in specific nuclear-interacting form of dark matter. Such candidate to DM is surprisingly close to Warm Dark Matter by its role in large scale structure formation. It catalyzes primordial heavy element production in Big Bang Nucleosynthesis and new types of nuclear transformations around us.

1. Introduction

The question about existence of new quarks and/or leptons is among the most important in the modern particle physics. Possibility of existence of new (meta)stable quarks which form new (meta)stable hadrons is of special interest. New stable hadrons can play the role of strongly interacting dark matter [1–3]. This question is believed to find solution in the framework of future Grand Unified Theory. A strong motivation for existence of new long-living hadrons comes from a possible solution [4] of the "doublet-triplet splitting" problem in supersymmetric GUT models. Phenomenology of string theory offers another motivation for new long lived hadrons.

A natural extension of the Standard model can lead in the heterotic string phenomenology to the prediction of fourth generation of quarks and leptons [5, 6] with a stable 4th neutrino [7–10]. The comparison between the rank of the unifying group E_6 ($r = 6$) and the rank of the Standard model ($r = 4$) can imply the existence of new conserved charges. These charges can be related with (possibly strict) gauge symmetries. New strict gauge U(1) symmetry (similar to U(1) symmetry of electrodynamics) is excluded for known particles but is possible, being ascribed to the fermions of 4th generation only. This provides theoretic motivation for a stability of the lightest fermion of 4th generation, assumed to be neutrino. Under the condition of existence of strictly conserved charge, associated to 4th generation, the lightest 4th generation quark Q (either U or D) can decay only to 4th generation leptons owing to GUT-type interactions, what makes it sufficiently long living.

Whatever physical reason was for a stability of new hypothetical particles, it extends potential for testing respective hypothesis due to its implications in cosmology. Especially rich in this sense is a hypothesis on (meta)stable quarks of new family. It defines the goal of current work.

As we will show, in the case when 4th generation possesses strictly conserved $U(1)$-gauge charge (which will be called y-charge), 4th generation fermions are the source of new interaction of Coulomb type (which we'll call further y-interaction). It can be crucial for viability of model with equal amounts of 4th generation quarks and antiquarks in Universe. The case of cosmological excess of 4th generation antiquarks offers new form of dark matter with a very unusual properties. Owing to strict conservation of y-charge, this excess should be compensated by excess of 4th generation neutrinos.

Recent analysis [11] of precision data on the Standard model parameters admits existence of the 4th generation particles, satisfying direct experimental constraints which put lower limit 220 GeV for the mass of lightest quark [12].

If the lifetime of the lightest 4th generation quark exceeds the age of the Universe, primordial Q-quark (and $\bar{Q}$-quark) hadrons should be present in the modern matter. If this lifetime is less than the age of the Universe, there should be no primordial 4th generation quarks, but they can be produced in cosmic ray interactions and be present in cosmic ray fluxes. The search for this quark is a challenge for the present and future accelerators.

In the present work we will assume that up-quark of 4th generation (U) is lighter than its down-quark (D). The opposite assumption is found to be virtually excluded, if D is stable. The reason is that D-quarks might form stable hadrons with electric charges ± 1 ($(DDD)^-$, $(\bar{D}u)^+$, $(\bar{D}\bar{D}\bar{D})^+$), which eventually form hydrogen-like atoms (hadron $(DDD)^-$ is combined with $^4He^{++}$ into $+1$ bound state), being strongly constrained in surrounding matter. It will become more clear from consideration of U-quark case, presented below.

The following hadron states containing (meta)stable U-quarks (U-hadrons) are expected to be (meta)stable and created in early Universe: "baryons" $(Uud)^+$, $(UUu)^{++}$, $(UUU)^{++}$; "antibaryons" $(\bar{U}\bar{U}\bar{U})^{--}$, $(\bar{U}\bar{U}\bar{u})^{--}$, meson $(\bar{U}u)^0$. The absence in the Universe of the states $(\bar{U}\bar{u}\bar{d})$, $(U\bar{u})$ containing light antiquarks are suppressed because of baryon asymmetry. Stability of double and triple U bound states (UUu), (UUU) and

$(\bar{U}\bar{U}\bar{u})$, $(\bar{U}\bar{U}\bar{U})$ is provided by the large chromo-Coulomb binding energy ($\propto \alpha_{QCD}^2 \cdot m_Q$) [13, 14]. Formation of these states in particle interactions at accelerators and in cosmic rays is strongly suppressed, but they can form in early Universe and cosmological analysis of their relics can be of great importance for the search for 4th generation quarks.

We analyze the mechanisms of production of metastable U (and $\bar{U}$) hadrons in the early Universe, cosmic rays and accelerators and point the possible signatures of their existence.

We'll show that in case of charge symmetry of U-quarks in Universe, a few conditions play a crucial role for viability of the model. An electromagnetic binding of $(\bar{U}\bar{U}\bar{U})^{--}$ with $^4He^{++}$ into neutral nucleus-size atom-like state (O-$Helium$) should be accompanied by a nuclear fusion of $(Uud)^+$ and $^4He^{++}$ into lithium-like isotope $[^4He(Uud)]$ in early Universe. The realization of such a fusion requires a marginal supposition concerning respective cross section. Furthermore, assumption of $U(1)$-gauge nature of the charge, associated to U-quarks, is needed to avoid a problem of overproduction of anomalous isotopes by means of an y-annihilation of U-relics ($[^4He(Uud)]$, (UUu), (UUU), $^4He(\bar{U}\bar{U}\bar{U})$, $^4He(\bar{U}\bar{U}\bar{u})$, $(\bar{U}u)$). Residual amount of U-hadrons with respect to baryons in this case is estimated to be less than 10^{-10} in Universe in toto and less than 10^{-20} at the Earth.

A negative sign charge asymmetry of U-quarks in Universe can provide a nontrivial solution for dark matter (DM) problem. For sctricly conserved charge such asymmetry in $\bar{U}$ implies corresponding asymmetry in leptons of 4th generation. In this case the most of $\bar{U}$ in Universe are contained in O-$Helium$ states $[^4He(\bar{U}\bar{U}\bar{U})]$ and minor part of them in mesons $\bar{U}u$. On the other hand the set of direct and indirect effects of relic U-hadrons existence provides the test in cosmic ray and underground experiments which can be decisive for this hypothesis. The main observational effects for asymmetric case do not depend on the existence of y-interaction.

The structure of this paper is as the following. Section 2 is devoted to the charge symmetric case of U-quarks. Cosmological evolution of U-quarks in early Universe is considered in subsection 2.1, while in subsection 2.2 the evolution and all possible effects of U-quarks existence in our Galaxy are discussed. The case of charge asymmetry of quarks of 4th generation in Universe is considered in Section 3. Section 4 is devoted to the questions of the search for U-quarks at accelerators. We summarize the results of our present study, developing earlier investigations [15, 16], in Conclusion.

2. Charge Symmetric Case of U-quarks

2.1. Primordial U-hadrons from Big Bang Universe

Freezing out of U-quarks

In the early Universe at temperatures highly above their masses fermions of 4th generation were in thermodynamical equilibrium with relativistic plasma. When in the course of expansion the temperature T falls down below the mass of the lightest U-quark, m, equilibrium concentration of quark-antiquark pairs of 4th generation is given by

$$n_4 = g_4 \left(\frac{Tm}{2\pi}\right)^{3/2} \exp(-m/T), \tag{1}$$

where $g_4 = 6$ is the effective number of their spin and colour degrees of freedom. We use the units $\hbar = c = k = 1$ throughout this paper.

The expansion rate of the Universe at RD-stage is given by the expression

$$H = \frac{1}{2t} = \sqrt{\frac{4\pi^3 g_{tot}}{45}}\frac{T^2}{m_{Pl}} \approx 1.66\, g_{tot}^{1/2}\frac{T^2}{m_{Pl}}, \tag{2}$$

where temperature dependence follows from the expression for critical density of the Universe

$$\rho_{crit} = \frac{3H^2}{8\pi G} = g_{tot}\frac{\pi^2}{30}T^4.$$

When it starts to exceed the rate of quark-antiquark annihilation

$$R_{ann} = n_4 \langle \sigma v \rangle, \tag{3}$$

in the period, corresponding to $T = T_f < m$, quarks of 4th generation freeze out, so that their concentration does not follow the equilibrium distribution Eq.(1) at $T < T_f$. For a convenience we introduce the variable

$$r_4 = \frac{n_4}{s}, \tag{4}$$

where

$$s = \frac{2\pi^2 g_{tot\,s}}{45}T^3 \approx 1.80\, g_{tot\,s} n_\gamma \approx 1.80\, g_{tot\,s}^{mod}\eta^{-1} n_B \tag{5}$$

is the entropy density of all matter. In Eq.(5) s was expressed through the thermal photon number density $n_\gamma = \frac{2\zeta(3)}{\pi^2}T^3$ and also through the baryon number density n_B, for which at the modern epoch we have $n_B^{mod}/n_\gamma^{mod} \equiv \eta \approx 6\cdot 10^{-10}$.

Under the condition of entropy conservation in the Universe, the number density of the frozen out particles can be simply found for any epoch through the corresponding thermal photon number density n_γ. Factors g_{tot} and $g_{tot\,s}$ take into account the contribution of all particle species and are defined as

$$g_{tot} = \sum_{i=bosons} g_i \left(\frac{T_i}{T}\right)^4 + \frac{7}{8}\sum_{i=fermions} g_i \left(\frac{T_i}{T}\right)^4$$

and

$$g_{tot\,s} = \sum_{i=bosons} g_i \left(\frac{T_i}{T}\right)^3 + \frac{7}{8}\sum_{i=fermions} g_i \left(\frac{T_i}{T}\right)^3,$$

where g_i and T_i are the number of spin degrees of freedom and temperature of ultrarelativistic bosons or fermions. For epoch $T \ll m_e \approx 0.5\,\mathrm{MeV}$ it is assumed that only photons and neutrinos with $T_\nu = (4/11)^{1/3}T$ give perceptible contribution into energy (until the end of RD-stage) and entropy (until now) densities so one has

$$g_{tot\,s}^{mod} \approx 3.91 \qquad g_{tot}^{mod} \approx 3.36. \tag{6}$$

For modern entropy density we have $s_{mod} \approx 2890\,\mathrm{cm}^{-3}$.

From the equality of the expressions Eq.(2) and Eq.(3) one gets

$$m/T_f \approx 42 + \ln(g_{tot}^{-1/2} m_p m \langle\sigma v\rangle)$$

with m_p being the proton mass and obtains, taking $\langle\sigma v\rangle \sim \frac{\alpha_{QCD}^2}{m^2}$ and $g_{tot}(T_f) = g_{tot\,s}(T_f) = g_f \approx 80 - 90$,

$$T_f \approx m/30$$

and

$$r_4 = \frac{H_f}{s_f \langle\sigma v\rangle} \approx \frac{4}{g_f^{1/2} m_{Pl} T_f \langle\sigma v\rangle} \approx 2.5 \cdot 10^{-14} \frac{m}{250\,\text{GeV}}. \quad (7)$$

Index "f" means everywhere that the corresponding quantity is taken at $T = T_f$. Note, that the result Eq.(7), obtained in approximation of "instantaneous" freezing out, coincides with more acurate one if $\langle\sigma v\rangle$ and g_f can be considered (as in given case) to be constant. Also it is worth to emphasize, that given estimation for r_4 relates to only 4th quark or 4th antiquark abundances, assumed in this part to be equal to each other.

Note that if $T_f > \Delta = m_D - m$, where m_D is the mass of D-quark (assumed to be heavier, than U-quark) the frozen out concentration of 4th generation quarks represent at $T_f > T > \Delta$ a mixture of nearly equal amounts of $U\bar{U}$ and $D\bar{D}$ pairs.

At $T < \Delta$ the equilibrium ratio

$$\frac{D}{U} \propto \exp\left(-\frac{\Delta}{T}\right)$$

is supported by weak interaction, provided that β-transitions ($U \to D$) and ($D \to U$) are in equilibrium. The lifetime of D-quarks, τ, is also determined by the rate of weak ($D \to U$) transition, and at $t \gg \tau$ all the frozen out $D\bar{D}$ pairs should decay to $U\bar{U}$ pairs.

At the temperature T_f annihilation of U-quarks to gluons and to pairs of light quarks $U\bar{U} \to gg, q\bar{q}$ terminates and $U\bar{U}$ pairs are frozen out. The frozen out concentration is given by Eq.(7). Even this value of primordial concentration of U-quarks with the mass $m = 250$ GeV would lead to the contribution into the modern density $2mr_4 s_{mod}$, which is by an order of magnitude less than the baryonic density, so that in the charge symmetric case U-quarks can not play a significant dynamical role in the modern Universe.

The actual value of primordial U-particle concentration should be much smaller due to QCD, hadronic and radiative recombination, which reduce the abundance of frozen out U-particles. y-Interaction can play essential role in successive evolution to be considered. It accounts for radiative recombination and plays crucial role in galactic evolution of U-hadrons. So, it will be included into further consideration which will be carried out for both sub-cases (with and without y-interaction).

QCD recombination

At

$$T \le I_1 = m\bar{\alpha}^2/4 = 3.2\,\text{GeV}\frac{m}{250\,\text{GeV}},$$

where $\bar{\alpha} = 0.23$ accounts for joint effect of Coulomb-like attraction due to QCD and y-interactions, formation of bound $(U\bar{U})$ states is possible, in which frozen out Heavy quarks and antiquarks can annihilate. Effect of y-interaction is not essential here.

Note that at $T \leq I_1$ rate of $(U\bar{U})$ annihilation in bound systems exceeds the rate of "ionization" of these systems by quark gluon plasma. So the rate of QCD recombination, given by [14, 15]

$$\langle \sigma v \rangle \approx \left(\frac{16\pi}{3^{5/2}}\right) \cdot \frac{\bar{\alpha}}{T^{1/2} \cdot m_U^{3/2}}, \tag{8}$$

is the rate, with which abundance of frozen out U-quarks decreases.

The decrease of U-hadron abundance owing to $U\bar{U}$ recombination is governed by the equation

$$\frac{dn_4}{dt} = -3Hn_4 - n_4^2 \cdot \langle \sigma v \rangle . \tag{9}$$

Using notation Eq.(4) and relation

$$-dt = \frac{dT}{HT}, \tag{10}$$

which follows from Eq.(2) and is true as long as $g_{tot} \approx const$, Eq.(9) is reduced to

$$dr_4 = r_4^2 \cdot sHT \langle \sigma v \rangle \, dT, \tag{11}$$

where $sHT = \sqrt{\pi g/45}$ with $g \equiv g_{tot\,s}^2 / g_{tot} =$ (for $T > m_e$) $= g_{tot\,s} = g_{tot}$.

At $T_0 = I_1 > T > T_{QCD} = T_1$, assuming in this period $g = const = g_f \approx 17$, the solution of Eq.(11) is given by

$$r_4 = \frac{r_0}{1 + r_0 \sqrt{\frac{\pi g_f}{45}}\, m_{Pl} \int_{T_1}^{T_0} \langle \sigma v \rangle \, dT} \approx 0.16 \left(\frac{m}{I_1}\right)^{1/2} \frac{m}{\bar{\alpha} m_{Pl}} \approx \tag{12}$$

$$\approx 1.6 \cdot 10^{-16} \frac{m}{250\ \mathrm{GeV}}.$$

It turns to be independent on the frozen out concentration r_0 given by Eq.(7).

At $T < I_{UU} \leq m\tilde{\alpha}^2/4 = 1.6\,\mathrm{GeV}\frac{m}{250\,\mathrm{GeV}}$, where effective constant $\tilde{\alpha} = C_F\alpha_s - \alpha_y \sim (4/3) \cdot 0.144 - 1/30 = 0.16$ accounts for repulsion of the same sign y-charges, reactions $U + U \to (UU) + g$ and $U + (UU) \to (UUU) + g$ can lead to formation (UU)-diquark and colorless (UUU) "hadron" (as well as similar $\bar{U}$ bound states) in quark gluon plasma [13, 14]. However, disruption of these systems by gluons in inverse reactions prevents their effective formation at $T \gtrsim I_{UU}/30$ [14]. Therefore, such systems of U quarks with mass $m < 700$ GeV are not formed before QCD phase transition.

Hadronic recombination

After QCD phase transition at $T = T_{QCD} \approx 150$MeV quarks of 4th generation combine with light quarks into U-hadrons. In baryon asymmetrical Universe only excessive valence quarks should enter such hadrons. Multiple U states formation can start only in processes of hadronic recombination for U-quark mass $m < 700$ GeV what is discussed below.

As it was revealed in [5, 6] in the collisions of such mesons and baryons recombination of U and $\bar{U}$ into unstable $(U\bar{U})$ "charmonium -like" state can take place, thus successively reducing the U-hadron abundance. Hadronic recombination should take place even in the absence of long range y-interaction of U-particles. So, we give first the result without the account of radiative recombination induced by this interaction.

There is a large uncertainties in the estimation of hadronic recombination rate. The maximal estimation for the reaction rate of recombination $\langle \sigma v \rangle$ is given by

$$\langle \sigma v \rangle \sim \frac{1}{m_\pi^2} \approx 6 \cdot 10^{-16} \, \frac{\mathrm{cm}^3}{\mathrm{s}} \tag{13}$$

or by

$$\langle \sigma v \rangle \sim \frac{1}{m_\rho^2} \approx 2 \cdot 10^{-17} \, \frac{\mathrm{cm}^3}{\mathrm{s}}. \tag{14}$$

The minimal realistic estimation gives [15]

$$\langle \sigma v \rangle \approx 0.4 \cdot (T_{eff} m^3)^{-1/2} (3 + \log{(T_{QCD}/T_{eff})}), \tag{15}$$

where $T_{eff} = \max{\{T, \alpha_y m_\pi\}}$.

Solution of Eq.(11) for $\langle \sigma v \rangle$ from the Eq.(13) is given by

Case A

$$r_4 = \frac{r_0}{1 + r_0 \cdot \sqrt{\frac{\pi g_{QCD}}{45}} \frac{m_{Pl}}{m_\pi} \frac{T_{QCD}}{m_\pi}} \approx 1.0 \cdot 10^{-20} \tag{16}$$

and it is $(\frac{m_\rho}{m_\pi})^2 \sim 30$ times larger for $\langle \sigma v \rangle$ from the Eq.(14):

Case B

$$r_4 = \frac{r_0}{1 + r_0 \cdot \sqrt{\frac{\pi g_{QCD}}{45}} \frac{m_{Pl}}{m_\rho} \frac{T_{QCD}}{m_\rho}} \approx 3.0 \cdot 10^{-19}. \tag{17}$$

For the minimal estimation of recombination rate (15) the solution of Eq.(11) has the form

$$r_4 = \frac{r_0}{1 + r_0 \cdot 2 \cdot \sqrt{\frac{\pi g_{QCD}}{45}} \frac{m_{Pl}}{m} \sqrt{\frac{T_{QCD}}{m}}} \tag{18}$$

where in all the cases r_0 is given by Eq.(12) and $g_{QCD} \approx 15$. We neglect in our estimations possible effects of recombination in the intermediate period, when QCD phase transition proceeds.

The solutions (16) and (17) are independent on the actual initial value of $r_4 = r_0$, if before QCD phase transition it was of the order of (12). For the minimal estimation of the recombination rate (15) the result of hadronic recombination reads

Case C

$$r_4 \approx 1.2 \cdot 10^{-16} \left(\frac{m}{250\ \mathrm{GeV}} \right)^{3/2}. \tag{19}$$

As we mentioned above, for the smallest allowed mass of U-quark, diquarks (UU), $(\bar{U}\bar{U})$ and the tripple U (and $\bar{U}$) states (UUU), $(\bar{U}\bar{U}\bar{U})$ can not form before QCD phase transition. Therefore U-baryonic states (UUu), (UUU) and their antiparticles should originate from single U (and $\bar{U}$) hadron collisions. The rate of their creation shares the same theoretical uncertainty as in the case of $(U\bar{U})$ formation, considered above. Moreover, while baryon (UUu) can be formed e.g. in reaction $(Uud) + (Uud) \to (UUu) + n$, having no energetic threshold, formation of antibaryon $(\bar{U}\bar{U}\bar{u})$ may be suppressed at smallest values of m by the threshold of nucleon production in reaction $(\bar{U}u) + (\bar{U}u) \to (\bar{U}\bar{U}\bar{u}) + p + \pi^+$, which can even exceed $\bar{U}\bar{U}$ binding energy. In further consideration we will not specify

$\bar{U}$-hadronic content, assuming that $(\bar{U}\bar{U}\bar{U})$, $(\bar{U}\bar{U}\bar{u})$ and $(\bar{U}u)$ can be present with appreciable fraction, while the content of residual U-hadrons is likely to be realized with multiple U-states and with suppressed fraction of single U-states. Nevertheless we can not ignore single U-baryonic states $(Uud)^+$ because only reliable inference on their strong suppression would avoid opposing to strong constraint on $+1$ heavy particles abundance which will be considered below.

Radiative recombination

Radiative $U\bar{U}$ recombination is induced by "Coulomb-like" attraction of U and $\bar{U}$ due to their y-interaction. It can be described in the analogy to the process of free monopole-antimonopole annihilation considered in [17]. Potential energy of Coulomb-like interaction between U and $\bar{U}$ exceeds their thermal energy T at the distance

$$d_0 = \frac{\alpha}{T}.$$

In the case of y-interaction its running constant $\alpha = \alpha_y \sim 1/30$ [5]. For $\alpha \ll 1$, on the contrary to the case of monopoles [17] with $g^2/4\pi \gg 1$, the mean free path of multiple scattering in plasma is given by

$$\lambda = (n\sigma)^{-1} \sim \left(T^3 \cdot \frac{\alpha^2}{Tm}\right)^{-1} \sim \frac{m}{\alpha^3 T} \cdot d_0,$$

being $\lambda \gg d_0$ for all $T < m$. So the diffusion approximation [17] is not valid for our case. Therefore radiative capture of free U and $\bar{U}$ particles should be considered. According to [17], following the classical solution of energy loss due to radiation, converting infinite motion to finite, free U and $\bar{U}$ particles form bound systems at the impact parameter

$$a \approx (T/m)^{3/10} \cdot d_0. \tag{20}$$

The rate of such binding is then given by

$$\langle \sigma v \rangle = \pi a^2 v \approx \pi \cdot (m/T)^{9/10} \cdot \left(\frac{\alpha}{m}\right)^2 \approx \tag{21}$$

$$\approx 6 \cdot 10^{-13} \left(\frac{\alpha}{1/30}\right)^2 \left(\frac{300\,\mathrm{K}}{T}\right)^{9/10} \left(\frac{250\,\mathrm{GeV}}{m}\right)^{11/10} \frac{\mathrm{cm}^3}{\mathrm{s}}.$$

The successive evolution of this highly excited atom-like bound system is determined by the loss of angular momentum owing to y-radiation. The time scale for the fall on the center in this bound system, resulting in $U\bar{U}$ recombination was estimated according to classical formula in [18]

$$\tau = \frac{a^3}{64\pi} \cdot \left(\frac{m}{\alpha}\right)^2 = \frac{\alpha}{64\pi} \cdot \left(\frac{m}{T}\right)^{21/10} \cdot \frac{1}{m} \tag{22}$$

$$\approx 4 \cdot 10^{-4} \left(\frac{300\,\mathrm{K}}{T}\right)^{21/10} \left(\frac{m}{250\,\mathrm{GeV}}\right)^{11/10} \mathrm{s}.$$

As it is easily seen from Eq.(30) this time scale of $U\bar{U}$ recombination $\tau \ll m/T^2 \ll m_{Pl}/T^2$ turns to be much less than the cosmological time at which the bound system was formed.

The above classical description assumes $a = \frac{\alpha}{m^{3/10}T^{7/10}} \gg \frac{1}{\alpha m}$ and is valid at $T \ll m\alpha^{20/7}$ [14].

Kinetic equation for U-particle abundance with the account of radiative capture on RD stage is given by Eq.(11).

At $T < T_{rr} = \alpha_y^{20/7} m \approx 10\,\mathrm{MeV}(m/250\,\mathrm{GeV})\left(\frac{\alpha_y}{1/30}\right)^{20/7}$ the solution for the effect of radiative recombination is given by

$$r_4 \approx \frac{r_0}{1 + r_0\sqrt{\frac{20 g_{rr}\pi^3}{9}\frac{\alpha^2 m_{Pl}}{m}}\left(\frac{T_{rr}}{m}\right)^{1/10}} \approx r_0 \qquad (23)$$

with r_0 taken at $T = T_{rr}$ equal to r_4 from Eqs.(16),(17) or (19).

Owing to more rapid cosmological expansion radiative capture of U-hadrons in expanding matter on MD stage is less effective, than on RD stage. So the result $r_4 \approx r_0$ holds on MD stage with even better precision, than on RD stage. Therefore radiative capture does not change the estimation of U-hadron pregalactic abundance, given by Eqs.(16),(17) or (19).

On the galactic stage in the most of astrophysical bodies temperature is much less than T_{rr} and radiative recombination plays dominant role in the decrease of U-hadron abundance inside dense matter bodies.

U-hadrons during Big Bang Nucleosynthesis and thereafter

One reminds that to the beginning of Big Bang Nucleosynthesis (BBN) there can be $(Uud)^+$, $(UUu)^{++}$, $(UUU)^{++}$, $(\bar{U}u)^0$, $(\bar{U}\bar{U}\bar{U})^{--}$, $(\bar{U}\bar{U}\bar{u})^{--}$ states in plasma. We do not specify here possible fractions of each of the U-hadron species (i) in U-hadronic matter, assuming that any of them can be appreciable ($r_i \lesssim r_4$).

After BBN proceeded, the states $(\bar{U}\bar{U}\bar{U})^{--}$, $(\bar{U}\bar{U}\bar{u})^{--}$ are combined with ${}^4He^{++}$ due to electromagnetic interaction. The binding energy of the ground state can be estimated with reasonable accuracy following Bohr formulas (for point-like particles)

$$I_b = \frac{(Z_A Z_X \alpha)^2 m_A}{2} \approx 1.5\,\mathrm{MeV}, \qquad (24)$$

where $Z_X = 2$, $Z_A = 2$ and $m_A \approx 3.7$ GeV are the charges of U-hadron and Helium and the mass of the latter. Cross section of this recombination is estimated as [19]

$$\langle\sigma v\rangle = \frac{2^8\pi\sqrt{2\pi}\alpha^3 Z_A^4 Z^2}{3\exp(4) m_A\sqrt{m_A T}} \approx \frac{3.06\cdot 10^{-4}}{m_A\sqrt{m_A T}}. \qquad (25)$$

Evolution of abundance of U-hadrons combining with He is described by equation

$$\frac{dn_{(\bar{U}\bar{U}\bar{U})}}{dt} = -3Hn_{(\bar{U}\bar{U}\bar{U})} - \langle\sigma v\rangle\, n_{(\bar{U}\bar{U}\bar{U})} n_{He}. \qquad (26)$$

The term corresponding to disintegration of $[(\bar{U}\bar{U}\bar{U})He]$ is neglected, since the energy of thermal photons is insufficient to disintegrate $[(\bar{U}\bar{U}\bar{U})He]$ (the same for $[(\bar{U}\bar{U}\bar{u})He]$) in the ground state in this period. Following procedure Eqs.(9-11), we get

$$r_{(\bar{U}\bar{U}\bar{U})} = r_{(\bar{U}\bar{U}\bar{U})0} \exp\left(-\sqrt{\frac{\pi g}{45}} m_{Pl} \int_0^{T_0} r_{He} \langle\sigma v\rangle\, dT\right) \approx, \tag{27}$$

$$\approx r_{(\bar{U}\bar{U}\bar{U})0} \exp\left(-0.6 \cdot 10^{12}\right),$$

where $r_{He} \equiv n_{He}/s = Y_p/4 \cdot \eta \cdot n_\gamma^{mod}/s_{mod} \approx 5.2 \cdot 10^{-12}$, g follows from Eq.(6) and $T_0 = 100$ keV was taken. As one can see, Eq.(27) gives in this case strong exponential suppression of free $(\bar{U}\bar{U}\bar{U})$ (the same for $(\bar{U}\bar{U}\bar{u})$), while neutral $[(\bar{U}\bar{U}\bar{U})He]$ and $[(\bar{U}\bar{U}\bar{u})He]$ states, being one of the forms of O-helium [16, 20–26], catalyze additional annihilation of free U-baryons and formation of primordial heavy elements [27]. New type of nuclear reactions, catalyzed by O-helium, seem to change qualitatively the results of BBN, however (see Sec. 3. and arguments in [16, 20–27]) it does not lead to immediate contradiction with the observations.

On the base of existing results of investigation of hyper-nuclei [28], one can expect that the isoscalar state $\Lambda_U^+ = (Uud)^+$ can form stable bound state with ^4He due to nuclear interaction. The change of abundance of U-hyperons Λ_U^+ owing to their nuclear fusion with ^4He is described by Eq.(26,27), substituting $(\bar{U}\bar{U}\bar{U}) \leftrightarrow \Lambda_U^+$. Disintegration of $[\Lambda_U He]$ is also negligible, since the period, when BBN is finished, is of interest ($T < T_0 \ll I([\Lambda_U He])$).

Cross section for nuclear reaction of question can be represented in conventional parameterization through the so called astrophysical S-factor

$$\sigma = \frac{S(E)}{E} \exp\left(-\frac{2\pi\alpha Z_X Z_A}{v}\right), \tag{28}$$

where $E = \mu v^2/2$ with μ being reduced mass of interacting particles and v being their relative velocity. The exponent in Eq.(28) expresses penetration factor, suppressing cross section, which reflects repulsive character of Coulomb force contrary to the case of $(\bar{U}\bar{U}\bar{U})$. S-factor itself is unknown, being supposed $S(E \to 0) \to const.$ Averaging σv over Maxwell velocity distribution gives, using saddle point method,

$$\langle\sigma v\rangle \approx \frac{4v_0 \cdot S(E(v_0))}{\sqrt{3}T} \exp\left(-\frac{3\mu v_0^2}{2T}\right), \tag{29}$$

where $v_0 = \left(\frac{2\pi\alpha Z_A Z_X T}{\mu}\right)^{1/3}$.

Calculation gives that suppression of free Λ_U^+ on more than 20 orders of magnitude is reached at $S(E) \gtrsim 2\,\mathrm{MeV}\cdot$ barn. S-factor for reaction of ^4He production is typically distinguished by high magnitudes from those of other reactions and lies around $5 - 30$ MeV$\cdot$ barn [29]. However, reactions with γ in final state, which is assumed in our case ($\Lambda_U + {}^4He \to [\Lambda_U He] + \gamma$), have as a rule S-factor in 10^4 times smaller. Special conditions should be demanded from unknown for sure physics of Λ_U-nucleus interaction to provide a large suppression of Λ_U abundance. Such suppression is needed, as we will see below, to avoid contradiction with data on anomalous hydrogen abundance in terrestrial matter. The experimental constraints on anomalous lithium are less restrictive and can be satisfied in this case.

y-plasma

The existence of new massless U(1) gauge boson (y-photon) implies the presence of primordial thermal y-photon background in the Universe. Such background should be in equilibrium with ordinary plasma and radiation until the lightest particle bearing y-charge (4th neutrino) freezes out. For the accepted value of 4th neutrino mass ($\geq$ 50 GeV) 4th neutrino freezing out and correspondingly decoupling of y-photons takes place before the QCD phase transition, when the total number of effective degrees of freedom is sufficiently large to suppress the effects of y-photon background in the period of Big Bang nucleosynthesis. This background does not interact with nucleons and does not influence the BBN reactions rate (its possible effect in formation and role of $[^4He(\bar{U}\bar{U}\bar{U})]$ "atom" is discussed in [15]), while the suppression of y-photon energy density leads to insignificant effect in the speeding up cosmological expansion rate in the BBN period. In the framework of the present consideration the existence of primordial y-photons does not play any significant role in the successive evolution of U-hadrons.

Inclusion of stable y-charged 4th neutrinos strongly complicate the picture. Condition of cancellation of axial anomalies requires relationship between the values of y-charges of 4th generation leptons (N, E) and quarks (U, D) as the following

$$e_{yN} = e_{yE} = -e_{yU}/3 = -e_{yD}/3.$$

In course of cosmological combined evolution of U and N and y, "y-molecules" of kind U-U-U-N, where different U-quarks can belong to different U-hadrons (possibly bound with nucleus) should form. Such y-neutral molecules can avoid effect of U-hadrons suppression in the terrestrial matter, relevant in charge symmetric case, and lead to contradiction with observations, analysis of which is started now. UUU-N-type states will be considered in section 3. devoted to the charge-asymmetric case.

2.2. Evolution and Manifestations of U-hadrons at Galactic Stage

In the period of recombination of nuclei with electrons the positively charged U-baryons recombine with electrons to form atoms of anomalous isotopes. The substantial (up to 10 orders of magnitude) excess of electron number density over the number density of primordial U-baryons makes virtually all U-baryons to form atoms. The cosmological abundance of free charged U-baryons is to be exponentially small after recombination.

Hadrons (UUu), (UUU) form atoms of anomalous He at $T \sim 2\,\mathrm{eV}$ together with recombination of ordinary helium. The states $[(\bar{U}\bar{U}\bar{U})He]$, $[(\bar{U}\bar{U}\bar{u})He]$, $(\bar{U}u)$ escape recombination with electrons because of their neutrality; hadrons (Uud), if they are not involved into chain of nuclear transitions, form atoms of anomalous hydrogen.

The formed atoms, having atomic cross sections of interaction with matter follow baryonic matter in formation of astrophysical objects like gas clouds, stars and planets, when galaxies are formed.

On the contrary, O-helium and $(\bar{U}u)$ mesons, having nuclear and hadronic cross sections, respectively, can decouple from plasma and radiation at $T \sim 1\,\mathrm{keV}$ and behave in Galaxy as collisionless gas. In charge asymmetric case, considered in the next Section 3., or in charge symmetric case without y-interaction O-helium and $(\bar{U}u)$ mesons behave on this reason as collisionless gas of dark matter particles. On that reasons one can expect suppression of their concentration in baryonic matter.

However, in charge symmetric case with y-interaction, the existence of Coulomb-like y-attraction will make them to obey the condition of neutrality in respect to the y-charge. Therefore owing to neutrality condition the number densities of U- and $\bar{U}$-hadrons in astrophysical bodies should be equal. It leads to effects in matter bodies, considered in this subsection.

U-hadrons in galactic matter

In the astrophysical body with atomic number density n_a the initial U-hadron abundance $n_{U0} = f_{a0} \cdot n_a$ can decrease with time due to $U\bar{U}$ recombination. Here and in estimations thereafter we will refer to U-quark abundance as U-hadron one (as if all U-hadrons were composed of single U-quarks), if it is not specified otherwise.

Under the neutrality condition

$$n_U = n_{\bar{U}}$$

the relative U-hadron abundance $f_{a0} = n_U/n_a = n_{\bar{U}}/n_a$ is governed by the equation

$$\frac{df_a}{dt} = -f_a^2 \cdot n_a \cdot \langle \sigma v \rangle \,. \tag{30}$$

Here $\langle \sigma v \rangle$ is defined by Eq.(21). The solution of this equation is given by

$$f_a = \frac{f_{a0}}{1 + f_{a0} \cdot n_a \cdot \langle \sigma v \rangle \cdot t}. \tag{31}$$

If

$$n_a \cdot \langle \sigma v \rangle \cdot t \gg \frac{1}{f_{a0}}, \tag{32}$$

the solution (31) takes the form

$$f_a = \frac{1}{n_a \cdot \langle \sigma v \rangle \cdot t}. \tag{33}$$

and, being independent on the initial value, U-hadron abundance decreases inversely proportional to time.

By definition $f_{a0} = f_0/A_{atom}$, where A_{atom} is the averaged atomic weight of the considered matter and f_0 is the initial U-hadron to baryon ratio. In the pregalactic matter this ratio is determined by r_4 from A) Eq.(16), B) Eq.(17) and C) Eq.(19) and is equal to

$$f = \frac{r_4}{r_b} = \begin{cases} 10^{-10} & \text{for the case A,} \\ 3 \cdot 10^{-9} & \text{for the case B,} \\ 1.2 \cdot 10^{-6} & \text{for the case C.} \end{cases} \tag{34}$$

Here $r_b \approx 10^{-10}$ is baryon to entropy ratio.

Taking for averaged atomic number density in the Earth $n_a \approx 10^{23}\ \mathrm{cm}^{-3}$, one finds that during the age of the Solar system primordial U-hadron abundance in the terrestrial matter should have been reduced down to $f_a \approx 10^{-28}$. One should expect similar reduction of U-hadron concentration in Sun and all the other old sufficiently dense astrophysical bodies.

Therefore in our own body we might contain just one of such heavy hadrons. However, as shown later on, the persistent pollution from the galactic gas nevertheless may increase this relic number density to much larger value ($f_a \approx 10^{-23}$).

The principal possibility of strong reduction in dense bodies for primordial abundance of exotic charge symmetric particles due to their recombination in unstable charmonium like systems was first revealed in [30] for fractionally charged colorless composite particles (fractons).

The U-hadron abundance in the interstellar gas strongly depends on the matter evolution in Galaxy, which is still not known to the extent, we need for our discussion.

Indeed, in the opposite case of low density or of short time interval, when the condition (32) is not valid, namely, at

$$n_a < \frac{1}{f_{a0}\left<\sigma v\right>t} = A_{atom} \cdot \frac{T}{300\,\mathrm{K}} \cdot \frac{t_U}{t}\,\mathrm{cm}^{-3} \begin{cases} 4 \cdot 10^4 \text{ for the case A}, \\ 10^3 \text{ for the case B}, \\ 1.2 \text{ for the case C}, \end{cases} \tag{35}$$

where $t_U = 4 \cdot 10^{17}$ s is the age of the Universe, U-hadron abundance does not change its initial value.

In principle, if in the course of evolution matter in the forming Galaxy was present during sufficiently long period ($t \sim 10^9$ yrs) within cold ($T \sim 10$ K) clouds with density $n_a \sim 10^3\,\mathrm{cm}^{-3}$ U-hadron abundance retains its primordial value for the cases A and B ($f_0 = f$), but falls down $f_0 = 5 \cdot 10^{-9}$ in the case C, making this case close to the case B. The above argument may not imply all the U-hadrons to be initially present in cold clouds. They can pass through cold clouds and decrease their abundance in the case C at the stage of thermal instability, when cooling gas clouds, before they become gravitationally bound, are bound by the external pressure of the hot gas. Owing to their large inertia heavy U-hadron atoms from the hot gas can penetrate much deeply inside the cloud and can be captured by it much more effectively, than ordinary atoms. Such mechanism can provide additional support for reduction of U-hadron abundance in the case C. The reduction of this abundance down to values, corresponding to the case A can be also provided by O-helium catalysis in the period of BBN.

However, in particular, annihilation of U-hadrons leads to multiple γ production. If U-hadrons with relative abundance f annihilate at the redshift z, it should leave in the modern Universe a background γ flux [15]

$$F(E > E_\gamma) = \frac{N_\gamma \cdot f \cdot r_b \cdot s_{mod} \cdot c}{4\pi} \approx 3 \cdot 10^3 f\,(\mathrm{cm}^2 \cdot \mathrm{s} \cdot \mathrm{ster})^{-1},$$

of γ quanta with energies $E > E_\gamma = 10\,\mathrm{GeV}/(1+z)$. The numerical values for γ multiplicity N_γ are given in table 1 [15].

So annihilation even as early as at $z \sim 9$ leads in the case C to the contribution into diffuse extragalactic gamma emission, exceeding the flux, measured by EGRET by three orders of magnitude. The latter can be approximated as

$$F(E > E_\gamma) \approx 3 \cdot 10^{-6} \left(\frac{E_0}{E_\gamma}\right)^{1.1} (\mathrm{cm}^2 \cdot \mathrm{s} \cdot \mathrm{ster})^{-1},$$

Table 1. Multiplicities of γ produced in the recombination of $(Q\bar{Q})$ pair with $m = 250$ GeV for different energy intervals.

Energy fraction	> 0	> 0.1 GeV	> 1 GeV	> 10 GeV	> 100 GeV
N_γ	69	62	28	2.4	0.001

where $E_0 = 451\,\mathrm{MeV}$.

The above upper bound strongly restricts ($f \leq 10^{-9}$) the earliest abundance because of the consequent impossibility to reduce the primordial U-hadron abundance by U-hadron annihilation in low density objects. In the cases A and B annihilation in such objects should not take place, whereas annihilation within the dense objects, being opaque for γ radiation, can avoid this constraint due to strong suppression of outgoing γ flux. However, such constraint nevertheless should arise for the period of dense objects' formation. For example, in the course of protostellar collapse hydrodynamical timescale $t_H \sim 1/\sqrt{\pi G \rho} \sim 10^{15}\,\mathrm{s}/\sqrt{n}$ exceeds the annihilation timescale [15]

$$t_{an} \sim \frac{1}{fn\langle\sigma v\rangle} \sim \frac{10^{12}\,\mathrm{s}}{fn}\left(\frac{1/30}{\alpha}\right)^2\left(\frac{T}{300\,\mathrm{K}}\right)^{9/10}\left(\frac{m}{250\,\mathrm{GeV}}\right)^{11/10}$$

at $n > 10^{14}\left(\frac{10^{-10}}{f}\right)^2\left(\frac{1/30}{\alpha}\right)^4\left(\frac{T}{300\,\mathrm{K}}\right)^{9/5}\left(\frac{m}{250\,\mathrm{GeV}}\right)^{11/5}$, where n is in cm^{-3}. We consider homogeneous cloud with mass M has radius $R \approx \frac{10^{19}\,\mathrm{cm}}{n^{1/3}}\left(\frac{M}{M_\odot}\right)^{1/3}$, where $M_\odot = 2\cdot 10^{33}$ g is the Solar mass. It becomes opaque for γ radiation, when this radius exceeds the mean free path $l_\gamma \sim 10^{26}\,\mathrm{cm}/n$ at $n > 3\cdot 10^{10}\,\mathrm{cm}^{-3}\left(\frac{M}{M_\odot}\right)^{1/2}$. As a result, for f as large as in the case C, rapid annihilation takes place when the collapsing matter is transparent for γ radiation and the EGRET constraint can not be avoided. The cases A and B are consistent with this constraint.

Note that at $f < 5\cdot 10^{-6}\left(\frac{M}{M_\odot}\right)^{2/9}$, i.e. for all the considered cases energy release from U-hadron annihilation does not exceed the gravitational binding energy of the collapsing body. Therefore, U-hadron annihilation can not prevent the formation of dense objects but it can provide additional energy source, e.g. at early stages of evolution of first stars. Its burning is quite fast (few years) and its luminosity may be quite extreme, leading to a short inhibition of star formation [15]. Similar effects of dark matter annihilation were recently considered in [31].

Galactic blowing of U-baryon atoms polluting our Earth

Since the condition Eq.(35) is valid for the disc interstellar gas, having the number density $n_g \sim 1\,\mathrm{cm}^{-3}$ one can expect that the U-hadron abundance in it can decrease relative to the primordial value only due to enrichment of this gas by the matter, which has passed through stars and had the suppressed U-hadron abundance according to Eq.(33). Taking the factor of such decrease of the order of the ratio of total masses of gas and stars in Galaxy $f_g \sim 10^{-2}$ and accounting for the acceleration of the interstellar gas by Solar gravitational force, so that the infalling gas has velocity $v_g \approx 4.2\cdot 10^6$cm/s in vicinity of Earth's orbit, one obtains that the flux of U-hadrons coming with interstellar gas should be of the order of [15]

$$I_U = \frac{f f_g n_g v_g}{8\pi} \approx 1.5 \cdot 10^{-7} \frac{f}{10^{-10}} \left(\mathrm{sm}^2 \cdot \mathrm{s} \cdot \mathrm{ster}\right)^{-1}, \tag{36}$$

where f is given by the Eq.(34).

Presence of primordial U-hadrons in the Universe should be reflected by their existence in Earth's atmosphere and ground. However, according to Eq.(33) (see discussion in Section 2.2.) primordial terrestrial U-hadron content should strongly decrease due to radiative recombination, so that the U-hadron abundance in Earth is determined by the kinetic equilibrium between the incoming U-hadron flux and the rate of decrease of this abundance by different mechanisms.

In the successive analysis we'll concentrate our attention on the case, when the U baryon has charge $+2$, and $\bar{U}$-hadrons are electrically neutral. In this case U baryons look like superheavy anomalous helium isotopes.

Searches for anomalous helium were performed in series of experiments based on accelerator search [32], spectrometry technique [33] and laser spectroscopy [34]. From the experimental point of view an anomalous helium represents a favorable case, since it remains in the atmosphere whereas a normal helium is severely depleted in the terrestrial environment due to its light mass.

The best upper limits on the anomalous helium were obtain in [34]. It was found by searching for a heavy helium isotope in the Earth's atmosphere that in the mass range 5 GeV – 10000 GeV the terrestrial abundance (the ratio of anomalous helium number to the total number of atoms in the Earth) of anomalous helium is less than $(2-3)\cdot 10^{-19}$. The search in the atmosphere is reasonable because heavy gases are well mixed up to 80 km and because the heavy helium does not sink due to gravity deeply in the Earth and is homogeneously redistributed in the volume of the World Ocean at the timescale of 10^3 yr.

The kinetic equations, describing evolution of anomalous helium and $\bar{U}$-hadrons in matter have the form [15]

$$\frac{dn_U}{dt} = j_U - n_U \cdot n_{\bar{U}} \cdot \langle \sigma v \rangle - j_{gU} \tag{37}$$

for $\bar{U}$-hadron number density n and

$$\frac{dn_{\bar{U}}}{dt} = j_{\bar{U}} - n_{\bar{U}} \cdot n_U \cdot \langle \sigma v \rangle - j_{g\bar{U}} \tag{38}$$

for number density of anomalous helium n_U. Here j_U and $j_{\bar{U}}$ take into account the income of, correspondingly, U-baryons and $\bar{U}$-hadrons to considered region, the second terms on the right-hand-side of equations describe $U\bar{U}$ recombination and the terms j_{gU} and $j_{g\bar{U}}$ determine various mechanisms for outgoing fluxes, e.g. gravitationally driven sink of particles. The latter effect is much stronger for $\bar{U}$-hadrons due to much lower mobility of U-baryon atoms. However, long range Coulomb like interaction prevents them from sinking, provided that its force exceeds the Earth's gravitational force.

In order to compare these forces let's consider the World's Ocean as a thin shell of thickness $L \approx 4 \cdot 10^5$ cm with homogeneously distributed y charge, determined by distribution of U-baryon atoms with concentration n. The y-field outside this shell according to Gauss' law is determined by

$$2E_y S = 4\pi e_y n S L,$$

being equal to

$$E_y = 2\pi e_y n L.$$

In the result y force, exerting on $\bar{U}$-hadrons

$$F_y = e_y E_y,$$

exceeds gravitational force for U-baryon atom concentration

$$n > 10^{-7} \frac{m}{250\,\mathrm{GeV}} \frac{30^{-1}}{\alpha_y}\,\mathrm{cm}^{-3}. \tag{39}$$

Note that the mobility of U-baryon atoms and $\bar{U}$ hadrons differs by 10 order of magnitude, what can lead to appearance of excessive y-charges within the limits of (39). One can expect that such excessive charges arise due to the effective slowing down of U-baryon atoms in high altitude levels of Earth's atmosphere, which are transparent for $\bar{U}$ hadrons, as well as due to the 3 order of magnitude decrease of $\bar{U}$ hadrons when they enter the Earth's surface.

Under the condition of neutrality, which is strongly protected by Coulomb-like y-interaction, all the corresponding parameters for $\bar{U}$-hadrons and U-baryons in the Eqs.(37)-(38) are equal, if Eq.(39) is valid. Provided that the timescale of mass exchange between the Ocean and atmosphere is much less than the timescale of sinking, sink terms can be neglected.

The stationary solution of Eqs.(37)-(38) gives in this case

$$n = \sqrt{\frac{j}{\langle \sigma v \rangle}}, \tag{40}$$

where

$$j_U = j_{\bar{U}} = j \sim \frac{2\pi I_U}{L} = 10^{-12} \frac{f}{10^{-10}}\,\mathrm{cm}^{-3}\mathrm{s}^{-1} \tag{41}$$

and $\langle \sigma v \rangle$ is given by the Eq.(21). For $j \le 10^{-12} \frac{f}{10^{-10}}\,\mathrm{cm}^{-3}\mathrm{s}^{-1}$ and $\langle \sigma v \rangle$ given by Eq.(21) one obtains in water

$$n \le \sqrt{\frac{f}{10^{-10}}}\,\mathrm{cm}^{-3}.$$

It corresponds to terrestrial U-baryon abundance

$$f_a \le 10^{-23} \sqrt{\frac{f}{10^{-10}}},$$

being below the above mentioned experimental upper limits for anomalous helium ($f_a < 10^{-19}$) even for the case C with $f = 2 \cdot 10^{-6}$. In air one has

$$n \le 10^{-3} \sqrt{\frac{f}{10^{-10}}}\,\mathrm{cm}^{-3}.$$

For example in a cubic room of 3m size there are nearly 27 thousand heavy hadrons.

Note that putting formally in the Eq.(40) the value of $\langle \sigma v \rangle$ given by the Eq.(13) one obtains $n \leq 6 \cdot 10^2\,\mathrm{cm}^{-3}$ and $f_a \leq 6 \cdot 10^{-20}$, being still below the experimental upper limits for anomalous helium abundance. So the qualitative conclusion that recombination in dense matter can provide the sufficient decrease of this abundance avoiding the contradiction with the experimental constraints could be valid even in the absence of gauge y-charge and Coulomb-like y-field interaction for U-hadrons. It looks like the hadronic recombination alone can be sufficiently effective in such decrease. However, if we take the value of $\langle \sigma v \rangle$ given by the Eq.(14) one obtains n by the factor of $\frac{m_\rho}{m_\pi} \sim 5.5$ larger and $f_a \leq 3.3 \cdot 10^{-19}$, what exceeds the experimental upper limits for anomalous helium abundance. Moreover, in the absence of y-attraction there is no dynamical mechanism, making the number densities of U-baryons and $\bar{U}$-hadrons equal each other with high accuracy. So nothing seems to prevent in this case selecting and segregating U-baryons from $\bar{U}$-hadrons. Such segregation, being highly probable due to the large difference in the mobility of U-baryon atoms and $\bar{U}$ hadrons can lead to uncompensated excess of anomalous helium in the Earth, coming into contradiction with the experimental constraints.

Similar result can be obtained for any planet, having atmosphere and Ocean, in which effective mass exchange between atmosphere and Ocean takes place. There is no such mass exchange in planets without atmosphere and Ocean (e.g. in Moon) and U-hadron abundance in such planets is determined by the interplay of effects of incoming interstellar gas, $U\bar{U}$ recombination and slow sinking of U-hadrons to the centers of planets.

Radiative recombination here considered is not able to save the case when free single positively charged U-hadrons (Uud) are present with noticeable fraction among U-hadronic relics. It is the very strong constraint on anomalous hydrogen ($f_a < 10^{-30}$) [35] what is hardly avoidable. So, the model with free stable $+1$ charged relics which are not bound in nuclear systems with charge $\geq +2$ can be discarded.

To conclude this section we present the obtained constraints on the Fig.(1).

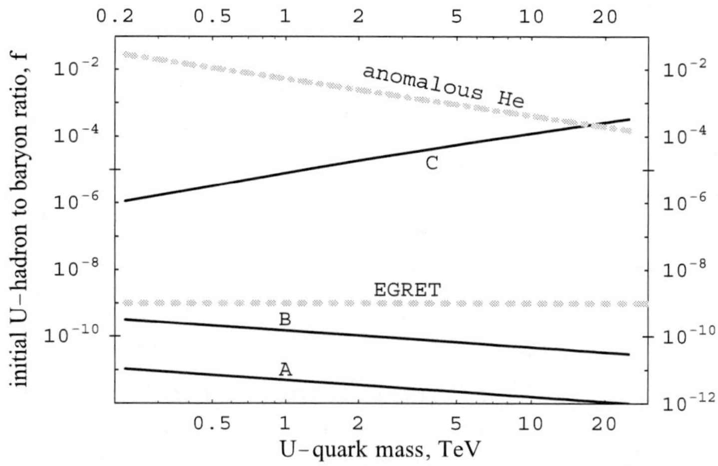

Figure 1. Upper constraints on relative abundances of U-hadrons (f) following from the search for anomalous helium in the Earth, from EGRET, constraining possible annihilation effect on pregalactic stage, in comparison with prediction f in three cases (A,B,C).

Cosmic rays experiments
In addition to the possible traces of U-hadrons existence in the Earth, they can be manifested in cosmic rays.

According to the arguments in previous section U-baryon abundance in the primary cosmic rays can be close to the primordial value f. It gives for case B

$$f = \frac{r_4}{r_b} \sim 3 \cdot 10^{-9}. \tag{42}$$

If U-baryons have mostly the form (UUU), its fraction in cosmic ray helium component can reach in this case the value

$$\frac{(UUU)}{^4He} \sim 3 \cdot 10^{-8},$$

which is accessible for cosmic rays experiments, such as RIM-PAMELA, being under run, and AMS 02 on International Space Station.

Similar argument in the case C would give for this fraction $\sim 2 \cdot 10^{-5}$, what may be already excluded by the existing data. However, it should be noted that the above estimation assumes significant contribution of particles from interstellar matter to cosmic rays. If cosmic ray particles are dominantly originated from the purely stellar matter, the decrease of U-hadron abundance in stars would substantially reduce the primary U-baryon fraction of cosmic rays. But in cosmic rays there are many different kind of particles. In order to differ U-hadrons from background events we can use the dependence of rigidity on velocity given on Fig.2.

The expected signal will strongly differ from background events in terms of relation between velocity and rigidity (momentum related to the charge). In the experiment PAMELA velocity is measured with a good accuracy, what can lead to the picture Fig.(2).

Correlation between cosmic ray and large volume underground detectors' effects
Inside large volume underground detectors (as Super Kamiokande) and in their vicinity U-hadron recombination should cause specific events ("spherical" energy release with zero total momentum or "wide cone" energy release with small total momentum), which could be clearly distinguished from the (energy release with high total momentum within "narrow cone") effects of common atmospheric neutrino - nucleon-lepton chain (as well as of hypothetical WIMP annihilation in Sun and Earth) [15].

The absence of such events inside 22 kilotons of water in Super Kamiokande (SK) detector during 5 years of its operation would give the most severe constraint

$$n < 10^{-3}\,\mathrm{cm}^{-3},$$

corresponding to $f_a < 10^{-26}$. For the considered type of anomalous helium such constraint would be by 7 order of magnitude stronger, than the results of present direct searches and 3 orders above our estimation in previous Section.

However, this constraint assumes that distilled water in SK does still contain polluted heavy hadrons (as it may be untrue). Nevertheless even for pure water it may not be the case for the detector's container and its vicinity. The conservative limit follows from the condition that the rate of U-hadron recombination in the body of detector does not exceed the rate

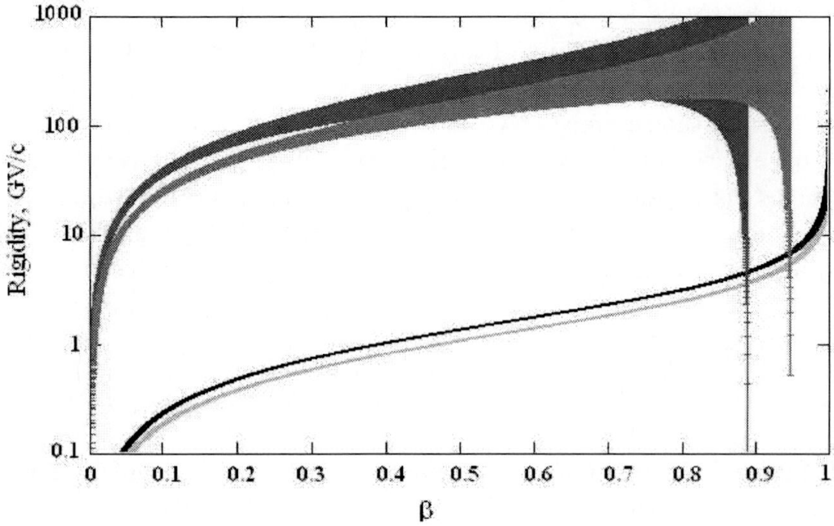

Figure 2. Lines show the region of expected signal, their thickness reflects expected accuricy in experiment PAMELA. Upper thick curves relate to anomalous helium of 500 and 750 GeV (a little upper and darker). The low thin curves relate to usual nuclei.

of processes, induced by atmospheric muons and neutrinos. The presence of clustered-like muons originated on the SK walls would be probably observed.

High sensitivity of large volume detectors to the effects of U-hadron recombination together with the expected increase of volumes of such detectors up to 1 km^3 offer the possibility of correlated search for cosmic ray U-hadrons and for effects of their recombination.

During one year of operation a 1 km^3 detector could be sensitive to effects of recombination at the U-hadron number density $n \approx 7 \cdot 10^{-6}\,\mathrm{cm}^{-3}$ and $f_a \approx 7 \cdot 10^{-29}$, covering the whole possible range of these parameters, since this level of sensitivity corresponds to the residual concentration of primordial U-hadrons, which can survive inside the Earth. The income of cosmic U-hadrons and equilibrium between this income and recombination should lead to increase of effect, expected in large volume detectors.

Even, if the income of anomalous helium with interstellar gas is completely suppressed, pollution of Earth by U-hadrons from primary cosmic rays is possible. The minimal effect of pollution by U-hadron primary cosmic rays flux I_U corresponds to the rate of increase of U-hadron number density $j \sim \frac{2\pi I_U}{R_E}$, where $R_E \approx 6 \cdot 10^8\,\mathrm{cm}$ is the Earth's radius. If incoming cosmic rays doubly charged U-baryons after their slowing down in matter recombine with electrons we should take instead of R_E the Ocean's thickness $L \approx 4 \cdot 10^5\,\mathrm{cm}$ that increases by 3 orders of magnitude the minimal flux and the minimal number of events, estimated below. Equilibrium between this income rate and the rate of recombination should lead to $N \sim jVt$ events of recombination inside the detector with volume V during its operation time t.

For the minimal flux of cosmic ray U-hadrons, accessible to AMS 02 experiment during 3 years of its operation ($I_{min} \sim 10^{-9} I_\alpha \sim 4 \cdot 10^{-11} I(E)$, in the range of energy per nucleon $1 < E < 10\,\mathrm{GeV}$) the minimal number of events expected in detector of volume V during time t is given by $N_{min} \sim \frac{2\pi I_{min}}{R_E} Vt$. It gives about 3 events per 10 years in

SuperKamiokande ($V = 2.2 \cdot 10^{10}\,\mathrm{cm}^3$) and about 10^4 events in the 1 km^3 detector during one year of its operation. The noise of this rate is one order and half below the expected influence of atmospheric ν_μ.

The possibility of such correlation facilitates the search for anomalous helium in cosmic rays and for the effects of U-hadron recombination in the large volume detectors.

The previous discussion assumed the lifetime of U-quarks τ exceeding the age of the Universe t_U. In the opposite case $\tau < t_U$ all the primordial U hadrons should decay to the present time and the cosmic ray interaction may be the only source of cosmic and terrestrial U hadrons.

3. The Case of a Charge-Asymmetry of U-quarks

The model [15] admits that in the early Universe an antibaryon asymmetry for 4th generation quarks can be generated [16,20,21]. Due to y-charge conservation $\bar{U}$ excess should be compensated by $\bar{N}$ excess. We will focus our attention here to the case of y-charged quarks and neutrinos of 4th generation and follow [16] in our discussion. All the main results concerning observational effects, presented here, can be generalized for the case without y-interaction.

$\bar{U}$-antibaryon density can be expressed through the modern dark matter density $\Omega_{\bar{U}} = k \cdot \Omega_{CDM} = 0.224$ ($k \leq 1$), saturating it at $k = 1$. It is convenient to relate the baryon (corresponding to $\Omega_b = 0.044$) and $\bar{U}$ ($\bar{N}$) excess with the entropy density s, introducing $r_b = n_b/s$ and $r_{\bar{U}} = n_{\bar{U}}/s = 3 \cdot n_{\bar{N}}/s = 3 \cdot r_{\bar{N}}$. One obtains $r_b \sim 8 \cdot 10^{-11}$ and $r_{\bar{U}}$, corresponding to $\bar{U}$ excess in the early Universe $\kappa_{\bar{U}} = r_{\bar{U}} - r_U = 3 \cdot (r_{\bar{N}} - r_N) = 10^{-12}(350\,\mathrm{GeV}/m_U) = 10^{-12}/S_5$, where $S_5 = m_U/350\,\mathrm{GeV}$.

Primordial composite forms of 4th generation dark matter

In the early Universe at temperatures highly above their masses $\bar{U}$ and $\bar{N}$ were in thermodynamical equilibrium with relativistic plasma. It means that at $T > m_U$ ($T > m_N$) the excessive $\bar{U}$ ($\bar{N}$) were accompanied by $U\bar{U}$ ($N\bar{N}$) pairs.

Due to $\bar{U}$ excess frozen out concentration of deficit U-quarks is suppressed at $T < m_U$ for $k > 0.04$ [21]. It decreases further exponentially first at $T \sim I_U \approx \bar{\alpha}^2 M_U/2 \sim 3S_5\mathrm{GeV}$ (where [15] $\bar{\alpha} = C_F\alpha_c = 4/3 \cdot 0.144 \approx 0.19$ and $M_U = m_U/2$ is the reduced mass), when the frozen out U quarks begin to bind with antiquarks $\bar{U}$ into charmonium-like state $(\bar{U}U)$ and annihilate. On this line $\bar{U}$ excess binds at $T < I_U$ by chromo-Coulomb forces dominantly into $(\bar{U}\bar{U}\bar{U})$ anutium states with electric charge $Z_\Delta = -2$ and mass $m_o = 1.05 S_5\mathrm{TeV}$, while remaining free $\bar{U}$ anti-quarks and anti-diquarks $(\bar{U}\bar{U})$ form after QCD phase transition normal size hadrons $(\bar{U}u)$ and $(\bar{U}\bar{U}\bar{u})$. Then at $T = T_{QCD} \approx 150\mathrm{MeV}$ additional suppression of remaining U-quark hadrons takes place in their hadronic collisions with $\bar{U}$-hadrons, in which $(\bar{U}U)$ states are formed and U-quarks successively annihilate.

Effect of $\bar{N}$ excess in the suppression of deficit N takes place at $T < m_N$ for $k > 0.02$ [21]. At $T \sim I_{NN} = \alpha_y^2 M_N/4 \sim 15\mathrm{MeV}$ (for $\alpha_y = 1/30$ and $M_N = 50\mathrm{GeV}$) due to y-interaction the frozen out N begin to bind with $\bar{N}$ into charmonium-like states $(\bar{N}N)$ and annihilate. At $T < I_{NU} = \alpha_y^2 M_N/2 \sim 30\mathrm{MeV}$ y-interaction causes binding of N with $\bar{U}$-hadrons (dominantly with anutium) but only at $T \sim I_{NU}/30 \sim 1\mathrm{MeV}$ this binding is not prevented by back reaction of y-photo-destruction.

To the period of Standard Big Bang Nucleosynthesis (SBBN) $\bar{U}$ are dominantly bound in anutium $\Delta^{--}_{3\bar{U}}$ with small fraction ($\sim 10^{-6}$) of neutral ($\bar{U}u$) and doubly charged ($\bar{U}\bar{U}\bar{u}$) hadron states. The dominant fraction of anutium is bound by y-interaction with $\bar{N}$ in ($\bar{N}\Delta^{--}_{3\bar{U}}$) "atomic" state. Owing to early decoupling of y-photons from relativistic plasma presence of y-radiation background does not influence SBBN processes [15, 16, 20].

At $T < I_o = Z^2 Z_{He}^2 \alpha^2 m_{He}/2 \approx 1.6$MeV the reaction $\Delta^{--}_{3\bar{U}} + ^4He \rightarrow \gamma + (^4He^{++}\Delta^{--}_{3\bar{U}})$ might take place, but it can go only after ^4He is formed in SBBN at $T < 100$keV and is effective only at $T \le T_{rHe} \sim I_o/\log{(n_\gamma/n_{He})} \approx I_o/27 \approx 60$keV, when the inverse reaction of photo-destruction cannot prevent it [14, 20, 22, 36]. In this period anutium is dominantly bound with $\bar{N}$. Since $r_{He} = 0.1 r_b \gg r_\Delta = r_{\bar{U}}/3$, in this reaction all free negatively charged particles are bound with helium [14, 20, 22, 36] and neutral Anti-Neutrino-O-helium (ANO-helium, $ANOHe$) ($^4He^{++}[\bar{N}\Delta^{--}_{3\bar{U}}]$) "molecule" is produced with mass $m_{OHe} \approx m_o \approx 1 S_5$TeV. The size of this "molecule" is $R_o \sim 1/(Z_\Delta Z_{He} \alpha m_{He}) \approx 2 \cdot 10^{-13}$ cm and it can play the role of a dark matter component and a nontrivial catalyzing role in nuclear transformations.

In nuclear processes ANO-helium looks like an α particle with shielded electric charge. It can closely approach nuclei due to the absence of a Coulomb barrier and opens the way to form heavy nuclei in SBBN. This path of nuclear transformations involves the fraction of baryons not exceeding 10^{-7} [20] and it can not be excluded by observations.

ANO-helium catalyzed processes

As soon as ANO-helium is formed, it catalyzes annihilation of deficit U-hadrons and N. Charged U-hadrons penetrate neutral ANO-helium, expel ^4He, bind with anutium and annihilate falling down the center of this bound system. The rate of this reaction is $\langle \sigma v \rangle = \pi R_o^2$ and an $\bar{U}$ excess $k = 10^{-3}$ is sufficient to reduce the primordial abundance of (Uud) below the experimental upper limits. N capture rate is determined by the size of ($\bar{N}\Delta$) "atom" in ANO-helium and its annihilation is less effective.

The size of ANO-helium is of the order of the size of ^4He and for a nucleus A with electric charge $Z > 2$ the size of the Bohr orbit for a ($Z\Delta$) ion is less than the size of nucleus A. This means that while binding with a heavy nucleus Δ penetrates it and effectively interacts with a part of the nucleus with a size less than the corresponding Bohr orbit. This size corresponds to the size of ^4He, making O-helium the most bound ($Z\Delta$)-atomic state.

The cross section for Δ interaction with hadrons is suppressed by factor $\sim (p_h/p_\Delta)^2 \sim (r_\Delta/r_h)^2 \approx 10^{-4}/S_5^2$, where p_h and p_Δ are quark transverse momenta in normal hadrons and in anutium, respectively. Therefore anutium component of ($ANOHe$) can hardly be captured and bound with nucleus due to strong interaction. However, interaction of the ^4He component of ($ANOHe$) with a ^A_ZQ nucleus can lead to a nuclear transformation due to the reaction $^A_ZQ + (\Delta He) \rightarrow ^{A+4}_{Z+2} Q + \Delta$, provided that the masses of the initial and final nuclei satisfy the energy condition $M(A, Z) + M(4, 2) - I_o > M(A + 4, Z + 2)$, where $I_o = 1.6$MeV is the binding energy of O-helium and $M(4, 2)$ is the mass of the ^4He nucleus. The final nucleus is formed in the excited $[\alpha, M(A, Z)]$ state, which can rapidly experience α- decay, giving rise to ($ANOHe$) regeneration and to effective quasi-elastic process of ($ANOHe$)-nucleus scattering. It leads to possible suppression of ANO-helium catalysis of nuclear transformations in matter.

ANO-helium dark matter

At $T < T_{od} \approx 1\,\mathrm{keV}$ energy and momentum transfer from baryons to ANO-helium $n_b \langle \sigma v \rangle (m_p/m_o)t < 1$ is not effective. Here $\sigma \approx \sigma_o \sim \pi R_o^2 \approx 10^{-25}\,\mathrm{cm}^2$. and $v = \sqrt{2T/m_p}$ is baryon thermal velocity. Then ANO-helium gas decouples from plasma and radiation and plays the role of dark matter, which starts to dominate in the Universe at $T_{RM} = 1\,\mathrm{eV}$.

The composite nature of ANO-helium makes it more close to warm dark matter. The total mass of (OHe) within the cosmological horizon in the period of decoupling is independent of S_5 and given by

$$M_{od} = \frac{T_{RM}}{T_{od}} m_{Pl} (\frac{m_{Pl}}{T_{od}})^2 \approx 2 \cdot 10^{42}\,\mathrm{g} = 10^9 M_{\odot}.$$

O-helium is formed only at $T_o = 60\,\mathrm{keV}$ and the total mass of OHe within cosmological horizon in the period of its creation is $M_o = M_{od}(T_o/T_{od})^3 = 10^{37}\,\mathrm{g}$. Though after decoupling Jeans mass in (OHe) gas falls down $M_J \sim 3 \cdot 10^{-14} M_{od}$ one should expect strong suppression of fluctuations on scales $M < M_o$ as well as adiabatic damping of sound waves in RD plasma for scales $M_o < M < M_{od}$. It provides suppression of small scale structure in the considered model. This dark matter plays dominant role in formation of large scale structure at $k > 1/2$.

The first evident consequence of the proposed scenario is the inevitable presence of ANO-helium in terrestrial matter, which is opaque for $(ANOHe)$ and stores all its infalling flux. If its interaction with matter is dominantly quasi-elastic, this flux sinks down the center of Earth. If ANO-helium regeneration is not effective and Δ remains bound with heavy nucleus Z, anomalous isotope of $Z-2$ element appears. This is the serious problem for the considered model.

Even at $k = 1$ ANO-helium gives rise to less than 0.1 [20, 21] of expected background events in XQC experiment [37], thus avoiding for all $k \leq 1$ severe constraints on Strongly Interacting Massive particles SIMPs obtained in [3] from the results of this experiment. In underground detectors $(ANOHe)$ "molecules" are slowed down to thermal energies far below the threshold for direct dark matter detection. However, $(ANOHe)$ destruction can result in observable effects. Therefore a special strategy in search for this form of dark matter is needed. An interesting possibility offers development of superfluid ^3He detector [38]. Due to high sensitivity to energy release above ($E_{th} = 1\,\mathrm{keV}$), operation of its actual few gram prototype can put severe constraints on a wide range of k and S_5.

At $10^{-3} < k < 0.02$ U-baryon abundance is strongly suppressed [16, 21], while the modest suppression of primordial N abundance does not exclude explanation of DAMA, HEAT and EGRET data in the framework of hypothesis of 4th neutrinos but makes the effect of N annihilation in Earth consistent with the experimental data.

4. Signatures for U-hadrons in Accelerator Experiments

Metastable U-quark within a wide range of expected mass can be searched on LHC and Tavatron. In spite on that its mass can be quite close to that of t-quark, strategy of their search should be completely different. U-quark in framework of considered model is metastable and will form metastable hadrons at accelerator contrary to t-quark.

Detailed analysis of possibility of U-quark search requires quite deep understanding of physics of interaction between metastable U-hadrons and nucleons of matter. However, strategy of U-quark search can be described in general outline, by knowing mass spectrum of U-hadrons, (differential) cross sections of their production. LHC certainly will provide a better possibility for U-quark search than Tevatron. Cross section of U-quark production in pp-collisions at the center mass energy 14 TeV is presented on the Fig. 3. For comparison, cross sections of 4th generation leptons are shown too. Cross sections of U- and D- quarks does not virtually differ.

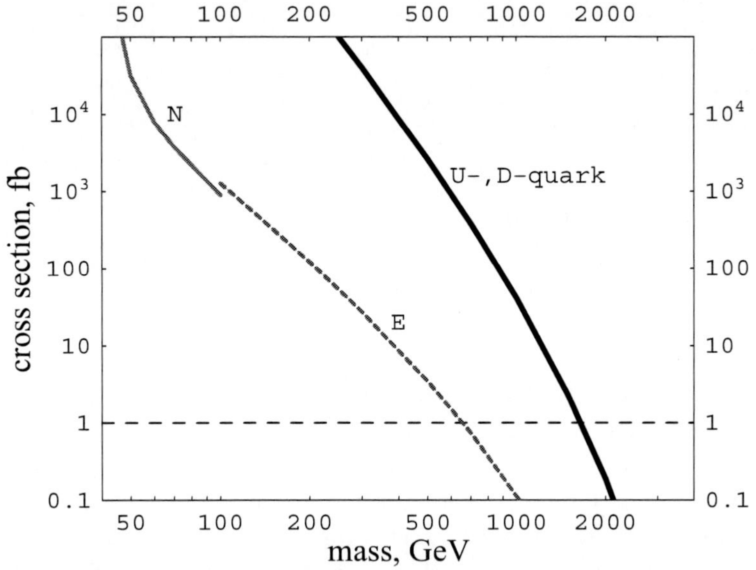

Figure 3. Cross sections of production of 4th generation particles (N, E, U, D) at LHC. Horisontal dashed line shows approximate level of sensitivity to be reached after first year of LHC operation.

Heavy metastable quarks will be produced with high transverse momentum p_T, velocity less than speed of light. In general, simultaneous measurement of velocity and momentum enables to find the mass of particle. Information on ionization losses is, as a rule, not so good thereto. All these features are typical for any heavy particle, while there can be subtle differences in the shapes of its angle-, p_T-distributions, defined by concrete model which it predicts.

It is peculiarities of long-lived hadronic nature what can be of special importance for clean selection of events of U-quarks creation. U-quark can form a whole class of U-hadronic states which can be perceived as stable in condition of experiment contrary to their relics in Universe. However, as we pointed out, double, triple U-hadronic states cannot be virtually created in collider. Many other hadronic states whose lifetime is $\gtrsim 10^{-7}$ s should look like stable. In the Table 2 expected mass spectrum of U-hadrons, obtained with the help of code Pythia [39], is presented.

The lower indeces in notation of U-hadrons in the Table 2 mean (iso)spin (I) of the light quark pair. From comparison of masses of different U-hadrons it follows that all $I = 1$ U-

Table 2. Mass spectrum and relative yields in LHC for U-hadrons. The same is for charged conjugated states.

	Difference between the masses of U-hadron and U-quark, GeV	Expected yields (in the right columns the yields of long-lived states are given)	
$\{U\bar{u}\}^0, \{U\bar{d}\}^+$	0.330	39.5(3)%, 39.7(3)%	
$\{U\bar{s}\}^+$	0.500	11.6(2)%	
$\{Uud\}^+$	0.579	5.3(1)%	7.7(1)%
$\{Uuu\}_1^{++}, \{Uud\}_1^+, \{Udd\}_1^0$	0.771	0.76(4)%, 0.86(5)%, 0.79(4)%	
$\{Usu\}^+, \{Usd\}^0$	0.805	0.65(4)%, 0.65(4)%	1.51(6)%
$\{Usu\}_1^+, \{Usd\}_1^0$	0.930	0.09(2)%, 0.12(2)%	
$\{Uss\}_{(1)}^0$	1.098	0.005(4)%	

hadrons decay quikly emitting π-meson or γ-quantum, except (Uss)-state. In the right column the expected relative yields are present. Unstable $I = 1$ U-hadrons decay onto respective $I = 0$ states, increasing their yields.

Firstly one makes a few notes. There are two mesonic states being quasi-degenerated in mass: $(U\bar{u})$ and $(U\bar{d})$ (we skip here discussion of strange U-hadrons). In interaction with medium composed of u and d quarks transformations of U-hadrons into those ones containing u and d are preferable (as it is the case in early Universe). From these it follows, that created pair of $U\bar{U}$ quarks will fly out from the vertex of pp-collision in form of U-hadron with positive charge in about 60% of such events and with neutral charge in 40% and in form of anti-U-hadron with negative charge in 60% and neutral in 40%. After traveling through detectors a few nuclear lengths from vertex, U-hadron will transform in (roughly) 100% to positively charged hadron (Uud) whereas anti-U-hadron will transform in 50% to negatively charged U-hadron $(\bar{U}d)$ and in 50% to neutral U-hadron $(\bar{U}u)$.

This feature will enable to discriminate the considered model of U-quarks from variety of alternative models, predicting new heavy stable particles.

Note that if the mass of Higgs boson exceeds $2m$, its decay channel into the pair of stable $Q\bar{Q}$ will dominate over the $t\bar{t}$, $2W$, $2Z$ and invisible channel to neutrino pair of 4th generation [40]. It may be important for the strategy of heavy Higgs searches.

5. Conclusion

To conclude, the existence of hidden stable or metastable quark of 4th generation can be compatible with the severe experimental constraints on the abundance of anomalous isotopes in Earths atmosphere and ground and in cosmic rays, even if the lifetime of such quark exceeds the age of the Universe. Though the primordial abundance $f = r_4/r_b$ of hadrons, containing such quark (and antiquark) can be hardly less than $f \sim 10^{-10}$ in case of charge symmetry, their primordial content can strongly decrease in dense astrophysical objects (in the Earth, in particular) owing to the process of recombination, in which hadron, containing quark, and hadron, containing antiquark, produce unstable charmonium-like quark-antiquark state.

To make such decrease effective, the equal number density of quark- and antiquark-containing hadrons should be preserved. It appeals to a dynamical mechanism, preventing segregation of quark- and antiquark- containing hadrons. Such mechanism, simultaneously providing strict charge symmetry of quarks and antiquarks, naturally arises, if the 4th generation possesses new strictly conserved U(1) gauge (y-) charge. Coulomb-like y-charge long range force between quarks and antiquarks naturally preserves equal number densities for corresponding hadrons and dynamically supports the condition of y-charge neutrality.

It was shown in the present paper that if U-quark is the lightest quark of the 4th generation, and the lightest free U-hadrons are doubly charged (UUU)- and (UUu)-baryons and electrically neutral $(U\bar{u})$-meson, the predicted abundance of anomalous helium in Earths atmosphere and ground as well as in cosmic rays is below the existing experimental constraints but can be within the reach for the experimental search in future. To realize this possibility nuclear binding of all the (Uud)-baryons with primordial helium is needed, converting potentially dangerous form of anomalous hydrogen into less dangerous anomalous lithium. Then the whole cosmic astrophysics and present history of these relics are puzzling and surprising, but nearly escaping all present bounds.

Searches for anomalous isotopes in cosmic rays and at accelerators were performed during last years. Stable doubly charged U baryons offer challenge for cosmic ray and accelerator experimental search as well as for increase of sensitivity in searches for anomalous helium. In particular, they seem to be of evident interest for cosmic ray experiments, such as PAMELA and AMS02. $+2$ charged U baryons represent the low Z/A anomalous helium component of cosmic rays, whereas -2 charged $\bar{U}$ baryons look like anomalous antihelium nuclei. In the baryon asymmetrical Universe the predicted amount of primordial single $(\bar{U}\bar{u}\bar{d})$ baryons is exponentially small, whereas their secondary fluxes originated from cosmic ray interaction with the galactic matter are predicted at the level, few order of magnitude below the expected sensitivity of future cosmic ray experiments. The same is true for cosmic ray $+2$ charged U baryons, if U-quark lifetime is less than the age of the Universe and primordial U baryons do not survive to the present time.

The models of quark interactions favor isoscalar (Uud) baryon to be the lightest among the 4th generation baryons (provided that U quark is lighter, than D quark, what also may not be the case). If the lightest U-hadrons have electric charge $+1$ and survive to the present time, their abundance in Earth would exceed the experimental constraint on anomalous hydrogen. This may be rather general case for the lightest hadrons of the 4th generation. To avoid this problem of anomalous hydrogen overproduction the lightest quark of the 4th generation should have the lifetime, less than the age of the Universe. Another possible solution of this problem, using double and triple $\bar{U}$ baryons $(\bar{U}\bar{U}\bar{U})$ and catalysis of (Uud) annihilation in atom-like bound systems ${}^4He(\bar{U}\bar{U}\bar{U})$ is considered in [15].

However short-living are these quarks on the cosmological timescale in a very wide range of lifetimes they should behave as stable in accelerator experiments. For example, with an intermediate scale of about 10^{11} GeV (as in supersymmetry models [41]) the expected lifetime of U-(or D-) quark $\sim 10^6$ years is much less than the age of the Universe but such quark is practically stable in any collider experiments.

First year operation of the accelerator LHC has good discovery potential for U(D)-quarks with mass up to 1.5 TeV. U-hadrons born at accelerator will distinguish oneself by high p_t, low velocity, by effect of a charge flipping during their propagation through the detectors. All these features enable strongly to increase efficiency of event selection from not only background but also from alternative hypothesis.

In the present work we studied effects of 4th generation having restricted our analysis by the processes with 4th generation quarks and antiquarks. However, as we have mentioned in the Introduction in the considered approach absolutely stable neutrino of 4th generation with mass about 50 GeV also bears y-charge. The selfconsistent treatment of the cosmological evolution and astrophysical effects of y-charge plasma of neutrinos, antineutrinos, quarks and antiquarks of 4th generation in charge symmetric case will be the subject of special studies. An attempt of such a treatment has been undertaken in the case of charge asymmetry, described in this paper.

We believe that a tiny trace of heavy hadrons as anomalous helium and stable neutral O−Helium and mesons[1] may be hidden at a low level in our Universe ($\frac{n_U}{n_b} \sim 10^{-10} - 10^{-9}$) and even at much lower level here in our terrestrial matter a density $\frac{n_U}{n_b} \sim 10^{-23}$ in case of charge symmetry. There are good reasons to bound the 4th quark mass below TeV energy. Therefore the mass window and relic density is quite narrow and well defined, open to a final test.

In case of charge asymmetry of 4th generation quarks, a nontrivial solution of the problem of dark matter (DM) can be provided due to neutral O−Helium-like U-hadrons states (ANO-helium in case of y-interaction existence). Such candidates to DM have many interesting implications in BBN, large scale structure of Universe and physics of DM [16, 20–26]. It should catalyze new types of nuclear transformations, reminding alchemists' dream on the philosopher's stone. It challenges direct search for species of such composite dark matter and its constituents. A very low probability for their existence is strongly compensated by the expectation value of their discovery.

Acknowledgements

We are grateful to G. Dvali for reading the manuscript and important recommendations

References

[1] Dimopoulos, S.; et al. *Phys Rev* 1990, vol D41, 2388.

[2] Dover, C.B.; Gaisser, T.K.; Steigman, G. *Phys Rev Lett* 1979, vol 42, 1117. Wolfram, S. *Phys Lett* 1979, vol B82, 65. Starkman, G.D.; et al. *Phys Rev* 1990, vol D41, 3594. Javorsek, D.; et al. *Phys Rev Lett* 2001, vol 87, 231804. Mitra, S. *Phys Rev* 2004, vol

[1] Storing these charged and neutral heavy hadrons in the matter might influence its e/m properties, leading to the appearance of apparent fractional charge effect in solid matter [15]. The present sensitivity for such effect in metals ranges from 10^{-22} to 10^{-20} .

D70, 103517; arXiv:astro-ph/0408341. Mack, G.D.; Beacom, J.F.; Bertone, G. *Phys Rev* 2007, vol D76, 043523; arXiv:0705.4298 [astro-ph].

[3] Wandelt, B.D.; et al. (2000). arXiv:astro-ph/0006344. McGuire, P.C.; Steinhardt, P.J. (2001). arXiv:astro-ph/0105567. Zaharijas, G.; Farrar, G.R. *Phys Rev* 2005, vol D72, 083502; arXiv:astro-ph/0406531.

[4] Dvali, G.R. *Phys Lett* 1992, vol B287, 101. Dvali, G.R. *Phys Lett* 1996, vol B372, 113; arXiv:hep-ph/9511237. Barbieri, R.; Dvali, G.R.; Strumia, A. *Nucl Phys* 1993, vol B391, 487.

[5] Khlopov, M.Yu.; Shibaev, K.I. Gravitation and Cosmology Suppl 2002, vol 8, 45.

[6] Belotsky, K.M.; Khlopov, M.Yu.; Shibaev, K.I. *Gravitation and Cosmology Suppl* 2000, vol 6, 140.

[7] Fargion, D.; et al. *JETP Letters* 1999, vol 69, 434; arXiv:astro/ph-9903086.

[8] Fargion, D.; et al. *Astropart Phys* 2000, vol 12, 307; arXiv:astro-ph/9902327.

[9] Belotsky, K.M.; Khlopov, M.Yu. *Gravitation and Cosmology Suppl* 2002, vol 8, 112.

[10] Belotsky, K.M.; Khlopov, M.Yu. *Gravitation and Cosmology* 2001, vol 7, 189.

[11] Maltoni, M.; et al. *Phys Lett* 2000, vol B476, 107. Ilyin, V.A.; et al. *Phys Lett* 2001, vol B503, 126. Novikov, V.A.; et al. Phys Lett 2002, vol B529, 111; *JETP Lett* 2002, vol 76, 119.

[12] Yao, W.M.; et al. (Particle Data Group) *Journal of physics G* 2006, vol 33, 1.

[13] Glashow, S.L. (2005). arXiv:hep-ph/0504287.

[14] Fargion, D.; Khlopov, M. (2005). arXiv:hep-ph/0507087.

[15] Belotsky, K.M.; Fargion, D.; Khlopov, M.Yu.; Konoplich, R.; Ryskin, M.; Shibaev, K. Gravitation and Cosmology 2005, vol 11, 3-15; arXiv:hep-ph/0411271.

[16] Belotsky, K.; Khlopov, M.; Shibaev, K. Gravitation and Cosmology 2006, vol 12, 1; arXiv:astro-ph/0604518.

[17] Zeldovich, Ya.B.; Khlopov, M.Yu. *Phys Lett* 1978, vol B79, 239.

[18] Dubrovich, V.K.; Fargion, D.; Khlopov, M.Yu. (2003). arXiv:hep-ph/0312105.

[19] Kohri, K.; Takayama F. Phys Rev 2007, vol D76 063507; arXiv:hep-ph/0605243.

[20] Khlopov, M.Yu. *JETP Lett* 2006, vol 83, 1 [Pisma Zh Eksp Teor Fiz 2006, vol 83, 3]; arXiv:astro-ph/0511796.

[21] Belotsky, K.; Khlopov, M.; Shibaev, K. (2006). arXiv:astro-ph/0602261.

[22] Fargion, D.; Khlopov, M.Yu.; Stephan, C.A. *Class Quantum Grav* 2006, vol 23, 7305; arXiv:astro-ph/0511789.

[23] Khlopov, M.Yu. (2006). arXiv:astro-ph/0607048.

[24] Khlopov, M.Yu.; Kouvaris, C. *Phys Rev* 2008, vol D77, 065002; arXiv:0710.2189 [astro-ph].

[25] Khlopov, M.Yu. *Bled Workshops in Physics* 2007, vol 8, 114.

[26] Khlopov, M.Yu. (2008). arXiv:0801.0167 [astro-ph].

[27] Khlopov, M.Yu. (2008). arXiv:0801.0169 [astro-ph].

[28] Shirkov, Yu.M.; Yudin, N.P. Yadernaya Fizika (in rus); Fizmatlit: Moscow, RU, 1980. Samanta, C.; et al. (2005). arXiv:nucl-th/0504085.

[29] Cyburt, R.; *Phys Rev* 2004, vol D70, 023505.

[30] Khlopov, M.Yu. *JETP Lett* 1981, vol 33, 162.

[31] Ascasibar, Y. (2006). arXiv:astro-ph/0612130. Freese, K.; Spolyar, D.; Aguirre, A. (2008). arXiv:0802.1724 [astro-ph]. Vasiliev, E.; Zelnikov, M. (2008). arXiv:0803.0002 [astro-ph].
Moskalenko, I.V.; Wai, L.L. (2006). arXiv:astro-ph/0608535; *Astrophys J* 2007, vol 659, L29–L32; arXiv:astro-ph/0702654; (2007). arXiv:0704.1324 [astro-ph]. Fairbairn, M.; Scott, P.; Edsjo, *J. Phys Rev* 2008, vol D77, 047301; arXiv:0710.3396 [astro-ph].
Spolyar, D.; Freese, K.; Gondolo, P. (2007). arXiv:0705.0521 [astro-ph]; (2007). arXiv:0709.2369 [astro-ph].

[32] Klein, J.; et al. P*roceedings of the Symposium on Accelerator Mass Spectrometry*; Argonne National Laboratory: Argonne, IL, 1981.

[33] Vandegriff, J.; et al. *Phys Lett* 1996, vol B365, 418.

[34] Mueller, P.; et al. *Phys Rev Lett* 2004, vol 92, 022501.

[35] Smith, P.F.; Bennet, J.R.J. *Nucl Phys* 1979, vol B149, 525. Smith, P.F.; et al. *Nucl Phys* 1982, vol B206, 333. Hemmick, T.K.; et al. *Phys. Rev.* 1990, vol D41, 2074. Verkerk, P.; et al. *Phys Rev Lett* 1992, vol 68, 1116. Yamagata, T.; Takamori, Y.; Utsunomiya, H. *Phys Rev* 1993, vol D47, 1231.

[36] Khlopov, M.Yu.; Stephan, C.A. (2006). arXiv:astro-ph/0603187.

[37] McCammon, D.; et al. *Nucl Instr Meth* 1996, vol A370, 266. McCammon, D.; et al. *Astrophys J* 2002, vol 576, 188; arXiv:astro-ph/0205012.

[38] Winkelmann, C.B.; Bunkov Y.M.; Godfrin, H. *Grav and Cosmol* 2005, vol 11, 87.

[39] Sjöstrand, T.; et al. *Computer Phys Commun* 2001, vol 135, 238; arXiv:hep-ph/0010017.

[40] Belotsky, K.M.; et al. *Phys Rev* 2003, vol D68, 054027; arXiv:hep-ph/0210153.

[41] Benakli, K.; *Phys Rev* 1999, vol D60, 104002. Burgess, C.P.; Ibánez, L.E.; Quevdo, F. *Phys Lett* 1999, vol B447, 257. Abel, S.A.; Allanach, B.C.; Quevdo, F.; Ibánez L.E.; Klein, M. (2000). arXiv:hep-ph/0005260.

Reviewed by Prof. Georgi Dvali

In: The Physics of Quarks: New Research
Editors: N.L. Watson and T.M. Grant, pp. 49-74
ISBN: 978-1-60456-802-8

Chapter 4

A BETHE-SALPETER FRAMEWORK UNDER COVARIANT INSTANTANEOUS ANSATZ WITH APPLICATIONS TO SOME HADRONIC PROCESSES

Shashank Bhatnagar*
Department of Physics, Addis Ababa University, P.O.Box 1148/1110,
Addis Ababa, Ethiopia

Abstract

Mesons are the simplest bound states in Quantum Chromodynamics (QCD). Their decays provide an important tool for understanding non-perturbative (long range) behavior of strong interactions which till date is not completely understood. Towards this end, we employ a Bethe-Salpeter framework under Covariant Instantaneous Ansatz for carrying out extensive studies on various processes in hadronic physics. We first derive the non-perturbative Hadron-quark vertex function which incorporates various Dirac covariants in accordance with our power counting scheme order-by-order in powers of inverse of meson mass since various studies have shown that the incorporation of various Dirac covariants is necessary to obtain quantitatively accurate observables. The power counting scheme we proposed gives us a lot of insight as to which of the covariants from their complete set are expected to contribute maximum to the calculation of various meson observables since all Dirac covariants do not contribute equally. Calculations employing this vertex function that have been done on leptonic decays of vector mesons and unequal mass pseudoscalar mesons along with the two photon decays of pions have yielded excellent agreements with experimental results and thus validating the power counting rule we have proposed.

---

*PACS: 11.10.St, 12.39.Ki, 13.20.-v, 13.20.Jf

1. Introduction

Before starting the discussion of our work, we first give a quick glimpse of the area of quark physics. This is a subject which has grown out of discoveries which rank in importance next only to two others in the last century, namely Quantum theory and Relativity, whose successful marriage gave birth to the concept of anti-matter and which was subsequently to be confirmed by experiment. In fact the discovery of quarks has shaken the very foundations of physical reality in a manner similar to what Quantum theory itself had done in the last century. This is due to their apparent invisibility by conventional yardsticks as have helped identify the existence of most other elementary particles in physics. Due to this fact, the very theory of quark interactions has had to be designed in such a way as to make it impossible to observe them as free particles, in the sense all other elementary particles have been observed!! This is an example of a case where a firm experimental fact was taken as a cornerstone of theoretical foundations. There exist two similar examples in history of physics where theoretical foundations were dictated by experimental evidence– one was the Theory of Relativity where the observed invariance of speed of light in various frames had led Einstein to incorporate this fundamental fact in his basic postulates, and the other was the development of Quantum Theory which was again motivated by some crucial experimental observations.

The quark picture has come to be accepted by the physics community mainly because of the external manifestations of two of its major attributes- colour and flavour. The underlying gauge theory of quark interactions is Quantum Chromodynamics (QCD). The eventual progress towards a correct gauge theory of strong interactions was made after the discovery of 3-fold colour charge as a basic non-abelian attribute of quark fields governed by the gauge group SU(3). The carrier gauge field of this attribute was identified as the massless gluon field. This is to be contrasted with the abelian U(1) Quantum Electrodynamics (QED)- the gauge theory for electromagnetic interactions, where the carrier of the gauge field, photon does not posses charge (is electrically neutral), though it acts as a carrier of charge field between particles having a single attribute- electric charge. On the other hand, the non-abelian SU(3) gluon field carries an octet of colour fields between particles having colour triplet (3) as well as colour anti-triplet ($\overline{3}$) attribute. An important feature of non-abelian gauge theories like QCD is that the gauge fields (gluons) self interact owing to the fact that the QCD lagrangian admits not only quark-gluon interactions, but also 3-gluon and 4-gluon interactions. This is in complete contrast to an abelian gauge theory like QED where the corresponding lagrangian does not admit interactions between gauge bosons (photons). In Table 1 we summarise the points of similarity and contrast between the gauge theories QED and QCD.

As we see from Table 1, an important distinction between abelian QED and non-abelian QCD lies in the variation of their respective "charges" as a function of momentum. The decrease of α_s with Q^2 has important consequence that the colour interactions become weaker at shorter distances, thus providing a natural mechanism for Asymptotic freedom, while the increase of α_s with small Q^2 provides a basis for confinement of colour attribute and hence of quarks and gluons (which possess colour) and thus preventing them to exist as free particles. Asymptotic freedom has been convincingly demonstrated through extensive calculations in perturbative QCD and their comparison with data in deep inelastic scattering experiments on protons and other nuclei. On the other hand, an equally convincing demonstration of confinement is vitiated by non-perturbative character of $\alpha_s(Q^2)$ operative at not so large Q^2 and the consequent unreliability of perturbative QCD calculations in the low-energy (long distance) regime.

Table 1. Comparison of gauge theories QED and QCD.

	QED	**QCD**
System	Atomic	Hadronic
Constituents	e^+, e^-	$q, \bar{q}$
Carrier field	photon	gluon
Attribute	charge	colour
Gauge group	U(1)	SU(3)
Conservation status	Exact	Exact
Coupling constant (asymptotic value)	$\alpha = \frac{1}{137}$	$\alpha \approx 1$
Q^2 variation	$\frac{d\alpha}{dQ^2} > 0$	$\frac{d\alpha_s}{dQ^2} < 0$
Short distance behavior	α increases at small r $\Rightarrow$ Asymptotic barrier	α_s decreases at small r $\Rightarrow$ Asymptotic freedom
Long distance behavior	α decreases at large r $\Rightarrow$ Infrared freedom	α_s increases at large r $\Rightarrow$ Infrared slavery
Perturbation theory status	Ok at large r, Not good at small r.	Ok at small r, Invalid at large r $\Rightarrow$ Confinement.

Though there is little doubt that QCD incorporates confinement feature within its conceptual premises, one of the major theoretical challenges of the day is the problem of how to bring out this remarkable feature explicitly from the QCD lagrangian. This has given birth to a number of non-perturbative approaches to deal with the long distance properties of QCD such as QCD sum rules, Lattice QCD and dynamical equation based approaches such as Schwinger-Dyson equation (SDE), Bethe-Salpeter equation (BSE) and potential models. All these approaches have both advantages as well as shortcomings. Among all these approaches, there is one approach that is model independent, namely Lattice QCD. However, since the task of calculating hadron structures from QCD alone is very difficult (as can be seen from studies in Lattice QCD), one relies on specific models of hadron dynamics to gain some understanding of hadronic structures at low energies (long distances). The dynamical equation based approach such as BSE is a conventional approach in dealing with two-body relativistic bound state problems. From solutions of BSE one can obtain useful information about the inner structure of hadrons. This equation which is firmly rooted in field theory provides a realistic description for analyzing mesons as composite objects. However one drawback while solving BSE can be traced to the fact that one has to use input kernel which is model dependent. However calculations have shown that BSE using phenomenological potentials can give satisfactory results.

In this work we have made use of Bethe-Salpeter equation under Covariant Instantaneous Ansatz (CIA) (see e.g. [1] and references therein) to study decays of various mesons. These studies are motivated by the fact that the investigation of bound states of hadrons is one of the most effective methods of studying the dynamics of interactions of its constituents. Meson decays provide an important tool for not only exploring the structure of these simplest bound states in QCD, but also for studying the non-perturbative (long distance) behavior of strong interactions which are not yet completely understood till date. This study can most effectively be accomplished by applying a particular framework of hadron dynamics to a diverse range of phenomena. Mesons have for a long time been a major focus of attention to understand the inner structure of hadrons from non-perturbative QCD. A number of such studies [1-6] dealing with decays of pseudo-scalar mesons and vector mesons at quark level of compositeness have been carried out recently. A realistic description of pseudoscalar and vector mesons at the quark level of compositeness would be an important element in our understanding of hadron dynamics and reaction processes at scales where QCD degrees of freedom are relevant. Such studies also offer a direct probe of hadron structure and help in revealing some aspects of the underlying quark-gluon dynamics.

Thus in this paper we study leptonic decays of vector mesons such as ρ, ω, ϕ , leptonic decays of pseudoscalar mesons such as π , K, D, D_S and B which proceed through the well known quark loop diagrams and two photon decays of a pion which proceeds through the famous quark triangle diagram. A relativistic framework for analyzing mesons as composite objects is provided by Bethe-Salpeter Equation. In this paper we employ a QCD oriented framework of Bethe-Salpeter Equation (BSE) under Covariant Instantaneous Ansatz (CIA) [7]. CIA is a Lorentz-invariant generalization of Instantaneous Approximation (IA). For $q\overline{q}$ system, CIA formulation [7] ensures an exact interconnection between 3D and 4D forms of the BSE. The 3D form of BSE serves for making contact with the mass spectra of hadrons, whereas the 4D form provides the $Hq\overline{q}$ vertex function $\Gamma(\hat{q})$ for evaluation of transition amplitudes. A BS framework under IA formulation similar to CIA formulation was also earlier suggested by Bonn group [8].

We had earlier employed the framework of BSE under CIA for calculation of decay constants [7,9] of heavy-light pseudoscalar mesons. We also evaluated the leptonic decays of vector mesons [10] in this framework. However, one of the simplified assumptions in all these calculations was that the hadron-quark vertex was restricted to have a single Dirac structure (e.g., γ_5 for pseudoscalar mesons, $\gamma.\varepsilon$ for vector mesons, etc.). However, recent studies [3,5,11] have revealed that various mesons have many different covariant structures in their wave functions whose inclusion was also found necessary to obtain quantitatively accurate observables [11] and it was further noticed that all Dirac covariants do not contribute equally and only some of them are relevant for calculation of meson mass spectrum and decay constants [3-5]. Such a Dirac structure of the BS wave function in fact was already indicated by Llewellyn Smith [12]. Hence it is necessary to introduce various Dirac structures into the $Hq\overline{q}$ vertex for different kinds of mesons. In the recent work [1], we developed a power counting rule for incorporating various Dirac covariants in the structure of hadron-quark vertex function, order by order in powers of inverse of meson mass, and calculated the leptonic decays of equal mass vector mesons such as ρ, ω, ϕ, taking into account the leading order covariants since they are expected to contribute maximum to observables according to our scheme. We very recently extended this work to pseudoscalar mesons [13] and studied their leptonic decay constants. In this paper we also study two photon decays of a pion and calculate the pion-photon coupling constant using the above structure of $Hq\bar{q}$ vertex function.

The organization of this paper is as follows: In section 2 we introduce the BSE framework under Covariant Instantaneous Ansatz and derive the generalized structure of Hadron-quark vertex function for the case of vector and pseudoscalar mesons which incorporates various Dirac covariants in its structure in accordance with a power counting rule [1] we have recently developed. The input BSE kernel we have employed is also introduced. In section 3, this generalized hadron-quark vertex function developed above is applied to: (i) Calculation of leptonic decay constants of vector mesons ρ, ω and ϕ, (ii) Calculation of leptonic decay constants of unequal mass pseudoscalar mesons π, K, D, D_S and B, and (iii) Calculation of pion-photon coupling constant in the process $\pi^0 \to \gamma\gamma$. Section 4 is devoted to Conclusion.

2.1. BSE Framework under Covariant Instantaneous Ansatz (CIA)

We first outline the BSE framework under Covariant Instantaneous Ansatz (CIA) that we have employed for the case of scalar quarks for simplicity. For a $q\bar{q}$ system with an effective kernel K and 4D wave function $\Phi(P,q)$, the 4D BSE takes the form,

$$i(2\pi)^4 \Delta_1 \Delta_2 \Phi(P,q) = \int d^4q' K(q,q') \Phi(P,q'), \tag{2.1}$$

where $\Delta_{1,2}$ are the inverse propagators of two scalar quarks are:

$$\Delta_{1,2} = m_{1,2}{}^2 + p_{1,2}{}^2, \tag{2.2}$$

where $m_{1,2}$ are (effective) constituent masses of quarks. The 4-momenta of the quark and anti-quark, $p_{1,2}$, are related to the internal 4-momentum q_μ and total momentum P of hadron of mass M as

$$p_{1,2\mu} = \hat{m}_{1,2} P_\mu \pm q_\mu \tag{2.3}$$

where $\hat{m}_{1,2} = [1 \pm (m_1^2 - m_2^2)/M^2]/2$ are the Wightman-Garding (WG) definitions [9] of masses of individual quarks. We now use an Ansatz on the BS kernel K in Eq. (2.1) which is assumed to depend on the 3D variables $\hat{q}_\mu, \hat{q}_\mu{}'$ i.e.

$$K(q,q') = K(\hat{q},\hat{q}') \tag{2.4}$$

where

$$\hat{q}_\mu = q_\mu - \frac{q.P}{P^2} P_\mu \tag{2.5}$$

is observed to be orthogonal to the total 4-momentum P (i.e., $\hat{q}.P = 0$) irrespective of whether the individual quarks are on-shell or off-shell (a similar form of the BS kernel was also suggested in ref. [8]). Hence, the longitudinal component of q_μ,

$$M\sigma = M\frac{q.P}{P^2} \tag{2.6}$$

does not appear in the form $K(\hat{q},\hat{q}')$ of the kernel. For reducing Eq. (2.1) to the 3D form, we define a 3D wave function $\phi(\hat{q})$ as

$$\phi(\hat{q}) = \int_{-\infty}^{+\infty} M d\sigma \Phi(P,q). \tag{2.7}$$

Substituting Eq.(2.7) in Eq.(2.1), with definition of kernel in Eq.(2.4), we get a covariant version of Salpeter equation,

$$(2\pi)^3 D(\hat{q})\phi(\hat{q}) = \int d^3\hat{q}' K(\hat{q},\hat{q}')\phi(\hat{q}') \tag{2.8}$$

where $D(\hat{q})$ is the 3D denominator function defined by

$$\frac{1}{D(\hat{q})} = \frac{1}{2\pi i} \int_{-\infty}^{+\infty} \frac{M d\sigma}{\Delta_1 \Delta_2} \tag{2.9}$$

whose value can be easily worked out by contour integration by noting positions of poles in the complex σ-plane (shown in detail in [10]). We can see that RHS of Eq.(2.8) is identical to RHS of Eq.(2.1) by virtue of Equations (2.4) and (2.7). We thus have an exact interconnection between 3D wave function $\phi(\hat{q})$ and 4D wave function $\Phi(P,q)$:

$$\Delta_1 \Delta_2 \Phi(P,q) = \frac{D(\hat{q})\phi(\hat{q})}{2\pi i} \equiv \Gamma(\hat{q}) \tag{2.10}$$

We thus also get the $Hq\bar{q}$ vertex function $\Gamma(\hat{q})$ under CIA for case of scalar quarks. Further in the process, an exact interconnection between 3D and 4D BSE is thus brought out where the 3D form serves for making contact with the mass spectrum of hadrons, whereas the 4D form provides the $Hq\overline{q}$ vertex function $\Gamma(\hat{q})$ which satisfies a 4D BSE with a natural off-shell extension over the entire 4D space (due to the positive definiteness of the quantity $\hat{q}^2 = q^2 - \frac{(q.P)^2}{P^2}$ throughout the entire 4D space) and thus provides a fully Lorentz-invariant basis for evaluation of various transition amplitudes through various quark loop diagrams. Due to these properties the framework can be profitably employed not only for mass spectral predictions but also for transition amplitudes all the way from low energies to high energies.

2.2. Generalized Forn of Hadron-Quark Vertex Function in BSE

To obtain the generalized form of Hadron-quark vertex function for the case of fermionic quarks constituting a particular meson, we first replace the scalar propagators Δ_i^{-1} in the above equation by the proper fermionic propagators S_F. Then for incorporation of relevant Dirac structures in vertex function $\Gamma(\hat{q})$, we take guidance from some of the recent studies [3,5,11] which have revealed that various mesons have many different covariant structures in their wave functions whose inclusion was also found necessary to obtain quantitatively accurate observables [11]. In our recent works we have attempted to incorporate other Dirac covariants from their complete set for a particular meson systematically.

For this purpose we make use of the power counting rule we developed in [1,13] for incorporating various Dirac covariants in the structure of $Hq\overline{q}$ vertex function for a particular meson (pseudoscalar, vector, scalar etc.) order-by-order in powers of inverse of meson mass M. In [1] we had thus written down the complete $Hq\overline{q}$ vertex function for a V-meson and applied it to the calculation of f_V values for ρ, ω and ϕ mesons. Our aim of that work was to find a "criterion" so as to systematically choose among various Dirac covariants from their complete set to write wave functions for different mesons (vector mesons, pseudoscalar mesons etc.):

(I)Vector Mesons

For a V-meson the $Hq\overline{q}$ vertex function is expressed as a linear combination of eight Dirac covariants $\Gamma_i^V\,(i = 0,...,7)$, each multiplying a Lorentz-scalar amplitude $F_i(q^2, q.P, P^2)$. However the choice of these covariants is not unique as can also be seen from the choice of covariants employed in [3,5,11] for a V-meson. It was noticed in earlier work that all covariants do not contribute equally and only five of the eight covariants ($\Gamma_1,...,\Gamma_5$) [3] were considered to be important for calculation of vector meson masses and decay constants and γ_μ was considered to be the most important covariant. Further in [3,5], calculations were also done for masses and decay constants for various other subsets of eight covariants. Thus as stated above, our motivation for this work was to find a criterion to systematically choose among various covariants from their complete set.

Thus for incorporating Dirac structures in the expression for $\Gamma(\hat{q})$, we take their forms as in [11]. We note that in the expression for CIA vertex function in equation (2.10), the factor $D(\hat{q})\phi(\hat{q})$ is nothing but the Lorentz-invariant momentum dependent scalar which depends on q^2, P^2 and $q.P$ and has a certain dimensionality of mass. However the Lorentz-scalar amplitudes multiplying various Dirac structures in [11] have different dimensionalities of mass. For adapting this decomposition to write the structure of $Hq\overline{q}$ vertex function $\Gamma(\hat{q})$ for a particular meson, we re-express this function by making these amplitudes dimensionless by weighing each covariant by an appropriate power M, the meson mass. Thus each term in the expansion of $\Gamma(\hat{q})$ is associated with a certain power of M. In detail we can express Hadron-quark vertex function for a V-meson as:

$$\Gamma^V.\varepsilon = (\Omega^V.\varepsilon)\frac{1}{2\pi i}N_V D(\hat{q})\phi(\hat{q});$$
$$\Omega^V.\varepsilon = i\not{\varepsilon}A_0 + \not{\varepsilon}\not{P}\frac{A_1}{M} + [q.\varepsilon - \not{\varepsilon}\not{q}]\frac{A_2}{M} + [\not{\varepsilon}\not{P}\not{q} - \not{\varepsilon}\not{q}\not{P} + 2iq.\varepsilon\not{P}]\frac{A_3}{M^2} + q.\varepsilon\frac{A_4}{M} + iq.\varepsilon\not{P}\frac{A_5}{M^2} - iq.\varepsilon\not{q}\frac{A_6}{M^2} \qquad (2.11)$$
$$+ q.\varepsilon[\not{P}\not{q} - \not{q}\not{P}]\frac{A_7}{M^3}.$$

where $A_i(i = 0,...,7)$ are eight dimensionless and constant coefficients to be determined. Now since we use constituent quark masses where the quark mass m is approximately half the hadron mass M, we can use the ansatz,

$$q << P \sim M \qquad (2.12)$$

in the hadron rest frame. Then each of the eight terms in Eq.(2.11) would receive suppression by different powers of 1/M. Thus we can arrange these terms as an expansion in powers of O(1/M). then we see that in the expansion of $\Omega.\varepsilon$, the structures associated with A_0, A_1 have magnitudes $0(1/M^0)$ and are of leading order. Those with A_2, A_3, A_4, A_5 are $O(1/M)$, while those with A_6, A_7 are $O(1/M^2)$. This naïve power counting rule suggests that the maximum contribution to calculation of any V-meson observable should come from the Dirac covariants $i\not{\varepsilon}$ and $\not{\varepsilon}\not{P}\frac{1}{M}$ associated with constant coefficients A_0 and A_1 respectively. As initial attempt [1], we took the form of $Hq\overline{q}$ vertex function incorporating only these leading order terms and in the perturbative expansion (2.11) and ignoring $O(1/M)$ and $O(1/M^2)$ terms and calculated V-meson decay constants using Leading order terms alone. Thus we take the modified form of $Hq\overline{q}$ vertex function as:

$$\Gamma^V(\hat{q}).\varepsilon = [i\not{\varepsilon} + \not{\varepsilon}\not{P}\frac{1}{M}]\frac{1}{2\pi i}N_V D(\hat{q})\phi(\hat{q}) \tag{2.13}$$

For flavourless vector mesons which are eigen states of charge parity, there is an extra restriction on the use of Dirac structures [1]. In general the coefficients of Dirac structures can be functions of $q.P$ and hence can be written as a Taylor's series in powers of $q.P$. However the coefficients used here are dimensionless. Hence various terms in the series should be powers of $q.P/M^2$ which is O(1/M). If we want to keep leading contributions, we should keep only the zero order terms in the Taylor's series. This justifies usage of A_0, A_1 as constants in the above equations. But now we choose only the C-even part of the coefficients since, since only odd powers of q are C-odd. Hence only proper C-value Dirac coefficients can be used. The Dirac structures used in Eq.(2.13) are consistent with the charge parity of flavourless vector mesons (the second term originates from $\sigma_{\mu\nu}P_\nu$). In this paper we will investigate numerical results to this order.

(II) Pseudoscalar Mesons

We proceed in a manner similar to the above case: The $Hq\bar{q}$ vertex function for a pseudoscalar meson can again be considered to have a certain dimensionality of mass [13] and can be expressed as a linear combination of four Dirac covariants $\Gamma_i^P (i = 1,2,3)$ [3,5,11], each multiplying a Lorentz scalar amplitude $F_i(q^2, q.P, P^2)$. The choice of the Dirac covariants is again not unique as can also be seen from the choice of covariants used in Ref. [3,11]. For adapting this decomposition to write the structure of vertex function $\Gamma(\hat{q})$, we again re-express the $Hq\bar{q}$ vertex function by again making the amplitudes $F_i(q^2, q.P, P^2)$ dimensionless, weighing each Dirac covariant with an appropriate power of M. Thus each term in the expansion of $\Gamma(\hat{q})$ is associated with a certain power of M and hence in detail we can express $\Gamma(\hat{q})$ as a polynomial in various powers of 1/M:

$$\Gamma^P(\hat{q}) = \Omega^P.\frac{1}{2\pi i}N_P D(\hat{q})\phi(\hat{q})\,; \tag{2.14}$$

$$\Omega^P = \gamma_5 B_0 - i\gamma_5\not{P}\frac{B_1}{M} - i\gamma_5\not{q}\frac{B_2}{M} - \gamma_5(\not{P}\not{q} - \not{q}\not{P})\frac{B_3}{M^2}$$

where B_i ($i = 0,...,3$) are four dimensionless and constant coefficients (which are taken to be constant on lines of [1]) to be determined. Using the ansatz in Eq.(2.12) in the rest frame of the hadron (however we wish to mention that among all the pseudoscalar mesons, pion enjoys the special status in view of its unusually small mass ($M < \Lambda_{QCD}$) and its case should be considered separately) , each of the four terms in Eq.(2.10) would again receive suppression by different powers of $1/M$. Thus we can arrange these terms as an expansion in powers of $O(\frac{1}{M})$. We can then see in the expansion of Ω^P, that the structures associated with the coefficients B_0, B_1 have magnitudes $O(\frac{1}{M^0})$ and are of leading order, while those with B_2, B_3 are $O(\frac{1}{M^1})$ and are next-to-leading-order. This naïve power counting rule suggests that the maximum contribution to the calculation of any pseudoscalar meson observable should come from the Dirac structures γ_5 and $i\gamma_5(\gamma.P)\frac{1}{M}$ associated with the constant coefficients B_0 and B_1 respectively. As a first application of this to P-meson sector and on lines similar to [1] (for V-meson case), we take the form of the $Hq\overline{q}$ vertex function incorporating these leading order terms in expansion (2.11) and ignoring $O(\frac{1}{M^1})$ terms for the moment and try to calculate the P-meson decay constants f_P taking only these leading order terms. Thus we take the modified form of $Hq\overline{q}$ vertex function as,

$$\Gamma^P(\hat{q}) = [\gamma_5 B_0 - i\gamma_5\gamma.P\frac{B_1}{M}].\frac{1}{2\pi i}N_P D(\hat{q})\phi(\hat{q}) \tag{2.15}$$

In general the coefficients B_i of the Dirac structures could be functions of $q.P$, and hence can be written as a Taylor series in powers of $q.P$. However the coefficients used here are dimensionless on lines of [1] employing the same reasoning. In a similar manner one can express the full hadron-quark vertex function for a scalar and axial vector mesons also in BSE under CIA, taking guidance from Ref. [11]. Then we can incorporate the Dirac structures (to various orders in 1/M) according to our power counting rule. At the same time, the restriction by charge parity on wave function of eigen state should also be respected. Further, to get the complete set of the Dirac structure for a certain kind of mesons, the restriction by the (space) Parity have been employed; and it is easy to see that the requirements of the space Parity and the charge Parity are the same for the vertex as well as the full wavefunction (see also [12]).

2.3. BSE Kernel

From the above analysis of the structure of $Hq\overline{q}$ vertex function, we notice that the structure of 3D wave function $\phi(\hat{q})$ as well as the form of the 3D BSE are left untouched and have the same form as in our previous works which justifies the usage of the same form of the input kernel we used earlier. Now we briefly mention some features of the BS formulation employed. The structure of BSE is characterized by a single effective kernel arising out of a four-fermion lagrangian in the Nambu-Jonalasino [14-15] sense. The formalism is fully consistent with Nambu-Jona-Lasino [15] picture of chiral symmetry breaking but is additionally Lorentz-invariant because of the unique properties of the quantity $\hat{q}^2$, which is positive definite throughout the entire 4D space. The input kernel $K(q,q')$ in BSE is taken as one-gluon-exchange like as regards color $(\frac{1}{2}\vec{\lambda}.\frac{1}{2}\vec{\lambda}_2)$ and spin$(\gamma_\mu^{(1)}\gamma_\mu^{(2)})$ dependence. The scalar function $V(q-q')$ is a sum of one-gluon exchange V_{OGE} and a confining term $V_{Conf.}$ [1,16]. Thus we can write

$$K(q,q') = \frac{1}{2}\vec{\lambda}^{(1)}.\frac{1}{2}\vec{\lambda}^{(2)}V_\mu^{(1)}V_\mu^{(2)}V(q-q');$$

$$V_\mu^{(1,2)} = \pm 2m_{1,2}\gamma_\mu^{(1,2)};$$

$$V(\hat{q}-\hat{q}') = \frac{4\pi\alpha_S(Q^2)}{(\hat{q}-\hat{q}')^2} + \frac{3}{4}\omega_{q\bar{q}}^2\int d^3\vec{r}[r^2(1+4a_0\hat{m}_1\hat{m}_2M^2r^2)^{-\frac{1}{2}} - \frac{C_0}{\omega_0^2}]e^{i(\hat{q}-\hat{q}').\vec{r}};$$
$$\alpha_S(Q^2) = \frac{12\pi}{33-2f}[\ln\frac{M_>^2}{\Lambda^2}]^{-1}; M_> = Max(M, m_1+m_2). \qquad (2.16)$$

The ansatz employed for the spring constant $\omega_{q\bar{q}}^2$ in Eq. (2.16) is [1,16]:

$$\omega_{q\bar{q}}^2 = 4\hat{m}_1\hat{m}_2M\omega_0^2\alpha_S(M_>^2) \qquad (2.17)$$

where $\hat{m}_1, \hat{m}_2$ are the Wightman-Garding definitions of masses of constituent quarks defined earlier. Here the proportionality of $\omega_{q\bar{q}}^2$ on $\alpha_S(Q^2)$ is needed to provide a more direct QCD motivation to confinement. This assumption further facilitates a flavour variation in $\omega_{q\bar{q}}^2$. And ω_0^2 in Eq.(2.16) and Eq.(2.17) is postulated as a universal spring constant which is common to all flavours. Here in the expression for $V(\hat{q}-\hat{q}')$, as far as the integrand of the confining term $V_{conf.}$ is concerned, the constant term C_0/ω_0^2 is designed to take account of the correct zero point energies, while a_0 term ($a_0 << 1$) simulates an effect of an almost linear confinement for heavy quark sectors (large m_1, m_2), while retaining the harmonic form for light quark sectors (small m_1, m_2) [14] as is believed to be true for QCD. Hence the term $r^2(1+4a_0\hat{m}_1\hat{m}_2M_>^2r^2)^{-\frac{1}{2}}$ in the above expression is responsible for effecting a smooth transition from harmonic ($q\bar{q}$) to linear ($Q\bar{Q}$) confinement. The basic input parameters in the kernel are just four- i.e. $a_0 = .028, C_0 = .29, \omega_0 = .158$ Gev and QCD length scale,

$\Lambda = .20$ Gev and quark masses $m_{u,d} = .265Gev$, $m_s = .415Gev$, $m_c = 1.530Gev$ and $m_b = 4.900Gev$ [1,9,10,14] which have been calibrated to fit the meson mass spectrum obtained earlier by solving the 3D BSE under Null-Plane Ansatz (NPA) (reported in [14]) (see also Ref.[16] and references therein). However due to fact that the 3D BSE under Covariant Instantaneous Ansatz (CIA) has a structure which is formally equivalent to the 3D BSE under Null-Plane Ansatz (NPA) near the surface P.q = 0 (see Ref.[1,10] for details), the mass spectral results in CIA formulation are exactly the same as the $q\overline{q}$ mass spectral results in NPA formulation predicted earlier in [14](see Ref.[1,10] for details). The details of this BS model under CIA in respect of spectroscopy are thus directly taken over from NPA formalism (see [1,10]for details) . Now comes to the problem of the 3D BS wave function. The ground state wave function $\phi(\widehat{q})$ satisfies the 3D BSE [14] on the surface P.q = 0, which is appropriate for making contact with O(3)-like mass spectrum [14]. Its fuller structure (described in Ref.[14]) is reducible to that of a 3D harmonic oscillator with coefficients dependent on the hadron mass M and the total quantum number N. The ground state wave function $\phi(\hat{q})$ deducible from this equation thus has a gaussian structure [1,9,10,14] and is expressible as:

$$\phi(\widehat{q}) \approx e^{-\widehat{q}^2/2\beta^2} \tag{2.18}$$

In the structure of $\phi(\hat{q})$ in Eq. (2.14), the parameter β is the inverse range parameter which incorporates the content of BS dynamics and is dependent on the input kernel $K(q,q')$. The structure of β is given in Section 3. We now give the calculation of leptonic decays constants of vector mesons and pseudoscalar mesons as well as the pion decay constant in the process $\pi^0 \rightarrow \gamma\gamma$ in the framework discussed in next sections.

3.1. Leptonic Decays of Vector Mesons

Vector meson decay proceeds through the quark loop diagram shown in [1]. The coupling of a vector meson to a photon is expressed by the dimensionless coupling constant g_V which can be described as

$$\frac{M^2}{g_V}\varepsilon_\mu(P) = <0|\overline{Q}\Theta\gamma_\mu Q|V(P)> \tag{3.1}$$

where Q is the flavour multiplet of quark field, and Θ is the quark electromagnetic charge operator, which can in turn be expressed by the loop integral,

$$\frac{M^2}{g_V}\varepsilon_\mu(P) = \sqrt{3}e_Q\int d^4q Tr[\Psi_V(P,q)i\gamma_\mu] \tag{3.2}$$

where $e_Q^2 = \frac{1}{2}, \frac{1}{9}, \frac{1}{18}$ for $V = \rho, \omega, \phi$ mesons respectively and the polarization vector ε_μ satisfies $\varepsilon.P = 0$. The Bethe-Salpeter wave function for a V-meson, $\Psi_V(P,q)$ is expressed as

$$\Psi_V(P,q) = S_F(p_1)\Gamma(\hat{q})S_F(-p_2);$$
$$S_F(p_1) = -i\frac{m_1 - \not{p}_1}{\Delta_1}, S_F(-p_2) = -i\frac{m_2 + \not{p}_2}{\Delta_2}. \tag{3.3}$$

For the structure of hadron-quark vertex function $\Gamma^V(\hat{q})$, we consider the form in Eq. (2.13). S_F are the fermionic propagators of the two constituent quarks of the hadron, and the non-perturbative aspects enter through the vertex function $\Gamma^V(\hat{q})$. Using $\Psi(P,q)$ from Eq.(3.3), vertex function from Eq.(2.13), evaluating trace over γ-matrices, and following the usual steps and finally carrying out integration over $d\sigma$ by the method of contour integration in the complex σ-plane, and noting that the V-meson decay constant $f_V = M/(e_Q g_V)$, we can write

$$f_V = \frac{4\sqrt{3}}{M} N_V \left[\int d^3\hat{q} \left\{ \left[\left(\frac{M^2}{6} + \frac{2}{3}m^2 \right) + \frac{1}{6} D_0(\hat{q}) \right] A_0 - mMA_1 \right\} \right] \tag{3.4}$$

where the relationship between the functions $D_0(\hat{q})$ and $D(\hat{q})$ is:

$$D(\hat{q}) = \frac{D_0(\hat{q})}{(\frac{1}{2\omega_1} + \frac{1}{2\omega_2})}; \tag{3.5}$$

$$D_0(\hat{q}) = (\omega_1 + \omega_2)^2 - M^2; \omega_{1,2}^2 = m_{1,2}^2 + \hat{q}^2. \tag{3.6}$$

In Eq.(3.4), $-mMA_1$ is the contribution to the integral for f_V due to presence of additional Dirac structure $\not{\varepsilon}\not{P}/M$ in the vertex function $\Gamma^V(\hat{q})$.

The structure of parameter β in the expression for the 3D wave function $\phi(\hat{q})$ is taken as [1,16]:

$$\beta^2 = (2\hat{m}_1\hat{m}_2 M\omega_{q\bar{q}}^2/\gamma^2)^{1/2}; \gamma^2 = 1 - \frac{2\omega_{q\bar{q}}^2 C_0}{M_> \omega_0^2} \tag{3.7}$$

where $\alpha_S(Q^2)$ is taken as in Eq. (2.16) and $\omega^2_{q\bar{q}}$ is expressed as in Eq. (2.17). To calculate the normalizer N_V in Eq.(3.5), we use the current conservation condition [1],

$$2iP_\mu = (2\pi)^4 \int d^4q Tr[\overline{\Psi}(P,q)(\frac{\partial}{\partial P_\mu} S_F^{-1}(p_1))\Psi(P,q)S_F^{-1}(-p_2)] + (1 \leftrightarrow 2), \quad (3.8)$$

We do not give here the details of BS normalizer calculation which can be similarly worked out (see [1] for details).

We have thus evaluated the general expression for f_V in BSE under Covariant Instantaneous Ansatz (CIA) with Dirac structure $\not{\varepsilon}\not{P}/M$ introduced into the hadron-quark vertex $\Gamma^V(\hat{q})$ besides the Dirac structure $i\not{\varepsilon}$. We see that so far the results are independent of any model for $\phi(\hat{q})$. However for calculating these decay constants, one needs to know the constant coefficients A_0, A_1 associated with Dirac structures $i\not{\varepsilon}$ and $\not{\varepsilon}\not{P}/M$ respectively. The relative value A_1/A_0 is a free parameter without any further knowledge of the meson structure in framework discussed above. In general, one can incorporate all the Dirac structures (i.e. associated with orders $O(1/M^1)$ besides the leading orders $O(1/M^0)$ along with their respective coefficients A_i in a Taylor's series of $\frac{q.P}{M^2}$ to express BS wave function and calculate decay constants f_V to the desired order. However we wish to mention that even if we wish to calculate decay constants to order $O(1/M^1)$, we introduce four additional parameters $A_2,...,A_5$. To fit these additional four parameters would lead to the necessity of taking data of more types of vector mesons which includes the heavy quark vector mesons such as J/Ψ. To have a global analysis and best fit to the spectra of both light and heavy quark mesons along with their decay constants, we may face the problem of reparametrizing the input BS kernel. However as a first step we vary this parameter A_1/A_0 at the lowest order to see the effect of introducing the Dirac structure $\not{\varepsilon}\not{P}/M$. Further from the expression for f_V (Eq.(3.5), with constituent quark mass $m \sim M/2$, we can note that $\frac{1}{6}A_0$ and $2A_1$ are of the same order if we consider both Dirac structures $i\not{\varepsilon}$ and $\not{\varepsilon}\not{P}/M$ give leading contributions. This suggests that as a rough estimate $A_1/A_0 \approx .083$. We calculate f_V for ρ, ω, ϕ mesons for values of A_1/A_0 in the range .06- .10. This range does not mean any preference but is adequate to show the dependence of f_V on A_1/A_0. The results are given in Table II along with other models and data. It is seen that numerical values of these decay constants in BSE under CIA improve considerably and come close to the experimental results when additional Dirac structure $\not{\varepsilon}\not{P}/M$ is introduced in the vertex function besides the structure $i\not{\xi}$ in comparison to decay constant values calculated when vertex function contains only the Dirac structure $i\not{\xi}$. Further Discussions are in Section 4.

Table II. Calculated values in MeV (along with results of other models) of leptonic decay constants $f_V (V = \rho, \omega, \phi)$ in BSE under CIA for range of values $A_1 / A_0 = .06 - .10$. The decay constants are calculated from data [20] from $\Gamma_{V \to e^+ e^-} = \frac{4\pi}{3} \frac{\alpha_{em}^2 e_q^2}{M} f_V^2$. The masses of hadrons are also used from [20]. The values of constituent quark masses used are: $m_{u,d} = 265 MeV, m_s = 450 MeV$.

	f_ρ	f_ω	f_ϕ
BSE-CIA using both covariants i $\not{\xi}$ and $\not{\xi}\not{P}/M$ for the ratio A_1 / A_0:			
.06	188.3	183.8	283.9
.07	197.4	192.7	295.2
.08	206.4	201.5	306.4
.09	215.3	210.3	317.1
.10	224.4	219.5	328.2
BSE-CIA using only the covariant i $\not{\xi}$	142.2	137.0	191.3
SDE [3,5] for parameter set $\omega = .4 GeV, D = .93 GeV^2$ and covariants:			
(i) $T_\mu^1 = \hat{\gamma}_\mu$	200.0		220.0
(ii) $T_\mu^1, ..., T_\mu^5$	199.0		250.0
(iii) $T_\mu^1, ..., T_\mu^8$	207.0		259.0
SDE [2]	163.0		
Experimental Data [20]	220.9 ± 1.7	194.6 ± 3.2	228.0 ± 2.87

3.2. Leptonic Decays of Pseudoscalar Mesons

Decay constants f_P can be evaluated through the loop diagram which gives the coupling of the two-quark loop to the axial vector current and can be evaluated as:

$$f_P P_\mu = < 0 | \overline{Q} i \gamma_\mu \gamma_5 Q | P(P) > \tag{3.9}$$

which can in turn be expressed as a loop integral,

$$f_P P_\mu = \sqrt{3} \int d^4 q Tr[\Psi_P(P, q) i \gamma_\mu \gamma_5] \tag{3.10}$$

Bethe-Salpeter wave function $\Psi_P(P,q)$ for a P-meson is expressed as in Eq. (3.3). In the following calculation, we again take the leading order terms in the structure of hadron-quark vertex function $\Gamma^P(\hat{q})$ for a pseudoscalar meson as in Eq. (2.15). However unlike the decay constant calculation of equal mass V-mesons studied in the last section, we have to use unequal mass kinematics for the case of calculation of decay constants of unequal mass pseudoscalar mesons. Using $\Psi_P(P,q)$ from Eq.(3.3) and the structure of $Hq\overline{q}$ vertex $\Gamma^P(\hat{q})$ from Eq.(2.15), evaluating trace over γ-matrices and multiplying both sides of Eq.(3.12) by $P_\mu/(-M^2)$, we can express f_P as:

$$f_P = \sqrt{3}N_P \int d^3\hat{q} D(\hat{q})\phi(\hat{q}).I \tag{3.11}$$

$$I = \int_{-\infty}^{+\infty} \frac{Md\sigma}{2\pi i\Delta_1\Delta_2}\{2B_0[m_{12}(1-\frac{\delta m^2}{M^2}) + 2\delta m\sigma]$$

$$+\frac{B_1}{M^3}[\{(-M^4 - 4m_1m_2M^2 + m_{12}^2\delta m^2) + 4M^2\hat{q}^2\} - 4M^2 m_{12}\delta m\sigma + 4M^4\sigma^2]\}$$

where using Eq.(2.3) and Eq.(2.5), we had expressed scalar products $p_1{\cdot}p_2$, $p_1.P$ and $p_2.P$ as:

$$p_1{\cdot}p_2 = -M^2(\hat{m}_1+\sigma)(\hat{m}_2-\sigma) - \hat{q}^2, \tag{3.12}$$

$$p_1.P = -M^2(\hat{m}_1+\sigma),$$
$$p_2.P = -M^2(\hat{m}_2-\sigma)$$

while $m_{12} = m_1 + m_2$ and $\delta m = m_2 - m_1$. Here we have employed unequal mass kinematics when the hadron constituents have different masses. We see that on the right hand side of the expression for f_P, each of the expressions multiplying the constant parameters B_0 and B_1 consist of two parts, of which only the second part explicitly involves the off-shell parameter σ (that is terms involving σ^1 multiplying B_0, and terms involving both σ^1 and σ^2 multiplying B_1). It is seen that the off-shell contribution which vanishes for $m_1 = m_2$ in case of f_P calculation in CIA [9] using only the leading covariant γ_5, now no longer vanishes for $m_1 = m_2$ in the above calculation for f_P (when another covariant $i\gamma_5\gamma.P/M$ is incorporated in $Hq\overline{q}$ vertex function besides the leading covariant γ_5) due to the term $4M^4$ multiplying σ^2 in B_1. This possibly implies that when other covariants besides the leading covariant γ_5 are incorporated into the vertex function, the off-shell part of

f_P does not arise from unequal mass kinematics alone (which is in complete contrast to the earlier CIA calculation of f_P employing only γ_5). This may also be a pointer to the fact that Dirac covariants other than γ_5 might also be important for the study of processes involving large q^2 (off-shell) (as suggested in [17]). Carrying out integration over $d\sigma$ by method of contour integration by noting the pole positions in the complex σ -plane:

$$\begin{aligned}\Delta_1 &= 0 \Rightarrow \sigma_1^{\pm} = \pm\frac{\omega_1}{M} - \hat{m}_1 \mp i\varepsilon, \omega_1^2 = m_1^2 + \hat{q}^2;\\ \Delta_2 &= 0 \Rightarrow \sigma_2^{\mp} = \mp\frac{\omega_2}{M} + \hat{m}_2 \pm i\varepsilon, \omega_2^2 = m_2^2 + \hat{q}^2\end{aligned} \tag{3.13}$$

we can express f_P as:

$$f_P = \sqrt{3}N_P \int d^3\hat{q} D(\hat{q})\phi(\hat{q}) \begin{bmatrix} 2B_0[m_{12}(1-\frac{\delta m^2}{M^2})\frac{1}{D(\hat{q})} + 2\delta m R_1] \\ +\frac{B_1}{M^3}[\{(-M^4 - 4m_1m_2M^2 + m_{12}^2\delta m^2)\frac{1}{D(\hat{q})} + \frac{4\hat{q}^2}{MD(\hat{q})}\} \\ +\{-4M^2 m_{12}\delta m.R_1 + 4M^4 R_2\}] \end{bmatrix} \tag{3.14}$$

where the relationship between the functions $D_0(\hat{q})$ and $D(\hat{q})$ (see Ref.[10]) is stated in Eq.(3.6) and the results of σ -integration on whether the contour is closed from above or below the real σ -axis is:

$$R_1 = \int_{-\infty}^{+\infty} \frac{Md\sigma}{2\pi i\Delta_1\Delta_2}\sigma = \frac{M^2(-\omega_1+\omega_2) + (m_1^2 - m_2^2)(\omega_1+\omega_2)}{4M^2\omega_1\omega_2(M^2 - (\omega_1+\omega_2)^2)}; \tag{3.15}$$

$$R_2 = \int_{-\infty}^{+\infty} \frac{Md\sigma}{2\pi i\Delta_1\Delta_2}\sigma^2 = \frac{(-M^4 - m_{12}^2\delta m^2 + 4M^2\omega_1\omega_2)(\omega_1+\omega_2) + 2M^2 m_{12}\delta m(\omega_2 - \omega_1)}{8M^4\omega_1\omega_2(M^2 - (\omega_1+\omega_2)^2)}$$

To calculate BS normalizer N_P for a pseudoscalar meson in the expression for f_P in Eq.(3.16), we again use the current conservation condition Eq. (3.8). However the calculation of N_P is extremely complex due to unequal mass kinematics and very lengthy and we do not present it here.

Table III. Leptonic decay constants (in MeV) of f_P values in BSE under CIA for range of values of ratio B_1 / B_0. The decay constants are calculated from data [20]. The masses of hadrons are also from [20]. The values of constituent quark masses and other input parameters used are fixed from hadron spectrum and given in the text. Comparisons with results obtained from other models are also provided. Comparisons are done with results obtained from other models [3,5,11,18,19] and the experimental data [20,21].

		f_π	f_K	f_D	f_{D_S}	f_B
BSE-CIA (using both the covariants γ_5 and $i(\gamma.\varepsilon)(\gamma.P)/M$) for the ratio B_1 / B_0:	.17	93.0	156.7	229.12	291.0	187.5
	.165	101.0	158.7	231.7	293.9	190.4
	.163	104.3	159.8	232.7	295.2	191.6
	.16	109.2	160.7	234.3	296	193
	.155	117.6	162.7	237	299	196.4
	.15	126.1	164	239.7	303	199.6
	.148	**130.7**	**165**	**240**	**304.2**	**201**
	.145	134.7	166.8	242.5	306	202.8
	.14	143.5	168.8	245.3	309.1	206.2
SDE [11] for parameter set $\omega = .45GeV$, $D = .25GeV^2$			164			
SDE [3,5] for parameter sets: $\omega = .3GeV, D = 1.25Ge$			154			
$\omega = .4GeV$, $D = .93GeV^2$			155			
$\omega = .5GeV, D = .79Ge$			157			
Lattice [18]				201 ± 3	249 ± 3.7	
QCD Sum Rule [19]				203 ± 20	233 ± 23	
Expt. Results [20]		130.7	159.8 ± 1.4	222.6 ± 16.7	294 ± 27	
Babar and Belle Collaboration [21]						237 ± 37

We have thus evaluated the general expressions for f_P in framework of BSE under CIA, with Dirac structure $(i\gamma_5)(\gamma.P)/M$ introduced in the $Hq\overline{q}$ vertex function besides γ_5 according to our power counting rule. We see that so far the results are independent of any model for $\phi(\hat{q})$. However, for calculating the numerical values of these decay constants one needs to know the constant coefficients B_0 and B_1 which are associated with the Dirac structures γ_5 and $(i\gamma_5)(\gamma.P)/M$ respectively. The relative value, $\frac{B_1}{B_0}$ is a free parameter without any further knowledge of the meson structure in the framework discussed above. As a first step in P-meson sector, we vary this parameter at the lowest order to see the effect of introducing Dirac structure $i\gamma_5\gamma.P/M$. We see that at $B_1/B_0 = .148$, we get $f_\pi = 130.7 MeV$ which is the experimental value of this quantity. We thus vary B_1/B_0 in the range .14-.16. This range does not mean any preference but is adequate to show the dependence of f_P on B_1/B_0 since we were interested in studying the effect of incorporating only the leading order covariants (γ_5 and $i\gamma_5\gamma.P/M$) in our power counting scheme on the pseudoscalar decay constants f_P for π, K, D, D_S, B mesons on lines similar to our earlier work [1]. However the present calculation of f_P is much more complex due to unequal mass kinematics employed.

The results are given in Table III along with those of other models [3,5,11,18,19] and experimental data [20-21]. It is seen that the numerical values of these decay constants in BSE under CIA improve when Dirac structure $\gamma_5(\gamma.P)/M$ is introduced in the vertex function in comparison to the values calculated with only γ_5. Further discussions are in Sec. 4.

3.3. Decay Constant of Pion in the Process $\pi^0 \to \gamma\gamma$

In this section, we were primarily interested in studying the application of our BSE framework with generalized vertex function described above to decays of a pion into two photons and calculating the pion-photon coupling constant F_π, and in the process studying the validity of our power counting rule which has recently given encouraging results in calculations of leptonic decay constants of pseudoscalar mesons [13] and vector mesons [1]. The invariant amplitude for the decay of pion into two photons (given by the famous triangle diagram) can be expressed by the equation [7],

$$A(\pi^0 \to 2\gamma) = \frac{e^2}{\sqrt{6}} \int d^4q Tr[\Psi(P,q) i\not{\epsilon}_1 S_F(q-Q) i\not{\epsilon}_2] + (1 \Leftrightarrow 2) \qquad (3.16)$$

where the Bethe-Salpeter wave function $\Psi(P,q)$ for a pion is expressed as in Eq.(3.3).

In the following calculation, we again take the leading order terms in the structure of hadron-quark vertex function $\Gamma^P(\hat{q})$ as in Eq. (2.15) due to reasons mentioned in section 2. In Eq.(3.1) after evaluating traces over $\gamma-$ matrices over various terms and expressing the dot products of momenta of constituent quarks p_1 and p_2 with hadron momentum P by means of relations,

$$p_1.P = \frac{1}{2}(\Delta_1 - \Delta_2 - M^2);$$
$$p_2.P = \frac{1}{2}(-\Delta_1 + \Delta_2 - M^2); \qquad (3.17)$$
$$p_1.p_2 = m^2 - \frac{1}{2}(\Delta_1 + \Delta_2 + M^2),$$

we can express the amplitude in Eq.(3.16) as,

$$A(\pi^0 \to 2\gamma) = F_\pi \varepsilon_{\mu\nu\rho\sigma} P_\mu \varepsilon_{2\nu} Q_\rho \varepsilon_{1\sigma}, \qquad (3.18)$$

where $P = p_1 + p_2$ is the total momentum of the pion, k_1, k_2 are the momenta of the two emitted photons with polarizations $\varepsilon_1, \varepsilon_2$ respectively and $Q = k_1 - k_2$ is the momentum difference of the emitted photons. The pion-photon coupling constant F_π is expressed as,

$$F_\pi = \frac{e^2}{\sqrt{6}} N_P \int d^3\hat{q} D(\hat{q}) \phi(\hat{q}) I;$$
$$I = \int_{-\infty}^{+\infty} \frac{M d\sigma}{2\pi i \Delta_1 \Delta_2 \Delta_3} \left[B_0[8m] + \frac{B_1}{M}[-16m^2 + 4(\Delta_1 + \Delta_2)] \right] \qquad (3.19)$$

where Δ_3 is the inverse propagator of the third quark in the triangle diagram, whereas $\Delta_{1,2}$ are the inverse propagators of the two constituent quarks forming the pion. These inverse propagators of the three quarks can in turn be expressed in terms of the off-shell parameter σ as,

$$\Delta_1 = \omega^2 - M^2(\frac{1}{2} + \sigma)^2;$$
$$\Delta_2 = \omega^2 - M^2(\frac{1}{2} - \sigma)^2; \qquad (3.20)$$
$$\Delta_3 = \omega^2 - M^2\sigma^2$$

In Eq. (3.19), carrying out integration over $d\sigma$ by method of contour integration by noting the pole positions in the complex σ -plane:

$$\Delta_1 = 0 \Rightarrow \sigma_1^\pm = \pm\frac{\omega}{M} - \frac{1}{2} \mp i\varepsilon,$$
$$\Delta_2 = 0 \Rightarrow \sigma_2^\pm = \pm\frac{\omega}{M} + \frac{1}{2} \mp i\varepsilon,$$

$$\Delta_3 = 0 \Rightarrow \sigma_3^{\pm} = \pm\frac{\omega}{M} \mp i\varepsilon\,, \omega^2 = m^2 + \hat{q}^2, \tag{3.21}$$

we can express F_π as,

$$F_\pi = \frac{e^2}{\sqrt{6}} N_P \int d^3\hat{q} D(\hat{q})\phi(\hat{q}) \left[B_0[8m]S_1 + \frac{B_1}{M}[-16m^2 S_1 + 4(S_2 + S_3)] \right] \tag{3.22}$$

where the results $S_{1,2,3}$ of $\sigma -$integrations in the above equation on whether the contour is closed from above the real axis or from below the real axis are:

$$S_1 = \int_{-\infty}^{+\infty} \frac{M d\sigma}{2\pi i \Delta_1 \Delta_2 \Delta_3} = \frac{12}{M^4\omega - 20M^2\omega^3 + 64\omega^5};$$

$$S_2 = \int_{-\infty}^{+\infty} \frac{M d\sigma}{2\pi i \Delta_1 \Delta_3} = \frac{4}{-M^2\omega + 16\omega^3};$$

$$S_3 = \int_{-\infty}^{+\infty} \frac{M d\sigma}{2\pi i \Delta_2 \Delta_3} = \frac{4}{-M^2\omega + 16\omega^3}. \tag{3.23}$$

In Eq.(3.22), the coefficient of parameter B_1 gives us the contribution of the other leading covariant $\gamma_5\gamma.P/M$ to the pion-photon coupling constant F_π. We have thus evaluated the general expressions for F_π in framework of BSE under CIA, with Dirac structure $(i\gamma_5)(\gamma.P)/M$ introduced in the $Hq\bar{q}$ vertex function besides γ_5 according to our power counting rule. We see that so far the results are independent of any model for $\phi(\hat{q})$. However, for calculating the numerical values of F_π, one again needs to know the constant coefficients B_0 and B_1 which are associated with the Dirac structures γ_5 and $(i\gamma_5)(\gamma.P)/M$ respectively. The relative value, $\frac{B_1}{B_0}$ is a free parameter without any further knowledge of the meson structure in the framework discussed above. The ratio $\frac{B_1}{B_0}$ was fixed by calibrating to the decay constants of P-mesons (π, K, D, D_S, B) [13]. It was found that the best fit of calculated values of decay constants of π, K, D, D_S, B mesons to their experimental values are obtained for the range of parameter values: $\frac{B_1}{B_0}$ =.14 - .16 (see Table

II). We see that at $B_1/B_0 = .148$, we get the pion decay constant, $f_\pi = 130.7MeV$ which is the experimental value of this quantity. Thus for making comparison of decay constants of other mesons to their experimental values, we fix the value, $\frac{B_1}{B_0} = .148$. At this value of .148, the values of pseudoscalar decay constants for various mesons in our model with use of leading order (LO) Dirac covariants alone were found to be (see Table III for details):

$$f_\pi = 130.7MeV(Expt. = 130.7 \pm .1)$$

$$f_K = 165MeV(Expt. = 159.8 \pm 1.4), \tag{3.26}$$

$$f_D = 240MeV(Expt. = 222.6 \pm 16.7),$$

$$f_{D_S} = 304.2MeV(Expt. = 294 \pm 27),$$

$$f_B = 201MeV\ (Expt. = 237 \pm 37)\ .$$

The experimental data (in brackets) for f_P for π, K, D, D_S mesons have been taken from Ref. [20], while for B meson has been taken from Ref.[21]. It is seen that the numerical values of these decay constants listed above in BSE under CIA improve dramatically when Dirac structure $\gamma_5(\gamma.P)/M$ is introduced in the vertex function in comparison to the values calculated with only γ_5 (see Table III).

However as regards the calculation of pion-photon decay constant F_π is concerned, we see that at $B_1/B_0 = .148$, the BS normalizer for pion works out at $N_\pi = .3198$. At this value of the normalizer, we get,

$$F_\pi = .026GeV^{-1}, \tag{3.27}$$

which is close to the corresponding experimental value of this quantity at $.02424GeV^{-1}$ obtained from the experimental value of decay width $\Gamma_{\pi^0 \to 2\gamma} = 8.02eV$ [20] by means of the relation [16], $\Gamma(\pi^0 \to \gamma\gamma) = \frac{F_\pi^2 M_\pi^3}{64\pi}$.

We compare our result with one of the recent relativistic calculation [22] of two photon decay width of π^0 meson in a Bethe-Salpeter formalism. The decay width in their model is worked out as $\Gamma_{\pi^0 \to 2\gamma} = 7.7eV$ which corresponds to $F_\pi = .023GeV^{-1}$ which is close to the result of our calculation.

4. Conclusion

In this paper we first postulated and discussed in detail a naïve power counting rule for incorporation of various Dirac covariants in the wave functions of different mesons (vector, pseudo-scalar etc.). Then in the framework of Bethe-Salpeter equation under Covariant Instantaneous Ansatz (CIA), we calculate (i) decay constants f_V for equal mass vector mesons ρ,ω and ϕ, (ii) decay constants f_P for unequal mass pseudoscalar mesons π,K,D,D_S and B and (iii) pion-photon coupling constant F_π in the process $\pi^0 \rightarrow \gamma\gamma$ using leading order (LO) covariants in Hadron-quark vertex. The values of decay constants f_V and f_P are listed in Tables II and III respectively along with data and the results of other models. It is seen that the decay constants f_V improve considerably when Dirac structure $\not{\epsilon}P/M$ is introduced in the vertex function with tuned parameter A_1/A_0 and come close to results of recent calculations [3,5,11,18,19] and experimental data [20,21], where it is to be mentioned that for rough comparison with experiment, we take f_V values for the parameter ratio A_1/A_0 in the range .08-.09 due to reasons mentioned in Section 3.1. However we wish to mention that though the results of f_V are quite close to data for ρ and ω mesons with use of leading order covariants alone, those of ϕ are a bit larger. This may suggest that whereas the leading order covariants $i\not{\epsilon}$ and $\not{\epsilon}P/M$ are sufficient for calculation of decay constants of lightest vector mesons ρ and ω, they are not sufficient for calculation of decay constant of heavier meson like ϕ, for which one may have to go to next-to-leading order terms (i.e. $O(1/M^1), O(1/M^2)$ etc.) in Eq.(2.11). The reason is that the relative momentum may not be small and the assumption $q << P \sim M$ may not be too good for the heavier of the light quark mesons.

As regards pseudo-scalar mesons are concerned, it is seen that the values of Decay constants f_P can improve considerably when the Dirac structure $\gamma_5 P/M$ is introduced in the vertex function with tuned parameter B_1/B_0 and come closer to the results of some recent calculations [3,5,11,18,19] as well as agree with Experimental results [20,21] within error though f_B is a bit lower (for details see Table III).

In our model when we include the other leading order Dirac covariant $\gamma_5 P/M$ besides γ_5, and get f_P as a function of the parameter ratio B_1/B_0. If we calibrate B_1/B_0 to reproduce the experimental value of Kaon decay constant, f_K=159.8MeV [20] we get B_1/B_0=.163. Using this value of tuned parameter, we then obtain f_D=232.78MeV (Expt.=222.6$\pm$ 16.7), f_{D_S} =295.18MeV (Expt.=294$\pm$ 27), which agree with data [20] within the errors. The decay constant for B-meson predicted in our framework is f_B=191.6MeV, which is not far from the recent experimental result for f_B [21]. As can be seen from Table

III, our results after calibrating to kaon data are as good as results obtained after calibrating to pion data (in Eq.(3.26)). Further our model predicts f_{D_S} value to be around 22% larger than f_D value which is roughly consistent with the prediction of most of the models which generally predict f_{D_S} to be 10% - 25% larger than f_D (as per recent studies in Ref.[23]). However for making comparison with results of various models and experimental data, we make use of calibration to pion data.

We further studied the two photon decays of a pion using the above generalized form of hadron-quark vertex function. The numerical value of pion-photon coupling constant using the value of the parameter ratio fixed at $\frac{B_1}{B_0} = .148$ after calibrating to the pion data is $F_\pi = .026 GeV^{-1}$ which is quite close to its corresponding experimental value of $.02424 GeV^{-1}$ [20]

The numerical result for F_π for the process $\pi^0 \rightarrow 2\gamma$, as well as the results for f_P (for π, K, D, D_S and B mesons) with use of leading order covariants γ_5 and $\gamma_5 \gamma.P/M$, along with those of f_V [1] confirm the validity of our power counting rule, according to which the leading order covariants should contribute maximum to any meson observable, and inspire the possibility to investigate higher order terms in order to get better agreement between calculations and data. This would in turn help in achieving a better understanding of hadron structure.

Note added in Proof: *We have recently completed the calculation of leptonic decay constants of pseudoscalar mesons with use of full Hadron-quark vertex after incorporation of both LO (γ_5 and $-i\gamma_5 P/M$) and NLO ($-i\gamma_5 q/M$ and $-\gamma_5(Pq - qP)/M^2$) Dirac covariants (see [24] for details) in Hadron-quark vertex. for P-meson in Eq. (2.14) and also studied the relevance of both LO and NLO Dirac covariants to f_P calculation. Decay constant f_P with full Hadron-quark vertex is then expressible in a general form as [24]:*

$$f_P = f_P^{(0)} + f_P^{(1)} + f_P^{(2)} + f_P^{(3)}; \tag{4.1}$$

where $f_P^{(0)}$ and $f_P^{(1)}$ are the contributions to f_P from LO Dirac covariants γ_5 and $-i\gamma_5 P/M$ respectively associated with constant coefficients B_0 and B_1, while $f_P^{(2)}$ and $f_P^{(3)}$ are the contributions from NLO Dirac covariants $-i\gamma_5 q/M$ and $-\gamma_5(Pq - qP)/M^2$ associated with constant coefficients B_2 and B_3. For details of the expression for f_P see Ref.[24]. Using the method of least square fitting of data, we find that the values of coefficients $B_0,..., B_3$ (with average error with respect to experimental data < 3.5%) should respectively be $B_0 = .7045, B_1 = .2626, B_2 = .0573$ and $B_3 = .0573$ to give decay constant values [24]:

$f_\pi = 130 Mev(Exp. = 130 \pm .1 Mev)$, $f_K = 164 Mev(Exp. = 159 \pm 1 Mev)$,

$f_D = 194 Mev(Exp. = 220 \pm 20 Mev)$, $f_{D_S} = 296 Mev(Exp. = 290 \pm 30 Mev)$

$f_B = 228 Mev(Exp. = 240 \pm 40 Mev)$

which are within the error bars of experimental data [20,21] depicted in Table III for these five mesons. As far as the contribution of LO and NLO covariants (identified in accordance with our power counting rule) to decay constant was concerned, for K *meson it was found that, the LO covariants contribute 60%, while NLO contribute 40%. However for heavy-light meson* D*, the LO contribution increases to 90%, while the NLO contribution was found to drop to 10%. For* D_s *meson, the the LO contribution is 91%, while the NLO contribution is 9%, while for* B *meson, the LO contribution is 96%, while NLO contribution drops to just 4% (see Ref.[24] for details). These findings are completely in accordance with the power counting rule proposed according to which the leading order (LO) covariants in BS wave function should contribute maximum to* f_P *followed by the next-to-leading order (NLO) covariants. For the lightest meson, pion the NLO contribution is quite significant. However, the numerical results for* f_P *obtained in our framework with use of LO and NLO covariants demonstrates the validity of our power counting rule which also provides a practical means of incorporating various Dirac covariants in BS wave function for a hadron. Work is currently in progress on a number of other hadronic processes with the incorporation of the full BS wave function after incorporation of all the Dirac covariants.*

Acknowledgement

The author is thankful to Dr.Shi-Yuan Li, Shandong University, China for valuable suggestions on the form of Hadron-quark vertex in BS wave functions which laid the basis for this work. This work has been carried out at Addis Ababa University and has been done within the framework of Associateship scheme of ICTP.

References

[1] Bhatnagar,S., Li ,S-Y. *J. Phys.G* **32**, 949 (2006).
[2] Ivanov M.A., Kalinovsky Yu.A., Roberts C.D., *Phys. Rev.* **D60**, 034018 (1999).
[3] Maris P., Tandy P.C., *Phys. Rev.* **C60**, 055214 (1999).
[4] Burden C.J., Roberts C.D., Thomson M.J., *Phys. Lett.* **B371**, 163 (1996).
[5] Alkofer R., Smekel L.V., *Phys. Rep.***353**, 281 (2001).
[6] Maris P., Roberts C.D., *Phys. Rev.* **C56**, 3369 (1997).
[7] Mitra A.N., Bhatnagar S., *Intl. J. Mod. Phys.* **A7**, 121(1992).
[8] Resag J., Muenz C.R., Metsch B.C. and H.R. Petry , *Nucl.Phys.* **A578**, 397 (1994).
[9] Bhatnagar S., Kulshreshtha D.S., Mitra A.N., *Phys. Lett.* **B263**, 485 (1991).
[10] Bhatnagar S., *Intl. J. Mod. Phys. E* **14**, 909 (2005).
[11] Alkofer R., Watson P., Weigel H., *Phys. Rev.* **D65**, 094026 (2002).
[12] Smith C.H.L., *Ann. Phys.* **53**, 521 (1969).

[13] Bhatnagar S., Li S-Y., Arxiv: hep-ph/ 0612084v4 (2007).
[14] Chakraborty S. et al., *Prog. Part. Nucl. Phys.* **22**, 43 (1989).
[15] Nambu Y., Jona Lasino G., *Phys. Rev.* **122**, 345 (1961).
[16] Mitra A.N., Sodermark B.M., *Nucl. Phys.* **A695**, 328 (2001).
[17] Bhagwat M.S., Pichowsky M.A., Tandy P.C., *Phys. Rev.* **D67**, 054019 (2003).
[18] Follana E. et al (HPQCD and UKQCD), arXiv:0706.1726[hep-lat].
[19] Narison S., hep-ph/0202200.
[20] Eidelman S.,et al., (Particle Data Group), *Phys. Lett.* **B592**, 1, (2004).
[21] BaBar and Belle Collaboration, http://utfit.roma1.infn.it/.
[22] Artuso M. et al. (CLEO Collaboration), *Phys. Rev. Lett.* **95**, 251801 (2005).
[23] Costa P. et al., *Phys. Rev.* **C70**, 048202 (2004).
[24] Bhatnagar S., Mahecha J., ICTP preprint IC/2008/081 (2008).

In: The Physics of Quarks: New Research
Editors: N.L. Watson and T.M. Grant, pp. 75-86
ISBN 978-1-60456-802-8

Chapter 5

PENTAQUARKS – STRUCTURE AND REACTIONS

Atsushi Hosaka
Research Center for Nuclear Physics (RCNP), Osaka University
Ibaraki 567-0047 Japan

1. Introduction

History of exotic hadrons is as old as the quark model [1, 2]. Yet the recent experimental evidence for the pentaquark baryon Θ^+ observed by the LEPS group [3] has triggered enormous amount of research activities both in experimental and theoretical studies [4]. Recent experimental situation is rather uncertain, after many negative results have been reported [5, 6]. Nevertheless, the virtue of the study of the exotic pentaquarks should remain the same, since the baryons containing five valence quarks are totally new form of hadrons.

In hadron physics, the understanding of five quark systems, if they exist and can be studied in laboratories, will give us more information on the dynamics of non-perturbative QCD, such as confinement of colors and chiral symmetry breaking. Many ideas have been proposed attempting to explain the unique features of Θ^+. As it has turned out and will be discussed in this note, however, the current situation is not yet settled, having revealed that our understanding of hadron physics would be poorer than we thought. Even an *ab initio* method of lattice QCD does not seem in a conclusive status, see for instance [7, 8, 9] We definitely need more ideas and methods to answer the related questions.

Turning to the specific interest in Θ^+, its would-be light mass and narrow width are the issues to be understood, together with the determination of its spin and parity [10, 11]. In particular, the information of parity is important, since it reflects the internal motion of the constituents.

In this note after a brief summary for experiments, we discuss mostly theoretical aspects of the recent research on the pentaquark baryon. For the discussion of the structure, we use quark models. We discuss the parity and decay properties in a simple framework. We then show the results of the recent serious calculation for the five-quark $uudd\bar{s}$ system for Θ^+. Finally, we discuss production reactions with some remarks on the recent experimental status.

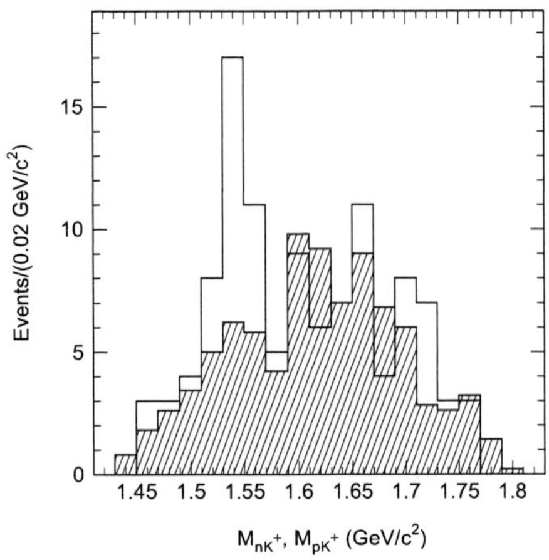

Figure 1. Invariant mass spectrum of the nK^+ extracted from the missing mass analysis of $\gamma n \rightarrow K^-K^+n$ (unhatched histogram) and that of pK^+ extracted from the analysis of $\gamma p \rightarrow K^-K^+p$ (hatched histogram) [3].

2. Experiment

The first observation was made by the LEPS group at SPring-8 [3]. The backward compton-scattered photon of energy 2.4 GeV produced at SPring-8 was used to hit a neutron target inside a carbon nucleus to produce a strangeness and antistrangeness pair (K^+ and K^-). The Fermi motion corrections were carefully done, and then a missing mass analysis was performed for the K^+n final state. They have seen an excess in the K^+n invariant mass spectrum over the background at 4.6σ level in the energy region 1.54 GeV as shown in Fig. 1. The width of the peak was as narrow as or less than the experimental resolution ($\sim$ 25 MeV). The peak was then identified with the exotic pentaquark state of strangeness $S = +1$. The absence of the similar peak structure in the K^+p system suggests the isospin of the state is likely to be $I = 0$.

Soon after the report of this result, many positive signals follow [4]. Major results of experiments so far are summarized in Table 1. However, after many positive signals were reported, negative results followed also, mostly from the analysis of high energy experiments. They are also summarized in Table 1. Recently, CLAS (g11) reported the null result in the reaction $\gamma p \rightarrow \bar{K}^0K^+n$ [12]. This has much larger statistics than the previous experiment performed at SAPHIA by about factor twenty. They extracted an upper limit of the Θ^+ production cross section, $\sigma \lesssim 1-4$ nb. The CLAS also reported the results for the neutron which did not see the prominent peak for Θ^+, but with the possible upper bound for the production cross section of several tens nb [13]. These results, however, do not immediately lead to the absence of Θ^+. This will be discussed in section 4.

Table 1. Brief summary of the previous experiments

Experiment	Reaction	Energy (GeV)	Mass (MeV)	Width (MeV)
LEPS	γn	$E_\gamma \sim 2.4$	$\Theta^+ \sim 1540$	$\Gamma \lesssim 25$
	γd		being analyzed	
Graal	$\gamma p \to \eta p$	$E_\gamma \lesssim 1.5$	$N_5^* \sim 1715$	
			$\Theta^+ \sim 1531$	
CLAS	γp	$1.6 \lesssim E_\gamma \lesssim 2.3$	$\Theta^+ \sim 1555 \pm 10$	$\Gamma \lesssim 20$
	γd	$1.5 \lesssim E_\gamma \lesssim 3.1$	$\Theta^+ \sim 1542 \pm 5$	
CLAS(g11)	γp	$1.6 \lesssim E_\gamma \lesssim 3.8$ GeV	–	
	γd		being analyzed	
HallA	ep	$E_\gamma \lesssim 5$	being analyzed	
HERMES	ed	$\sqrt{s} \sim 10$	$\Theta^+ \sim 1527 \pm 2.3$	$\Gamma = 17 \pm 9 \pm 3$
ZEUS	ep	$300 \lesssim \sqrt{s} \lesssim 319$	$\Theta^+ \sim 1521.5 \pm 1.5$	$\Gamma = 6.1 \pm 1.6^{+2}_{-1.4}$
H1	ep	$300 \lesssim \sqrt{s} \lesssim 319$	$\Theta_C^+ \sim 3100$	
NA49	pp	$\sqrt{s} \sim 17.2$	$\Xi^{--} \sim 1680$	
			$\Theta^+ = 1526 \pm 2$	
COSY-TOF	pp	$p \sim 2.95$	$\Theta^+ = 1530 \pm 5$	$\Gamma \leq 18 \pm 4$
E522(KEK)	$\pi^- p$	$p \sim 1.95$	$\Theta^+ = 1530 \pm 5$	
E690(Fermilab)	pp	$\sqrt{s} = 800$	–	
CDF(Fermilab)	$\bar{p}p$	$\sqrt{s} = 1960$	–	
Babar	e^+e^-		–	
Belle	e^+e^-	$e^+(3.5)e^-(8)$	–	

3. Theoretical Models

In this section, we discuss the structure of Θ^+ using quark models. Although the pioneering work of Diakonov *et. al.* was performed in the chiral soliton model [14], it is always instructive and intuitively understandable to work in the quark model [15, 17].

Quark model

The constituent quark model has been successfully applied to the description of the conventional mesons and baryons [16]. In this model, a confining potential for quarks is introduced to prepare basis states as single particle states for valence quarks. Then quark-quark interactions such as the spin-color interaction of one gluon exchange or the spin-flavor one of meson (the Nambu-Goldstone boson) exchange is introduced as a residual interaction. The interaction hamiltonian is then treated either perturbatively or diagonalized within a given model space. The role of various interactions for Θ^+ has been investigated in the literatures [15, 18, 19, 20].

In the naive quark model, the five quarks of $uudd\bar{s}$ occupy the lowest s-orbit which forms a $(0s)^5$ configuration. Two spin values ($J = 1/2$ or 3/2) are possible, but the parity is uniquely determined as negative. The $J^P = 1/2^-$ state of the $(0s)^5$ state is, however, rearranged to be the KN scattering states. Hence it can not be the narrow state as observed in experiments. On the other hand, $J^P = 3/2^-$ state can be rearranged as $K^*(J^P = 1^-)N$ state which, however, can not be a decay channel of Θ^+ due to energy conservation. Furthermore, a kinematically allowed KN mode from the $J^P = 3/2^-$ state should be in d-state, to which the $(0s)^5$ state can not decay due to the conservation of the orbital angular momentum. Hence, $J^P = 3/2^-$ state could be one of the candidates for a narrow resonance. That the mass of such a state takes 1.5 GeV is a dynamical question and is not yet solved.

The role of the pion

For $J^P = 1/2^+$, a quark model state requires an excitation of a quark in a p-orbit, $(0s)^4 0p$ configuration, with an energy excess of $\hbar\omega \sim 500$ MeV. For this state to be the ground state of about 1.5 GeV for Θ^+, there should be a mechanism to make this state lower than the $(0s)^5$ state. It was shown that the chiral interaction mediated by the pion may be responsible for this [21, 22].

Here we illustrate it using the chiral bag model, where quark single particle orbits are considered as functions of the chiral angle at the bag surface $F(R)$ (see Fig. 2) [22, 23]. The quark states are labeled by the grand angular momentum $\vec{K} = \vec{J} + \vec{I}$, the sum of the total angular momentum and isospin, due to the strong spin-isospin correlation of the hedgehog pion field. An important observation in the chiral bag model is that as the pion-quark interaction increases (as F gets larger), the level crossing between states of different parities occur. For instance, at $F(R) \gtrsim 0.3\pi$, the 0^+ level is no longer the lowest, which is replaced by the 1^- orbit. Hence the ground state configuration of the five quarks changes from $(0^+)^4 1^+ \sim (0s)^5$ to $(0^+)^4 1^- \sim (0s)^4 1p$. The latter configuration carries the positive parity. If such a mechanism dominates the five quark state may appear with relatively light mass with the positive parity.

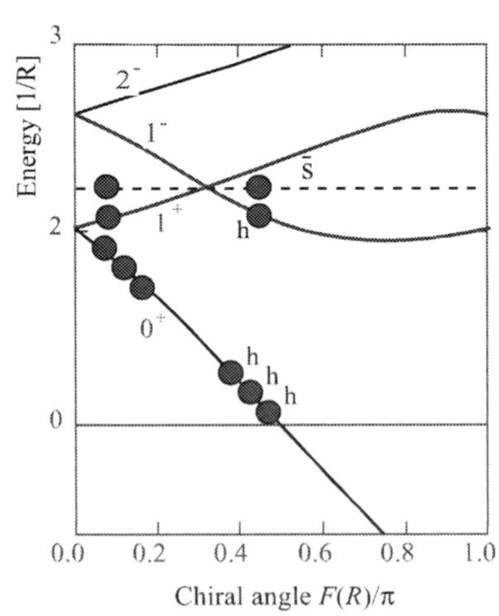

Figure 2. Quark energy levels of the chiral bag model as functions of the chiral angle F. For u, d hedgehog quarks, energy levels vary as functions of $F(R)$ as denoted by h, while the s quark is not affected by the interaction. The blobs indicate how five quarks occupy the levels.

Table 2. Spectroscopic factors and decay widths (in MeV) of Θ^+ for $J^P = 1/2^\pm$. For notations SF, SC and JW, see text.

	$J^P = 1/2^-$		$1/2^+$	
		SF	SC	JW
S-factor	$1/2\sqrt{2}$	$\sqrt{5/96}$	$\sqrt{5/192}$	$\sqrt{5/576}$
Γ (MeV)	890	63	32	11

Decay of Θ^+

In the quark model, decay of the pentaquark Θ^+ occurs through the fall-apart process as shown in Fig. 3 (left). In order to calculate such a process, one should first construct a state for Θ^+ by solving a model hamiltonian and then compute the transition amplitudes. However, in order to simplify the problem, we perform an analysis by assuming several five-quark configurations [24]. This is useful to see the relation of the decay and the internal structure.

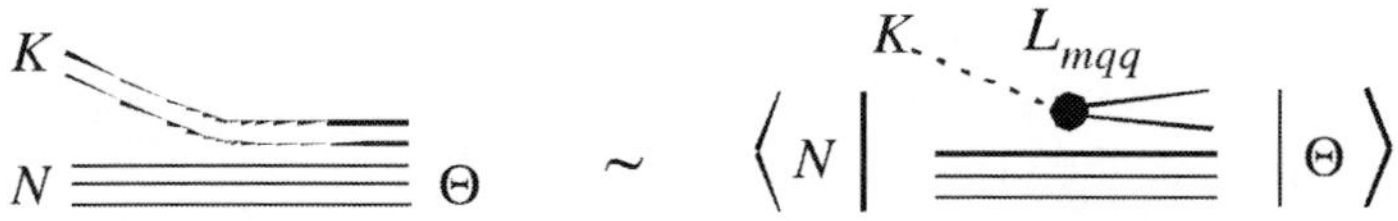

Figure 3. Decay of the pentaquark state.

We compute the matrix element of the standard quark-meson interaction of Yukawa type

$$L_\pm = g_{KN\Theta}\bar{\psi}_N\gamma_\pm\psi_\Theta K\,, \tag{1}$$

between the five-quark state for Θ^+ and the three-quark state for the nucleon. The interaction is responsible for the annihilation of the $u\bar{s}$ pair going to the kaon in the final state, see Fig. 3 (right). We have considered the following different cases, $(0s)^5$ states of $J^P = 1/2^-$ and $(0s)^4 0p$ of $J^P = 1/2^+$. The calculation for $J^P = 3/2^+$ is the same as the case of $J^P = 1/2^+$. The decay of $(0s)^5$ states of $J^P = 3/2^-$ is forbidden due to the reason that we discussed before. For $(0s)^4 0p$ states, we study three different configurations: the one minimizing the spin-flavor interaction of the type $\sum_{i>j}(\sigma_i\sigma_j)(\lambda_i^f\lambda_j^f)$ (SF), the one minimizing the spin-color interaction $\sum_{i>j}(\sigma_i\sigma_j)(\lambda_i^c\lambda_j^c)$ (SC), and the one of the strong diquark correlations as proposed by Jaffe and Wilczek (JW) [25].

Here we summarize the results as follows. For the negative parity state of $(0s)^5$, the decay width turns out to be of order of several hundreds MeV or more, typically $0.5 \sim 1$ GeV. This is an expected result since $(0s)^5$ $1/2^-$ configuration is equivalent to an s-wave KN state in the same oscillator basis. For the positive parity state, we obtain $\Gamma = 63$ MeV, 32 MeV and 11 MeV, for the SF, SC and JW configurations, respectively. The diquark correlation of (JW) develops a spin-flavor-color wave function having a small overlap with

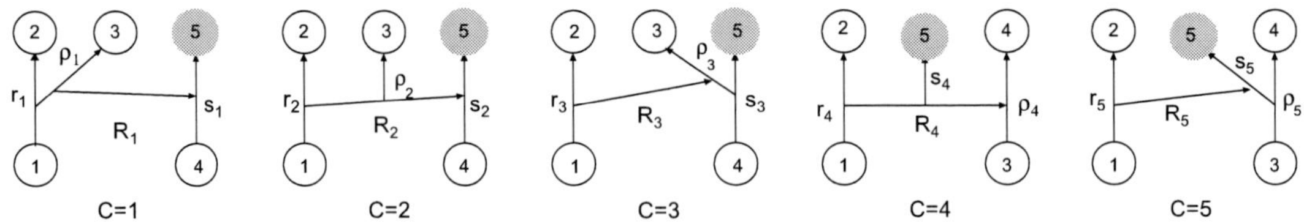

Figure 4. Five sets of Jacobi coordinates among five quarks. Four u, d quarks, labeled by particle $1-4$, are to be antisymmetrized, while particle 5 stands for $\bar{s}$ quark. Sets $c = 4, 5$ contain two qq correlations, while sets $c = 1-3$ do both qq and $q\bar{q}$ correlations. Sets $c = 4, 5$ describe molecular configurations and sets $c = 1-3$ does connected ones, as shown in the text. The NK scattering channel is treated with $c = 1$.

the decaying channel of the nucleon and kaon. Hence, as anticipated, $(0s)^5 1/2^-$ state can not be identified with a narrow resonance, while a $1/2^+$ state may be possible if a strongly correlated wave function is developed such as the JW configuration.

Solving the five-body system

Recently, we have considered a full treatment of the five-body system for the pentaquark $uudd\bar{s}$ [26]. The previous theoretical studies must have been qualitative since the arguments were based on some assumptions which looked reasonable without solving the five-body system, as we have done in the previous paragraph.

The method consists of two steps. First, an accurate bound state solution is obtained, by the variational Gaussian expansion method [27, 28] where the wave function is expanded by a set of Gauss functions with various combinations of internal coordinates. For the present study, we consider five combinations of the internal Jacobi coordinates as shown in fig. 4. The bound state solution is then coupled with scattering states between possible sub-particles forming $uudd\bar{s} \sim \Theta^+$.

Since such a study is new in the pentaquark study, we consider a simple quark model hamiltonian by Isgur-Karl. The validity of different model hamiltonians should be discussed later. Therefore, we take,

$$H = \sum_i \left(m_i + \frac{\mathbf{p}_i^2}{2m_i} \right) - T_G + V_{\text{Conf}} + V_{\text{CM}} , \tag{2}$$

where m_i and $\mathbf{p}_i$ are the mass and momentum of ith quark and T_G is the kinetic energy of the center-of-mass system. In what follows, u and d quarks are labeled by $i = 1-4$, and $\bar{s}$ by $i = 5$. The confining potential V_{Conf} is of harmonic oscillator type:

$$V_{\text{Conf}} = -\sum_{i<j} \sum_{\alpha=1}^{8} \frac{\lambda_i^\alpha}{2} \frac{\lambda_j^\alpha}{2} \left[\frac{k}{2} (\mathbf{x}_i - \mathbf{x}_j)^2 + v_0 \right] , \tag{3}$$

where $\mathbf{x}_i$ is the position vector of the ith quark, v_0 is a mass shift parameter, and λ_i^α are the Gell-Mann matrices for color. The color-magnetic potential V_{CM} is given by

$$V_{\text{CM}} = \sum_{i<j} \sum_{\alpha=1}^{8} \frac{\lambda_i^\alpha}{2} \frac{\lambda_j^\alpha}{2} \frac{\xi_\sigma}{m_i m_j} e^{-(\mathbf{x}_i - \mathbf{x}_j)^2/\beta^2} \boldsymbol{\sigma}_i \cdot \boldsymbol{\sigma}_j . \tag{4}$$

For parameters, first we take standard values of quark masses $m_u = m_d = 330$ MeV and $m_{\bar{s}} = 500$ MeV. As for k and β we take $k = 455.1$ MeV$\cdot$ fm^{-2} and $\beta = 0.5$ fm [16]. The remaining parameters ξ_σ and v_0 are determined such that the three-body calculation reproduces $m_N = 939$ MeV and $m_\Delta = 1232$ MeV [29]. The resulting values are $\xi_\sigma/m_u^2 = -474.9$ MeV and $v_0 = -428.3$ MeV.

The solution to the $uudd\bar{s}$ system for the hamiltonian (2) can be written as

$$\Psi_{J^\pi M}(E) = \Psi^{(NK)}_{J^\pi M}(E) + \sum_{\nu=1}^{\nu_{\max}} b_\nu(E)\widehat{\Phi}_{J^\pi M}(E_\nu). \tag{5}$$

where the first term represents the KN scattering states and the second term the would-be $uudd\bar{s}$ bound states.

We have searched for solutions for $J^\pi = \frac{1}{2}^-$ and $\frac{1}{2}^+$. First we consider a solution only including the first term of (5). In this case, N and K behave as inert particles. The resulting phase shifts are shown in Fig. 5 as dash-dotted lines showing little structure. Now by including the second terms of (5), the results changes drastically, as shown by the solid lines. No resonance is seen in the energy region $0 - 500$ MeV above the NK threshold ($1.4 - 1.9$ GeV in mass), while we find two resonances around 530 MeV; one is a sharp $J^\pi = \frac{1}{2}^-$ resonance with a width of $\Gamma = 0.12$ MeV located at $E - E_{\rm th} = 540$ MeV and the other is a broad $\frac{1}{2}^+$ resonance at ~ 520 MeV with $\Gamma \sim 110$ MeV.

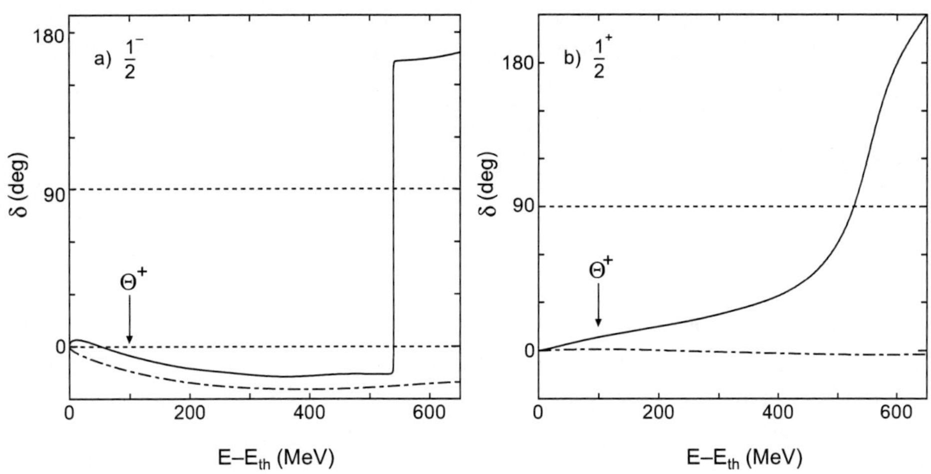

Figure 5. Calculated phase shifts for a) $J^\pi = \frac{1}{2}^-$ and b) $J^\pi = \frac{1}{2}^+$ states. The solid lines are given by the full-fledged calculation, while the dash-dotted lines are by the approximate calculation with the elastic NK channel alone (see (5)). Energies are measured from the NK threshold. The arrow indicates the energy of $\Theta^+(1540)$ in $E - E_{\rm th}$.

The structure of the two resonances are interesting to look at. In particular, $1/2^-$ state has a large contribution from the second term of (5). It is not, however, dominated by a single term, but many higher states contribute. This makes the coupling to the KN scattering state relatively weaker leading to the narrow resonance. On the other hand, for $1/2^+$, there is a significant contribution from the first term in (5). In both cases, the second term

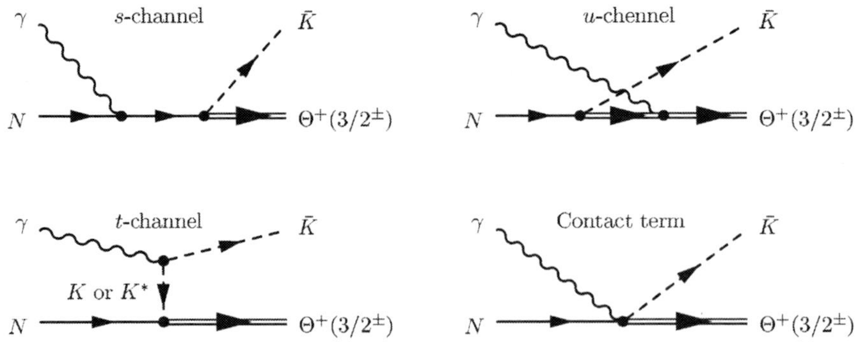

Figure 6. Born diagrams of the effective lagrangian approach.

is not dominated by components of 4 and 5 of Fig. 4, but there is a significant contribution from the configurations of 1-3. This follows from the fact that the color singlet correlation is strong. In fact, in the color exchange interaction such as (4), the $\bar{q}q$ interaction is twice as strong as the qq interaction. The formation of diquark correlation is thus a dynamical question depending on the nature of the underlying quark interaction.

The present study showed that if we naively apply the quark model Hamiltonian which has been used widely for three quark systems to the five-body system, we can not produce a state with a low mass and a narrow decay width. We do not know whether this indicates the absence of the narrow pentaquark state or an artifact of the incomplete knowledge of the low energy QCD.

4. Photoproductions

As discussed in the preceding sections, recent situation for the pentaquark Θ^+ is rather uncertain due to many negative results against the existence of the exotic states [5, 6]. Among others, recent data from CLAS experiments of Jefferson laboratory have been taken seriously [12, 13]. What was unexpected was, having about ten times more events, an originally observed peak of Θ^+ has disappeared [13]. Yet, the LESP has observed a pronounced peak corresponding to Θ^+ in the same mass region as before under an interesting kinematics of the $\Lambda(1520)$ production from the deuteron [31].

In this report, we propose a scenario which could provide one explanation for the null result of CLAS [30], and is still consistent with the existence of Θ^+. Our analysis is based on an effective lagrangian approach and is tested once again for the photoproduction of Θ^+ associated with the kaon ($\bar{K}$) production [32]. Our consideration is minimal in that baryon resonances coupling to $\bar{K}\Theta^+$ are not included. However, we take into account the gauge and chiral symmetries seriously, where we adopt a gauge invariant form factor in a covariant manner [33] and a method of low energy expansion is extended to the kaon sector together with Θ^+. Hence our consideration includes the four Born (tree) diagrams as shown in Fig. 6.

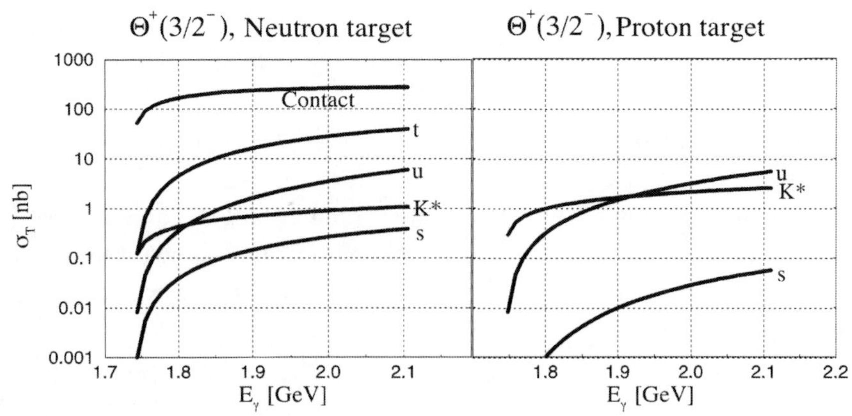

Figure 7. Various components of the total cross section for $J^P = 3/2^-$ as functions of the photon energy in the laboratory frame.

Among them, particularly important is the role of the contact diagram, which is the analogue of the Kroll-Ruderman term for the pion photoproduction. This term is present only for the charge exchange process, i.e. $\gamma n \to K^-\Theta^+$, as required by the gauge symmetry when minimal substitution is made for derivative couplings of meson-baryon interaction lagrangians. The relative importance of the contact term depends on J^P of Θ^+ and on the contributions from other terms. Here we discuss three cases of $J^P = 1/2^+, 3/2^\pm$. In quark models, a low-lying $1/2^-$ state is not likely compatible with a narrow resonance, while $1/2^+$ and $3/2^+$ states can be so if strong diquark correlation is developed [24, 26]. The $3/2^-$ state is an interesting possibility due to the d-wave nature of the final state KN which suppresses the decay of Θ^+.

As shown in Fig. 7, the contact term provides a dominant contribution when $J^P = 3/2^-$, partly due to the suppression of other terms by the form factor. A similar tendency is observed for the case of $J^P = 3/2^+$. Consequently, the values of total cross sections for the neutron target becomes large, while that of the proton target is as small as few nb. Also we have verified that the angular distribution for $J = 3/2$ is highly forward peaking. In the CLAS, detector setup is designed such that the efficiency in a large angle region is more achieved. Therefore, a large amount of kaons produced in a forward angle region are not detected. On the other hand, in the LEPS the detection is optimized in a forward angle region.

In Table 1, we have summarized total cross section values for several cases. We have used $KN\Theta^+$ coupling values corresponding to $\Gamma_{\Theta^+\to KN} = 1$ MeV. Also we have used the cutoff parameter in the form factor $\Lambda \sim 700$ MeV, as corresponding to a baryon size of $\sim \sqrt{6}/\Lambda \sim 0.7$ fm. For Θ^+ this size may be too small, hence a use of a smaller Λ may be preferred, which reduces the cross sections. Therefore, the actual values of cross sections may be smaller than those listed in Table 1.

We can make several remarks

- There is large asymmetry in cross section values between the cases of proton and neutron targets for $J = 3/2$. The importance of the contact term is enhanced partly due to the form factor in the kinematical region of the present reaction.

- Cross section values of the proton target are of only few nb, which is compatible with the upper bound estimated by using the recent high statistics data of CLAS [12].
- The results do not depend very much on the unknown magnetic coupling of Θ^+.
- The results, however, depend significantly on the $K^* N \Theta^+$ coupling. In the present calculation we have used a value compatible with a quark model estimation. The use of a larger value of $K^* N \Theta^+$ coupling reduces the difference between the proton and neutron.

Table 3. Main results of the Θ^+ photoproduction, where all results are for the case of finite $g_{K^* N \Theta}$.

J^P	$3/2^+$		$3/2^-$		$1/2^+$	
Target	n	p	n	p	n	p
σ	$\sim$ 25 nb	$\sim$ 1 nb	$\sim$ 200 nb	$\sim$ 4 nb	$\sim$ 1 nb	$\sim$ 1 nb
$\frac{d\sigma}{d\cos\theta}$	Forward	$\sim 60°$	Forward	–	$\sim 45°$	$\sim 45°$

5. Conclusions

In this note, we have briefly looked at the current situation of the research of the pentaquark baryon. The Θ^+ contains minimally $uudd\bar{s}$ quarks, and is a truly exotic hadron. Hadrons beyond what have so far observed open a door to the new field which we have not experienced before. This is the primary reason that the subject has been very much investigated. The situation, however, has turned out to be not very simple; the first signal seems to be now overtaken by the many negative signals.

Experimentally, as we have emphasized, the setup of the LEPS at Spring-8 is very much different from the others; their detector system is designed to be suited to the observation of particles at forward angles. It is then extremely important to check at the LEPS to perform observation in large angle regions. Theoretically, many models and methods are still not converging into an agreement. This situation is very much different from the case of the ground state hadrons, where many models and methods of QCD can make reasonable descriptions. Could this be an indication of the lack of our understanding of QCD, or non-existence of the pentaquark baryon under normal conditions? We do not have the definite answer yet.

The authors thank T. Nakano, H. Toki, T. Hyodo, A. Titov and M. Oka for useful discussions. A.H. is supported in part by the Grant for Scientific Research [(C) No.17959600] from the Ministry of Education, Culture, Science and Technology, Japan.

References

[1] M. Gell-Mann, *Phys. Lett.* **8** (1964) 214.

[2] For researches before the recent activities, see, for instance, H. Lipkin, *Nucl. Phys.* **A625** (1997) 207-219.

[3] T. Nakano *et al.* [LEPS Collaboration], *Phys. Rev. Lett.* **91**, 012002 (2003).

[4] For a recent review, see for instance, the *proceedings of the international workshop PENTAQUARK04*, held at SPring-8, Japan, July (2004); edited by A. Hosaka and T. Hotta, World Scientific (2005).

[5] K. H. Hicks, *Prog. Part. Nucl. Phys.* **55**, 647 (2005) [arXiv:hep-ex/0504027].

[6] R. A. Schumacher, arXiv:nucl-ex/0512042.

[7] N. Ishii, T. Doi, H. Iida, M. Oka, F. Okiharu and H. Suganuma, *Phys. Rev. D* **71**, 034001 (2005).

[8] T. T. Takahashi, T. Umeda, T. Onogi and T. Kunihiro, arXiv:hep-lat/0503019.

[9] K. Holland and K. J. Juge [BGR (Bern-Graz-Regensburg) Collaboration], arXiv:hep-lat/0504007.

[10] A.W. Thomas, K. Hicks and A. Hosaka, *Prog. Theor. Phys.* **111**, 291 (2004).

[11] S. I. Nam, A. Hosaka and H. C. Kim, *Phys. Lett. B* **602**, 180 (2004)

[12] M. Battaglieri, R. De Vita, V. Kubarovsky, L. Guo, G. S. Mutchler, P. Stoler and D. P. Weygand [CLAS Collaboration], arXiv:hep-ex/0510061.

[13] K. H. Hicks [CLAS Collaboration], arXiv:hep-ex/0510067.

[14] D. Diakonov, V. Petrov and M. V. Polyakov, *Z. Phys. A* **359**, 305 (1997).

[15] For a recent review, see for instance: M. Oka, hep-ph/0406211 and references therein.

[16] N. Isgur and G. Karl, *Phys. Rev. D* **20**, 1191 (1979).

[17] A. Hosaka and H. Toki, *Quarks, Baryons and Chiral Symmetry*, World Scientific (2001).

[18] B.K. Jennings and K. Maltman, *Phys. Rev. D* **68**, 094020 (2004).

[19] T. Shinozaki, M. Oka and S. Takeuchi, *Phys. Rev. D* 71, 074025 (2005).

[20] S. Takeuchi and K. Shimizu, arXiv:hep-ph/0411016.

[21] Fl. Stancu, D.O. Riska, *Phys. Lett. B* **575**, 242 (2003); Fl. Stancu, *Phys. Lett. B* **595**, 269 (2004).

[22] A. Hosaka, *Phys. Lett. B* **571**, 55 (2003).

[23] A. Hosaka and H. Toki, *Phys. Reports* **277**, 65 (1996), and references therein.

[24] A. Hosaka, M. Oka and T. Shinozaki, *Phys. Rev. D* **71**, 074021 (2005) [arXiv:hep-ph/0409102].

[25] R. L. Jaffe and F. Wilczek, *Phys. Rev. Lett.* **91**, 232003 (2003).

[26] E. Hiyama, M. Kamimura, A. Hosaka, H. Toki and M. Yahiro, *Phys. Lett. B* **633**, 237 (2006)

[27] M. Kamimura, *Prog. Theor. Phys. Suppl.* **62**, 236 (1977).

[28] E. Hiyama, Y. Kino and M. Kamimura, *Prog. Part. Nucl. Phys.* **51**, 223 (2003).

[29] E. Hiyama, K. Suzuki, H. Toki and M. Kamimura, *Prog. Theor. Phys.* **112**, 99 (2004).

[30] S. I. Nam, A. Hosaka and H. C. Kim, *Phys. Lett. B* **633**, 483 (2006) [arXiv:hep-ph/0505134].

[31] T. Nakano, For instance, talk given at the workshop Pentaquark05, J-Lab, October 20-22, 2005.

[32] S. I. Nam, A. Hosaka and H. C. Kim, *Phys. Lett. B* **579**, 43 (2004).

[33] H. Haberzettl, C. Bennhold, T. Mart and T. Feuster, *Phys. Rev. C* **58** (1998) 40.

In: The Physics of Quarks: New Research
Editors: N.L. Watson and T.M. Grant, pp. 87-138
ISBN 978-1-60456-802-8

Chapter 6

HEAVY QUARK DIFFUSION AS A PROBE OF THE QUARK-GLUON PLASMA

Ralf Rapp and Hendrik van Hees
Cyclotron Institute and Physics Department, Texas A&M University
College Station, Texas 77843-3366, U.S.A.

Abstract

In this article we report on recent research on the properties of elementary particle matter governed by the strong nuclear force, at extremes of high temperature and energy density. At about 10^{12} Kelvin, the theory of the strong interaction, Quantum Chromodynamics (QCD), predicts the existence of a new state of matter in which the building blocks of atomic nuclei (protons and neutrons) dissolve into a plasma of quarks and gluons. The Quark-Gluon Plasma (QGP) is believed to have prevailed in the Early Universe during the first few microseconds after the Big Bang. Highly energetic collisions of heavy atomic nuclei provide the unique opportunity to recreate, for a short moment, the QGP in laboratory experiments and study its properties. After a brief introduction to the basic elements of QCD in the vacuum, most notably quark confinement and mass generation, we discuss how these phenomena relate to the occurrence of phase changes in strongly interacting matter at high temperature, as inferred from first-principle numerical simulations of QCD (lattice QCD). This will be followed by a short review of the main experimental findings at the Relativistic Heavy Ion Collider (RHIC) at Brookhaven National Laboratory. The data taken in collisions of gold nuclei thus far provide strong evidence that a QGP has indeed been produced, but with rather remarkable properties indicative for an almost perfect liquid with unprecedentedly small viscosity and high opacity. We then discuss how heavy quarks (charm and bottom) can be utilized to quantitatively probe the transport properties of the strongly-coupled QGP (sQGP). The large heavy-quark mass allows to set up a Brownian motion approach, which can serve to evaluate different approaches for heavy-quark interactions in the sQGP. In particular, we discuss an implementation of lattice QCD computations of the heavy-quark potential in the QGP. This approach generates "pre-hadronic" resonance structures in heavy-quark scattering off light quarks from the medium, leading to large scattering rates and small diffusion coefficients. The resonance correlations are strongest close to the critical temperature (T_c), suggesting an intimate connection to the hadronization of the QGP. The implementation of heavy-quark transport into Langevin simulations of an expanding QGP fireball at

RHIC enables quantitative comparisons with experimental data. The extracted heavy-quark diffusion coefficients are employed for a schematic estimate of the shear viscosity, corroborating the notion of a strongly-coupled QGP in the vicinity of T_c.

1. Introduction

1.1. Elementary Particles and Forces

The quest for the elementary constituents from which the matter around us is built has always fascinated mankind. In the fifth century B.C., Greek philosophers introduced the notion of an *indivisible* entity of matter, the $\alpha\tau o\mu o\sigma$ (atom). More than 2000 years passed before this concept was systematized in the nineteenth century in terms of the chemical elements as the building blocks of the known substances. The large variety of the chemical elements, however, called for a deeper substructure within these atoms, which were soon revealed as bound states of negatively charged electrons (e^-) and positively charged atomic nuclei, held together by their mutual attraction provided by the electromagnetic force. The nuclei, while very small in size (but carrying about 99% of the atom's mass), were found to further decompose in positively charged protons (p) and uncharged neutrons (n), both of approximately equal mass, $M_{p,n} \simeq 0.94$ GeV/c^2. This was a great achievement, since at this point all matter was reduced to 3 particles: p, n, e^-. There was still the problem of the stability of the atomic nucleus, since packing together many positive charges (protons) in a small region of space obviously implies a large electric repulsion. The solution to this problem triggered the discovery of the Strong Nuclear Force acting between nucleons (protons and neutrons); it turned out to be a factor of $\sim$100 stronger than the electromagnetic one, but with a very short range of only a few femtometer (1 fm=10^{-15} m). In the 1950's and 1960's, rapid progress in particle accelerator technology opened new energy regimes in collision experiments of subatomic particles (e.g., p-p collisions). As a result, many more particles interacting via the Strong Force (so-called hadrons) were produced and discovered, including "strange" hadrons characterized by a for strongly interacting particles untypically long lifetime. Again, this proliferation of states (the "hadron zoo") called for yet another simplification in terms of hadronic substructure. Gell-Mann introduced three types of "quarks" [1] as the elementary constituents from which all known hadrons could be built; they were dubbed up (u), down (d) and strange (s) quarks, with fractional electric charges +2/3, -1/3 and -1/3, respectively. In this scheme, hadrons are either built from 3 quarks (forming baryons, e.g., $p = (uud)$, $n = (udd)$), or a quark and an antiquark (forming mesons, e.g., $\pi^+ = (u\bar{d})$ or $K^0 = (d\bar{s})$). Three more heavy quark "flavors", carrying significantly larger masses than the light quarks (u, d, s), were discovered in the 1970's – charm (c) and bottom (b) – as well as in 1995 – the top (t) quark.

The discovery of an increasingly deeper structure of the fundamental matter particles is intimately related to the question of their mutual forces which, after all, determine how the variety of observed composite particles is built up. The understanding of fundamental *interactions* is thus of no less importance than the identification of the matter constituents. The modern theoretical framework to provide a unified description are Quantum Field Theories (QFTs), which combine the principles of Quantum Mechanics with those of Special Relativity. In QFTs, charged matter particles (fermions of half-integer spin) interact via the

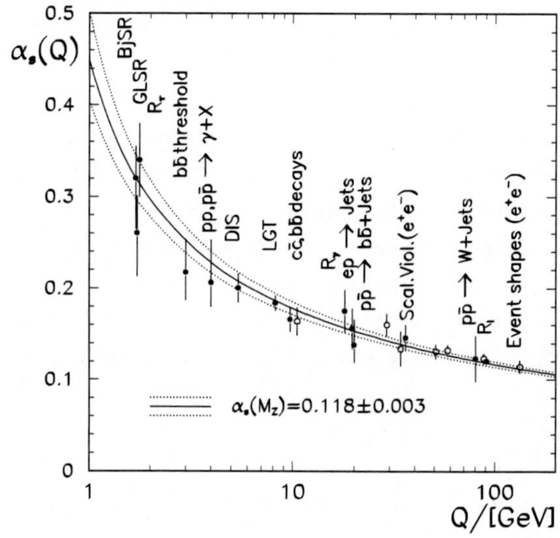

Figure 1. Dependence of the QCD coupling constant, $\alpha_s = g^2/4\pi$, on the momentum transfer, Q (or inverse distance $1/r \sim Q$), of the interaction. Figure taken from Ref. [5].

exchange of field quanta (bosons with integer spin). The QFT of the Electromagnetic Force is Quantum Electro-Dynamics (QED), where the associated field quantum is the photon (γ) coupling to electric charges (positive) and anticharges (negative). The coupling constant (or charge) of QED is rather small, $\alpha_{\rm em} = e^2/4\pi$=1/137, which allows to organize theoretical calculations in a series of terms characterized by increasing powers of $\alpha_{\rm em}$, so-called perturbation theory. The smallness of $\alpha_{\rm em}$ then allows for precise perturbative calculations of electromagnetic observables with an accuracy exceeding ten significant digits for select quantities, rendering QED one of the most successful theories in physics.

The QFT of the Strong Nuclear Force, Quantum Chromo-Dynamics (QCD), has been developed in the early 1970's [2, 3]. The chromo (=color) charge of quarks comes in three variants: red, green and blue (plus their anticharges), rendering QCD a mathematically more involved theory. In particular, the force quanta ("gluons") themselves carry a nonzero (color-) charge [4], giving rise to gluon self-interactions. The latter are closely related to another remarkable property of QCD, namely the "anti-screening" of its charges in the vacuum: quantum fluctuations, i.e., the virtual quark-gluon cloud around a color charge, induces an increase of the effective charge with increasing distance, a phenomenon known as asymptotic freedom which manifests itself in the running coupling constant (or charge) of QCD, $\alpha_s(Q)$, cf. Fig. 1. On the one hand, the interactions at small distances, r (which, by means of Heisenberg's uncertainty principle, corresponds to large momentum transfers, $Q \sim 1/r$, in a scattering process), are comparatively weak and perturbation theory is applicable (much like in QED). In this regime, QCD is well tested, being in excellent agreement with experiment (albeit not at the same level of precision as QED; even at very large Q, α_s is still a factor of $\sim$10 larger than $\alpha_{\rm em}$). On the other hand, the coupling constant grows toward small Q entering the realm of "strong QCD" where new *nonperturbative* phenomena occur. Most notably these are the Confinement of color charges and the "Spontaneous

Breaking of Chiral Symmetry" (SBCS). The former refers to the fact that quarks and gluons have never been observed as individual particles, but only come in "colorless" baryons (where the 3 quarks carry an equal amount of the 3 different color charges) or mesons (where quark and antiquark carry color charge and anticharge). Spontaneous Chiral Symmetry Breaking is closely related to the complex structure of the QCD vacuum; the latter is filled with various condensates of quark-antiquark and gluon fields. In particular, the scalar quark condensate of up and down quarks can be quantified by a vacuum expectation value, $\langle 0|\bar{q}q|0\rangle \simeq (-250\ \mathrm{MeV})^3$, translating into a total pair density of about 4 per fm^3.[1] Thus, the QCD vacuum is a rather dense state, and the quarks inside the hadrons propagating through it acquire an effective mass, $m^*_{u,d} \simeq 350$ MeV, which is much larger than their bare mass, $m^0_{u,d} \simeq 5\text{-}10$ MeV. The QCD condensates are thus the main source of the visible (baryonic) mass in the Universe. The theoretical understanding of the mechanisms underlying Confinement and SBCS, and their possible interrelation, constitutes a major challenge in contemporary particle and nuclear physics research. Currently, the only way to obtain first-principle information on this nonperturbative realm of QCD is through numerical lattice-discretized computer simulations (lattice QCD). However, even with modern day computing power, the numerical results of lattice QCD computations for observable quantities are often hampered by statistical and systematic errors (e.g., due to finite volume and discretization effects); the use of effective models is thus an indispensable tool for a proper interpretation and understanding of lattice QCD results, and to provide connections to experiment.

1.2. Elementary Particle Matter and the Quark-Gluon Plasma

A particularly fascinating aspect of the Strong Force is the question of what kind of matter (or phases of matter) it gives rise to. The conventional phases of matter (such as solid, liquid and gas phases of the chemical elements and their compounds, or even electron-ion plasmas) are, in principle, entirely governed by the electromagnetic force. Matter governed by the Strong Force is "readily" available only in form of atomic nuclei: the understanding of these (liquid-like) droplets of nuclear matter (characterized by a mass density of $\sim 1.67 \cdot 10^{15}\mathrm{g/cm^3}$) is a classical research objective of nuclear physics. But what happens to nuclear matter under extreme compression and/or heating? What happens to the (composite) nucleons? Is it possible to produce truly elementary-particle matter where nucleons have dissolved into their quark (and gluon) constituents (as may be expected from the asymptotic freedom, i.e., small coupling constant, of the quark and gluon interactions at short distance)? Does the condensate structure of the QCD vacuum melt, similar to the condensate of Cooper pairs in a superconductor at sufficiently high temperature? Are there phase transitions associated with these phenomena? The investigation of these questions not only advances our knowledge of strong QCD (including the fundamental problems of confinement and mass generation), but also directly relates to the evolution of the early universe, as well as to the properties of extremely compact stellar objects (so-called neutron stars). Lattice-QCD computations at finite temperature indeed predict that hadronic

[1] In nuclear and particle physics it is common practice to use units of $\hbar$ (Planck's constant), c (speed of light) and k (Boltzmann constant); in these units, energies, e.g., can be converted into inverse distance by division with $\hbar c \simeq 197.33$ MeV fm; energies are also equivalent to temperature.

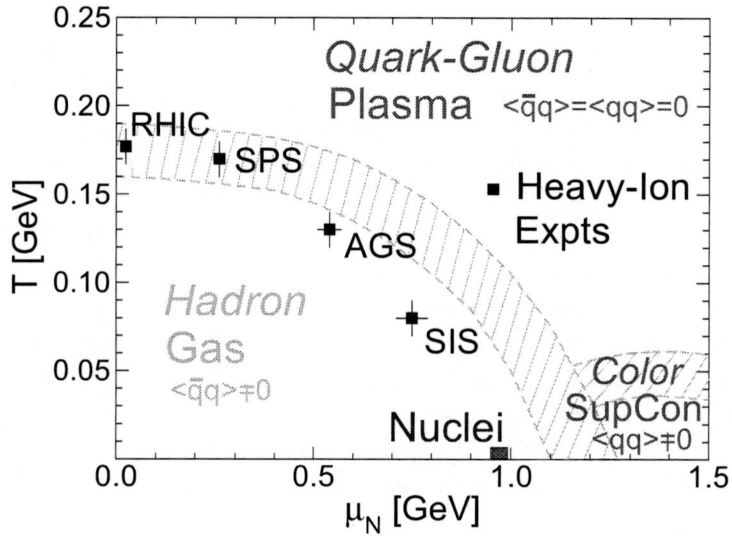

Figure 2. Schematic view of the phase diagram of strongly interacting matter, in terms of the nucleon (or baryon) chemical potential, μ_N, and temperature, T. The former determines the net baryon density in the system. The shaded bands are schematic dividers of the different phases as expected from theoretical model calculations. At vanishing μ_N, current lattice QCD calculations indicate a crossover transition from hot hadronic matter to the QGP at a (pseudo-) critical temperature of T_c=160-190 MeV [6, 7]. Normal nuclear matter (as present in atomic nuclei) is located on the T=0 axis at $\mu_N \simeq 970$ MeV (corresponding to a nucleon density of $\varrho_N \simeq 0.16$ fm^{-3}, the nuclear saturation density). At larger μ_N (and small $T\leq$ 50-100 MeV), one expects the formation of a Color-Superconductor [8, 9], i.e., cold quark matter with a BCS-type condensate of quark Cooper pairs, $\langle 0|qq|0\rangle \neq 0$. The "data" points are empirical extractions of (μ_N, T)-values from the observed production ratios of various hadron species (π, p, K, Λ, etc.) in heavy-ion experiments at different beam energies [10].

matter undergoes a transition to a state of matter where quarks and gluons are no longer confined into hadrons. The temperature required to induce this transition into the "Quark-Gluon Plasma" (QGP) is approximately $kT \sim 0.2$ GeV $= 2 \cdot 10^8$ eV, or $T \sim 10^{12}K$. The Universe is believed to have passed through this transition at about $10\mu s$ (=0.00001s) after its birth. This was, however, $\sim$ 15 billion years ago, and the question arises how one can possibly study the QGP, or more generally the phase diagram of QCD matter, today. Clearly, without input from experiment, this would be a hopeless enterprise.

It turns out that by colliding heavy atomic nuclei at high energies, one can create highly excited strongly interacting matter in the laboratory. The incoming kinetic energy of the colliding nuclei is largely converted into compression and thermal energy, and by varying the collision energy one is able to produce a wide range of different matter types as characterized by their baryon density, ϱ_B and temperature, T. This is illustrated in a schematic phase diagram of strongly interacting matter in Fig. 2. In the present article we will mainly focus on heavy-ion collisions at the highest currently available energies. These experiments are being conducted at the Relativistic Heavy-Ion Collider (RHIC) at Brookhaven National Laboratory (BNL, Upton, New York): gold (Au) nuclei, fully stripped of their electrons, are accelerated in two separate beam pipes to an energy of $E = 100$ GeV per nucleon, before being smashed together head on at four collision points where two large (PHENIX and STAR) and two smaller (BRAHMS and PHOBOS) detector systems have been positioned. With each gold nucleus consisting of A=197 nucleons (as given by the atomic mass number of gold), a total (center-of-mass) energy of $E_{\rm cm} \simeq 200\ A$ GeV $\simeq 40$ TeV $= 4 \cdot 10^{13}$ eV is brought into the collision zone. Note that the energy of the accelerated nuclei exceeds their rest mass by more than a factor of 100 (recall that the rest mass of the nucleon is $M_N \simeq 0.94$ GeV/c^2). In a central Au-Au collision at RHIC approximately 5000 particles are produced (as observed in the detectors), emanating from the collision point with velocities not far from the speed of light. Most of these particles are pions, but essentially all known (and sufficiently long-lived) hadrons made of u, d and s quarks are observed. The key challenge is then to infer from the debris of produced particles the formation and properties of the matter - the "fireball" - that was created in the immediate aftermath of the collision. While the typical lifetime of the fireball is only about $\sim 10^{-22}$s, it is most likely long enough to form locally equilibrated strongly interacting matter which allows for a meaningful analysis of its properties in terms of thermodynamic concepts, and thus to study the QCD phase diagram as sketched in Fig. 2. This has been largely deduced from the multiplicities and momentum spectra of produced hadrons, which allow to determine typical temperatures and collective expansion velocities of the exploding fireball, at least in the later (hadronic) phases of its evolution. A possibly formed QGP will, however, occur in the earlier (hotter and denser) phases of a heavy-ion reaction. The identification and assessment of suitable QGP signatures is at the very forefront of contemporary research. Hadrons containing heavy quarks (charm and bottom, $Q = c, b$) have been identified as particularly promising probes of the QGP. The basic idea is as follows: since charm- and bottom-quark masses, $m_c \simeq 1.5$ GeV/c^2 and $m_b \simeq 4.5$ GeV/c^2, are much larger than the typical temperatures, $T \simeq T_c \simeq 0.2\text{GeV}$, of the medium formed in a heavy-ion collision, they are (i) only produced very early in the collision (upon first impact of the colliding nuclei) and, (ii) not expected to thermalize during the lifetime of the fireball. Furthermore, the largest changes of their momentum spectra occur when the collision rate and momentum transfer are the highest. This is facilitated by a large density and temperature (i.e., in the early phases of a heavy-ion reaction), but is crucially dependent also on the interaction strength. Both aspects are embodied into the notion of transport coefficients. The main objective of this article is to provide a theoretical description of heavy-quark transport in the QGP, and to test the results in applications to RHIC data.

The remainder of this article is organized as follows: In Sec. 2. we present a general overview of the physics of the Strong Force and the facets of its different matter phases. We start in Sec. 2.1. by introducing basic features of Quantum Chromodynamics (QCD), the quantum field theory describing the strong interactions between quarks and gluons, the elementary building blocks of hadrons. In Sec. 2.2. we briefly review our current understanding of strongly interacting matter and its phase diagram as theoretically expected from both numerical lattice QCD computations and model analysis. In Sec. 2.3. we elucidate the main ideas and achievements of the experimental high-energy heavy-ion programs as conducted at the Relativistic Heavy Ion Collider (RHIC) at the Brookhaven National Laboratory (New York), as well as at the Super Proton Synchrotron (SPS) and the future Large Hadron Collider (LHC) at the European Organization for Nuclear Research (CERN, Geneva, Switzerland). We summarize what has been learned about hot and dense strongly interacting matter thus far, and which questions have emerged and/or remained open. This leads us to the central part of this article, Sec. 3., where we discuss in some detail the theoretical developments and phenomenological applications in using heavy quarks (charm and bottom) as a probe of the Quark-Gluon Plasma. In Sec. 3.1. we concentrate on the theoretical understanding of heavy-quark (HQ) interactions in the QGP. We mostly address elastic scattering, within both perturbative and nonperturbative approaches, where the latter are divided into a resonance model and T-matrix calculations utilizing potentials extracted from lattice QCD. The HQ interactions in the medium are used to compute pertinent selfenergies and transport coefficients for drag and diffusion. In Sec. 3.2. the latter are implemented into a Brownian motion framework of a Fokker-Planck equation which is particularly suitable for describing the diffusion of a heavy particle in a heat bath. In Sec. 3.3. these concepts are applied to heavy-ion collisions, by implementing a Langevin simulation of HQ transport into realistic QGP fireball expansions for Au-Au collisions at RHIC. To make contact with experiment, the quarks have to be hadronized which involves a quark coalescence approach at the phase transition, in close connection to successful phenomenology in light hadron spectra. This is followed by an analysis of transverse momentum spectra of heavy mesons and their electron decay spectra for which experimental data are available. In Sec. 3.4. we recapitulate on the ramifications of the theoretical approach in a broader context of Quark-Gluon Plasma research and heavy-ion phenomenology, and outline future lines of investigation. Sec. 4. contains a brief overall summary and conclusions.

2. The Quark-Gluon Plasma and Heavy-Ion Collisions

2.1. The Strong Force and Quantum Chromodynamics

The basic quantity which, in principle, completely determines the theory of the strong interaction, is the Lagrangian of QCD,

$$\mathcal{L}_{\rm QCD} = \bar{q}\,({\rm i}\,\not{D} - \hat{m}_q)\,q - \frac{1}{4} G_{\mu\nu} G^{\mu\nu} \ , \tag{1}$$

where q and $\bar{q}$ denote the elementary matter fields, quarks and antiquarks. The quark fields are specified by several quantum numbers: (i) color charge (red, green or blue), (ii) flavor (up, down, strange, charm, bottom and top) and (iii) spin ($\pm\frac{1}{2}\hbar$). The mass matrix $\hat{m}_q = {\rm diag}(m_u, m_d, m_s, m_c, m_b, m_t)$ is a simple diagonal matrix in flavor space. It roughly separates QCD into a light-flavor (u, d with bare masses $m_{u,d} \simeq 0.005$ GeV/c^2) and a heavy-flavor sector (c, b, t with $m_c \simeq 1.3$ GeV/c^2, $m_b \simeq 4.5$ GeV/c^2, $m_t \simeq 175$ GeV/c^2), while the strange-quark mass is somewhat in between ($m_s \simeq 0.12$ GeV/c^2). The interactions of the quarks are encoded in the covariant derivative, $\not{D} = \not{\partial} - {\rm i}g\,\not{A}$, where A denotes the gluon field, the carrier of the Strong Force, and the gauge coupling g quantifies the interaction strength as referred to in Fig. 1. A pictorial representation of these interactions can be given in terms of Feynman diagrams: quarks interact via the exchange of gluons, cf. Fig. 3. The

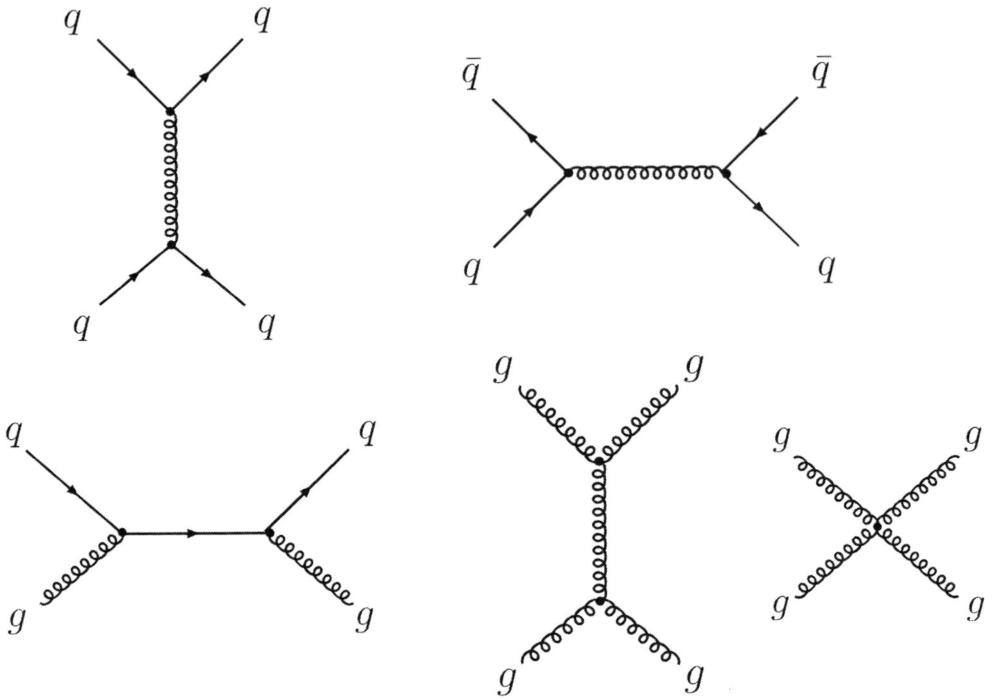

Figure 3. Feynman diagrams for quark-quark scattering via the exchange of a gluon (top left) quark-antiquark scattering via annihilation into a gluon (top right), quark-gluon scattering and gluon-gluon scattering (bottom panels).

gluons are massless particles with spin 1; most notably, they also carry color charge (which comes in 8 different charge-anticharge combinations, e.g., red-antigreen, red-antiblue, etc.), giving rise to gluon self-interactions. These are encoded in the last term of Eq. (1) where $G_{\mu\nu} = \partial_\mu A_\nu - \partial_\nu A_\mu + g[A_\mu, A_\nu]$; the commutator term, $[A_\mu, A_\nu] = A_\mu \cdot A_\nu - A_\nu \cdot A_\mu$ is a consequence of the 3×3 matrix structure in color-charge space and generates 3- and 4-gluon interaction vertices (as depicted in the bottom right panel of Fig. 3). Each interaction vertex brings in a factor of g into the calculations of a given diagram (except for the 4-gluon vertex which is proportional to g^2). All diagrams (more precisely, the pertinent scattering amplitudes) shown in Fig. 3 are thus proportional to $\alpha_s = g^2/4\pi$, while more complicated diagrams (involving additional quark/gluon vertices) are of higher order in α_s. This forms the basis for perturbation theory: for small α_s, higher order diagrams are suppressed, ensuring a rapid convergence of the perturbation series. Perturbative QCD (pQCD) indeed works very well for reactions at large momentum transfer, Q, where $\alpha_s(Q)$ is small, cf. Fig. 1, while it breaks down for small Q. One possibility to obtain information on the nonperturbative interactions is to investigate the potential between two static quark charges, i.e., the potential between a heavy quark and antiquark. In the color neutral channel, a phenomenological ansatz can be written as a combination of a Coulombic attraction at short distances which merges into a linearly rising potential at large distance (signifying confinement),

$$V_{Q\bar{Q}}(r) = -\frac{4}{3}\frac{\alpha_s}{r} + \sigma r \,, \tag{2}$$

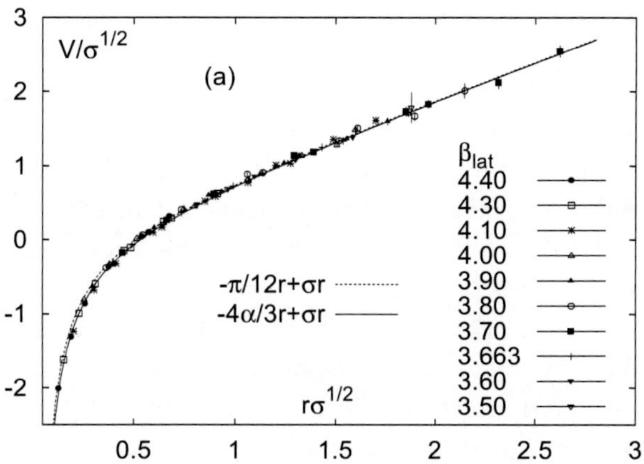

Figure 4. The static potential between a heavy quark and antiquark in vacuum, as a function of their distance, as computed in lattice QCD [12]. Potential and distance are given in units of the string tension, $\sqrt{\sigma} \simeq 0.42$ GeV $\simeq$ 1/(2.12 fm).

where $\sigma \simeq 1$ GeV/fm denotes the "string tension". Such a potential provides a good description of the observed spectra of heavy quarkonium states, i.e., charm-anticharm and bottom-antibottom quark bound states. In recent years, the heavy-quark potential has been computed with good precision in lattice QCD, which fully confirmed the phenomenological ansatz, Eq. (2), (see, e.g., Fig. 4). Subsequently, the potential approach, in combination with expansions organized in powers of the inverse HQ mass, $m_{c,b}$ (rather than the coupling constant), has been developed into an effective theory of low-energy QCD, cf. Ref. [13] for a review.

Besides the (external) quark masses, QCD has only one (intrinsic) dimensionful scale which is generated by quantum effects (loop corrections). The latter give rise to the running coupling constant,

$$\alpha_s(Q) = \frac{1}{(11N_c - 2N_f)\ln(Q^2/\Lambda_{\rm QCD}^2)}, \tag{3}$$

where N_c=3 is the number of color charges and N_f the number of active quark flavors at given Q (i.e., the number of flavors with $m_q \leq Q$). It is tempting to interpret the value of $\Lambda_{\rm QCD} = 0.2$ GeV as the dividing line between perturbative and nonperturbative regimes of QCD. In practice, however, the scale for the onset of nonperturbative effects is significantly larger, typically given by the hadronic mass scale of $\sim$1 GeV. To understand the emergence of this scale, it is important to realize that the QCD vacuum structure is rather rich, characterized by quark and gluon condensates. E.g., in the light scalar quark-antiquark channel ($\bar{u}u$ and $\bar{d}d$) a strongly attractive force leads to the spontaneous formation and condensation of $\bar{q}q$ pairs (reminiscent to a Bose condensate)[2]. An important consequence of the condensate formation is that the light quarks acquire an effective mass when propagating through

[2]The origin of this force is most likely not the perturbative exchange of gluons, but nonperturbative gluon configurations - so-called instantons - which correspond to tunneling events between topologically different vacua and are characterized by a 4-dimensional "radius" of $\rho \simeq (0.6\ \text{GeV})^{-1}$, cf. Ref. [14] for a review

the condensed vacuum, which is given by the condensate as $m_q^* \simeq G\langle 0|\bar{q}q|0\rangle \simeq 0.4$ GeV, where G is an (instanton-induced) effective quark coupling constant. Note that this mass exceeds the bare quark masses by about a factor of $\sim$100, being the major source of the proton mass, $M_p \sim 3m_q^*$, and thus of the visible mass in the Universe. Formally, the presence of the constituent quark mass (and quark condensate) is closely related to the phenomenon of "Spontaneous Breaking of Chiral Symmetry" (SBCS): in the limit of vanishing bare-quark masses (which is a good approximation for the very light u and d quarks), the QCD Lagrangian is invariant under rotations in isospin and chirality (handedness), i.e., transformations that change u into d quarks and left-handed into right-handed quarks. This invariance is equivalent to the conservation of isospin and chiral quantum numbers of a quark. However, the constituent quark mass breaks the chiral symmetry (i.e., massive quarks can change their chirality so that it is no longer conserved). SBCS not only manifests itself in the QCD ground-state, but also in its excitation spectrum, i.e., hadrons. Therefore, hadronic states which transform into each other under "chiral rotations" (so-called chiral multiplets or partners) are split in mass due to SBCS. Prominent examples in the meson spectrum are π(140)-σ(400-1200) and ρ(770)-a_1(1260), or N(940)-N^*(1535) in the nucleon spectrum.

In the following Section, we will discuss how the presence of strongly interacting matter affects the nonperturbative structure of the QCD vacuum and the interactions therein.

2.2. Strongly Interacting Matter and the QCD Phase Diagram

When heating a condensed state, the general expectation is that the condensate eventually "melts" (or "evaporates"), and that the interactions are screened due to the presence of charged particles in the medium. Transferring this expectation to the QCD vacuum implies that, at sufficiently large temperature, the condensates should vanish and the quarks and gluons are released from their hadronic bound states (deconfinement), forming the Quark-Gluon Plasma (QGP). Numerical lattice QCD (lQCD) computations of the thermodynamic partition function at finite temperature have substantially quantified this notion over the last two decades or so. The pertinent equation of state (EoS), i.e., the pressure, energy and entropy density, indeed exhibits a rather well defined transition, as shown in Fig. 5 for a lQCD calculation with close to realistic input for the bare light and strange quark masses. At high temperatures ($T > 3T_c$) the equation of state is within $\sim$15% of the values expected for an ideal gas, known as the Stefan-Boltzmann (SB) limit. The SB values are given by $\varepsilon_{q\bar{q}} = \frac{7}{8} d_{q\bar{q}} \frac{\pi^2}{30} T^4$ and $\varepsilon_g = d_g \frac{\pi^2}{30} T^4$ with degeneracies $d_{q\bar{q}} = \frac{7}{8} N_c N_s N_{\bar{q}} N_f = 10.5 N_f$ for quarks plus antiquarks (N_c=3 for red, green and blue colors, N_s=2 for spin up and down and $N_{\bar{q}}$=2 for antiquarks; N_f is the number of massless flavors), and $d_g = N_s N_c = 16$ for gluons (N_c=8 color-anticolor combinations, N_s=2 transverse spin polarizations; the relative factor of 7/8 is due to the difference of Fermi vs. Bose distribution functions for quarks vs. gluons). For N_f=3 one finds $\varepsilon_{\rm SB} = \varepsilon_{q\bar{q}} + \varepsilon_g \simeq 15.6\, T^4$, as indicated by the "$\varepsilon_{\rm SB}$" limit in the upper right corner of the left panel in Fig. 5. Likewise, using $P = sT - \varepsilon$, one finds for the SB limit of the entropy density $s = (10.5N_f + 16)\frac{4\pi^2}{90} T^3 \simeq 20.8\, T^3$, cf. right panel of Fig. 5.

Returning to the phase transition region, the rapid change in ε is accompanied by comparably sudden changes in the quark condensate, and the expectation value of the so-called Polyakov loop, an order parameter of deconfinement, cf. Fig. 6. The latter is, roughly

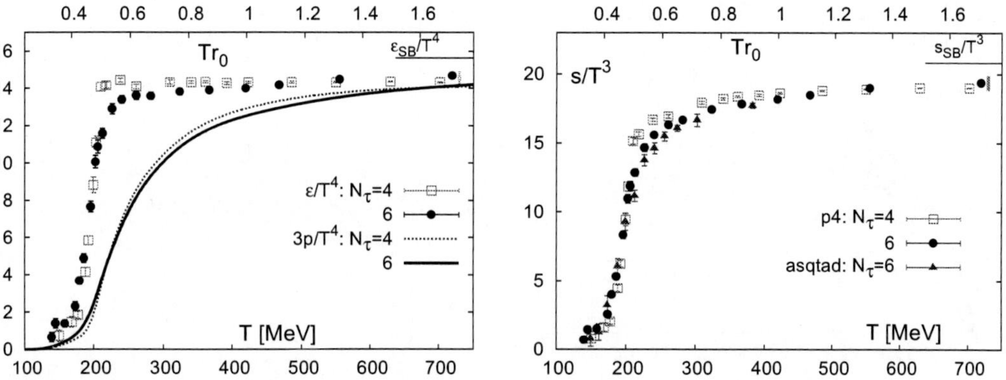

Figure 5. The equation of state of strongly interacting matter as computed in lattice QCD [11] in terms of the energy density (left panel) and entropy density (right panel) as a function of temperature (at $\mu_B = 0$). The computations use bare light and strange quark masses close to their physical values.

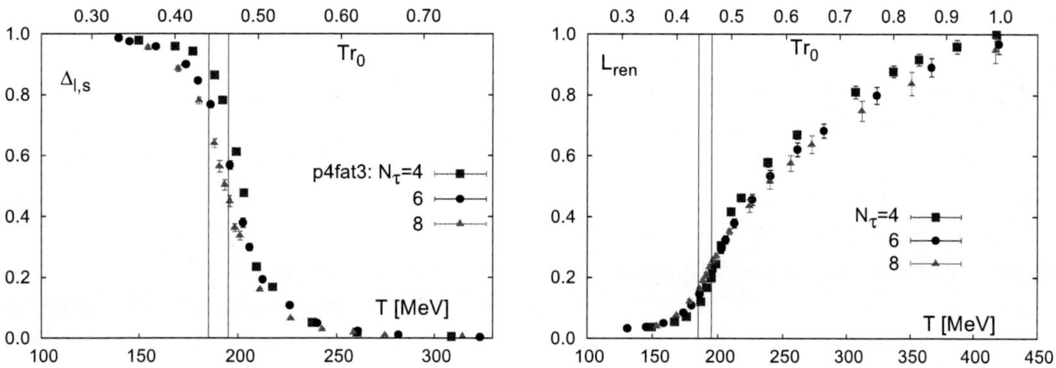

Figure 6. Lattice QCD computations of the (subtracted and normalized) light-quark chiral condensate (left panel) and the (renormalized) Polyakov loop expectation value as a function of temperature [11, 15].

speaking, proportional to the exponent of the heavy-quark free energy at large distance, $e^{-F^{\infty}_{Q\bar{Q}}/T}$, which vanishes in the confined phase (or at least becomes very small since $F^{\infty}_{Q\bar{Q}} \equiv F_{Q\bar{Q}}(r \to \infty)$ is large), but is finite in the deconfined QGP.

Detailed studies of the HQ free energy as a function of the relative distance, r, of the Q-$\bar{Q}$ pair have also been conducted in finite-T lattice QCD, see, e.g., Fig. 7. One finds the qualitatively expected behavior that the interaction is increasingly screened with increasing temperature, penetrating to smaller distances, as is characteristic for a decreasing color-Debye screening length (or, equivalently, increasing Debye mass, μ_D). However, the implications of these in-medium modifications for the binding of quarkonium states are quite subtle. The first problem is that, unlike in the vacuum case, the identification of the free energy with an interaction potential is no longer straightforward, due to the appearance

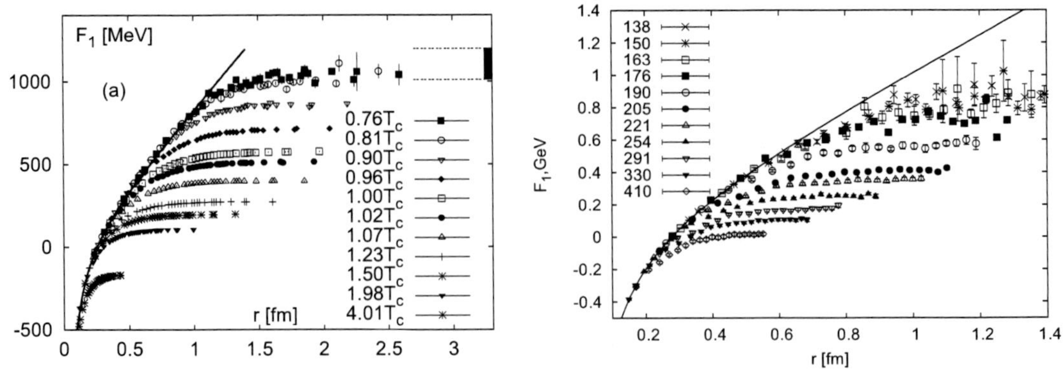

Figure 7. The heavy-quark free energy as a function of size, r, of the HQ pair, for various temperatures as computed in lattice QCD [12, 16].

of an entropy term, $TS_{Q\bar{Q}}$,

$$F_{Q\bar{Q}}(r,T) = U_{Q\bar{Q}}(r,T) - TS_{Q\bar{Q}}(r,T), \tag{4}$$

where $U_{Q\bar{Q}}$ denotes the internal energy. It is currently an open problem whether $F_{Q\bar{Q}}$ [17] or $U_{Q\bar{Q}}$ [18–22] (or even a combination thereof [23]) is the most suitable quantity to be inserted into a potential model calculation for quarkonium states in the medium (typically carried out using a Schrödinger equation). The different choices lead to considerably different results for the quarkonium binding in the QGP. On the one hand, when directly using the free energy $F_{Q\bar{Q}}$, the charmonium ground state (the S-wave J/ψ or η_c states) dissolves not far above the critical temperature, at about 1.2 T_c [17]. On the other hand, the use of $U_{Q\bar{Q}}$ implies deeper potentials and thus stronger binding, and the ground-state charmonium survives up to temperatures of ~2-2.5 T_c. The stronger bound bottomonia are more robust and may not dissolve until ~4 T_c.

An alternative way to address the problem of quarkonium dissociation in the QGP is the computation of Q-$\bar{Q}$ correlation functions, which are basically thermal Green's functions (or Q-$\bar{Q}$) propagators. However, the oscillating nature of the propagator for physical energies (i.e., in real time) renders numerical lQCD evaluations intractable. However, one can apply a trick by transforming the problem into "imaginary time", where the propagators are exponentially damped. The price one has to pay is an analytical continuation back into the physical (real time) regime of positive energies. This transformation is particularly problematic at finite temperature, where the imaginary-time axis in the statistical operator (free energy) is given by the inverse temperature, which severely limits the accuracy in the analytic continuation (especially for a limited (discrete) number of points on the imaginary-time axis). However, probabilistic methods, in particular the so-called Maximum Entropy Method (MEM), have proven valuable in remedying this problem [25]. LQCD computations of charmonium correlation functions in the QGP, followed by a MEM analysis to extract pertinent spectral functions, support the notion that the ground state (S-wave) charmonium survives up to temperatures of ~1.5-2 T_c [26], cf., e.g., the left panel of Fig. 8. Even in the light-quark sector lQCD computations indicate the possibility that mesonic resonance states persist above the phase transition [27, 28], as illustrated in the right panel of

Fig. 8.

To summarize this section, first principle lattice-QCD calculations at finite temperature have confirmed that hadronic matter undergoes a transition into a Quark-Gluon Plasma. This transition is characterized by rapid changes in the equation of state around a temperature of T_c=0.15-0.20 GeV, which is accompanied by variations in order parameters associated with deconfinement and the restoration of chiral symmetry (i.e., vanishing of the chiral quark condensate). While thermodynamic state variables are within 15% of the ideal gas limit at temperatures of $\geq 3\ T_c$, the analysis of the heavy-quark potential and mesonic spectral functions indicate substantial nonperturbative effects at temperatures below $\sim 2\ T_c$. This suggests that up to these temperatures the QGP is quite different from a weakly interacting gas of quarks and gluons. Substantial progress in our understanding of hot and dense QCD matter has emerged on a complementary front, namely from experiments using ultrarelativistic heavy-ion collisions. The following Section gives a short overview of the key observations and pertinent theoretical interpretations.

2.3. Relativistic Heavy Ion Collisions and the Quest for the QGP

The first years of experiments at the Relativistic Heavy-Ion Collider have indeed provided convincing evidence that a thermalized medium is produced in $\sqrt{s} = 200$ AGeV collisions. In this section we give a brief summary of the basic observations and pertinent interpretations [29][3]. A schematic pictorial sketch of the main stages of the evolution of a head-on collision of heavy nuclei is displayed in Fig. 9. The main observables are momentum spectra of various hadron species. We will concentrate on particles with zero longitudinal

[3] For an assessment of the earlier CERN-SPS experiments at lower energies, cf. Ref. [30].

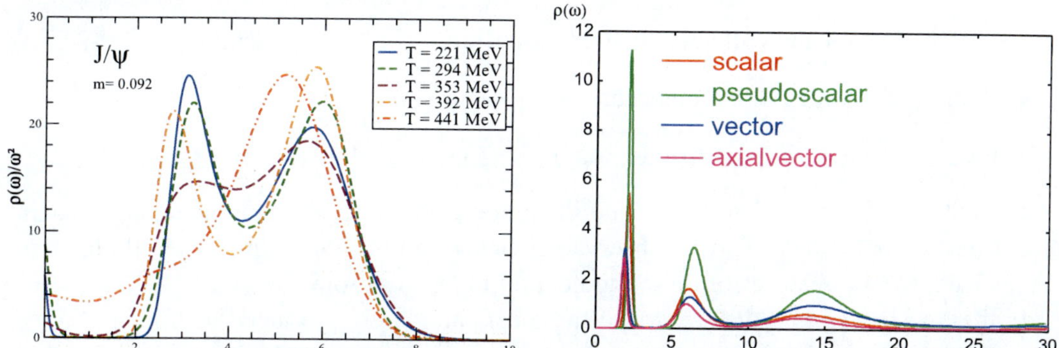

Figure 8. Meson spectral functions in the Quark-Gluon Plasma, based on imaginary-time correlation functions computed in finite-T lattice QCD, followed by a transformation in the physical regime using the Maximum Entropy Method. Left panel: S-wave charm-anticharm channel (spin-parity $J^P = 1^-$, corresponding to the J/ψ meson) in a QGP with $N_f = 2$ flavors [26]; the critical temperature in these calculations is about $T_c \simeq 200$ MeV. Right panel: quark-antiquark channels with a bare quark mass corresponding to strange quarks [27], in the scalar ($J^P = 0^+$), pseudoscalar ($J^P = 0^-$), vector ($J^P = 1^-$) and axialvector ($J^P = 1^+$) channels in a pure gluon plasma at T=1.4 T_c.

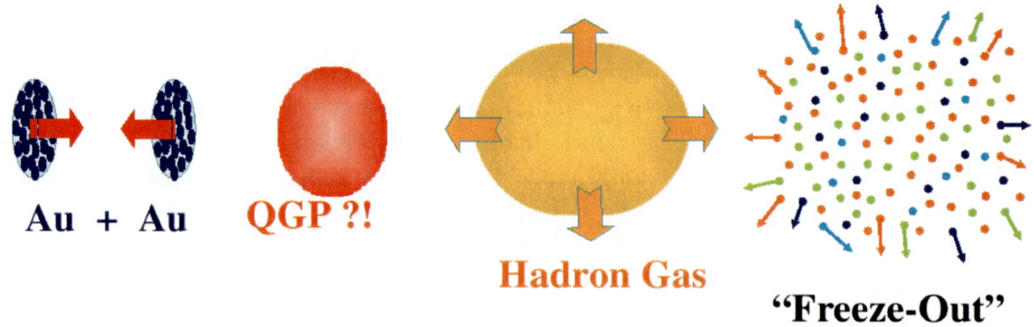

Figure 9. Schematic representation of the various stages of a heavy-ion collision. From left to right: incoming nuclei at highly relativistic energies moving at close to the speed of light (which induces a substantial Lorentz contraction relative to their transverse size; at full RHIC energy of 100+100 GeV, the Lorentz contraction is a factor of ~1/100); upon initial impact of the nuclei, primordial ("hard") nucleon-nucleon collisions occur; further reinteractions presumably induce the formation of a Quark-Gluon Plasma (after $\tau_0 = 0.5-1$ fm/c), whose pressure drives a collective expansion and cooling (for a duration of $\tau_{\rm QGP} \simeq 5$fm/c), followed by hadronization and further expansion in the hadronic phase (for a duration of $\tau_{\rm HG} \simeq 5 - 10$fm/$c$); at thermal freezeout the (short-range) strong interactions cease (after approximately 15 fm/c total fireball lifetime).

momentum (p_z=0) in the center-of-mass frame of a nucleus-nucleus collision, the so-called mid rapidity (y=0) region, where one expects the largest energy deposition of the interpenetrating nuclei. The main kinematic variable is thus the transverse momentum (p_T) of a particle. Three major findings at RHIC thus far may be classified by their p_T regime (see Fig. 10 for 3 representative measurements):

- Thermalization and collective matter expansion in the low-p_T regime $p_T \simeq$ 0-2 GeV;
- Quark coalescence in the intermediate-p_T regime, $p_T \simeq$ 2-5 GeV;
- Jet quenching in the high-p_T regime, $p_T \geq 5$ GeV.

It turns out that all of the 3 regimes, and the associated physical phenomena, are relevant for our discussion of heavy-quark observables below. In the following, we will elaborate on the characteristic features of these momentum regimes in some detail.

In the low-p_T regime, the spectra of the most abundantly produced hadrons (π, K, p, Λ, etc.) are well described by hydrodynamic simulations of an exploding fireball [34–37]. At its breakup (or thermal freezeout), where the (short-range) interactions of the hadrons cease rather abruptly, the fireball matter is expanding at an average collective velocity of about 60% of the speed of light and has cooled down to a temperature of about $T_{\rm fo} \simeq$ 100 MeV. The hadro-chemistry of the fireball, characterized by the thermal ratios of the various hadron species [10], is "frozen" at a higher temperature of about $T_{\rm ch} \simeq$ 170 MeV (as represented by the "data" points in Fig. 2). The separation of chemical and thermal freezeout is naturally explained by the large difference in the inelastic and elastic reaction rates in a hadronic gas. Elastic scattering among hadrons is dominated by strong resonances (e.g., $\pi\pi \to \rho \to \pi\pi$ or $\pi N \to \Delta \to \pi N$) with large cross sections, $\sigma_{\rm res} \simeq$ 100 mb,

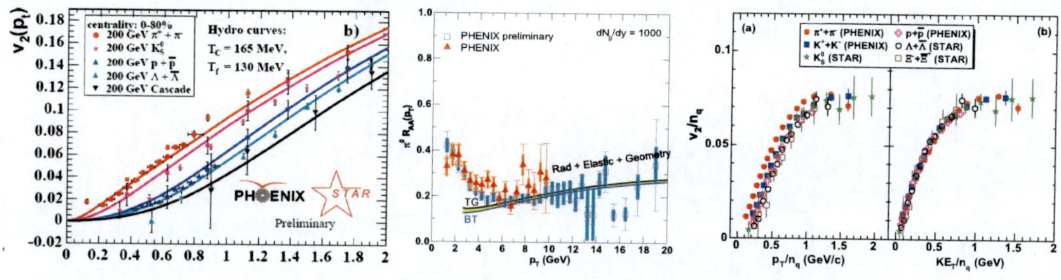

Figure 10. Key experimental measurements in Au-Au collisions at the Relativistic Heavy Ion Collider in its first years of operation, as reflected in transverse-momentum spectra of hadrons [29]. Left panel: Elliptic flow compared to hydrodynamic calculations [31] in the low-p_T regime; middle panel: nuclear modification factor of neutral pions compared to perturbative jet-quenching calculations [32] in the high-p_T regime; right panel: universal scaling of the hadronic v_2 interpreted as an underlying quark v_2 [33].

implying a thermal relaxation time of about $\tau_{\rm therm} \simeq \langle \sigma \varrho_h v_{\rm rel}\rangle \simeq 1$ fm/c (assuming a typical hadron density of $0.2\,{\rm fm}^{-3}$ and a relative velocity of $v_{\rm rel} = 0.5c$). Inelastic cross sections, on the other hand, are typically around 1 mb, implying chemical relaxation times in a hadron gas of $\sim$100 fm/c, well above its lifetime of $\sim$10 fm/c.

A hadronic observable which is sensitive to the early expansion phases of the fireball is the elliptic flow, v_2, which characterizes the azimuthal asymmetry in the emitted hadron spectra (i.e., the second harmonic) according to

$$\frac{{\rm d}N_h}{{\rm d}^2p_T{\rm d}y} = \frac{{\rm d}N_h}{\pi {\rm d}p_T^2{\rm d}y}\left[1 + 2v_2(p_T)\cos(2\phi) + \ldots\right] \tag{5}$$

(at mid rapidity, the system is mirror symmetric in the transverse x-y plane and odd Fourier components in the azimuthal angle ϕ vanish). In a noncentral Au-Au collision, the initial nuclear overlap zone in the transverse plane is "almond"-shaped, cf. Fig. 11. Once the system thermalizes, the pressure gradients along the short axis of the medium are larger than along the long axis. As a result, hydrodynamic expansion will be stronger along the x-axis relative to the y-axis, and thus build up an "elliptic flow" in the collective matter expansion, which eventually reflects itself in the final hadron spectra via a positive v_2 coefficient. The important point is that a large v_2 can only be generated if the thermalization of the medium is rapid enough: an initial period of (almost) free streaming will reduce the spatial anisotropy and thus reduce the system's ability to convert this spatial anisotropy into a momentum anisotropy (i.e., v_2). In this way, the magnitude of v_2 (and its p_T dependence) is, in principle, a quantitative "barometer" of the thermalization time, τ_0. Applications of ideal relativistic hydrodynamics have shown that the experimentally measured $v_2(p_T)$ for various hadrons (π, K, p, Λ) is best described when implementing a thermalization time of τ_0=0.5-1 fm/c. The agreement with data extends from $p_T = 0$ to $\sim$2-3 GeV, which (due to the exponentially falling spectra) comprises more than 95% of the produced hadrons. Not only do the interactions for thermalizing the matter rapidly have to be very strong, but, for ideal hydrodynamics to build up the observed v_2, the viscosity of the formed medium must remain very small (it is zero in ideal hydrodynamics): the typical timescale for build-up

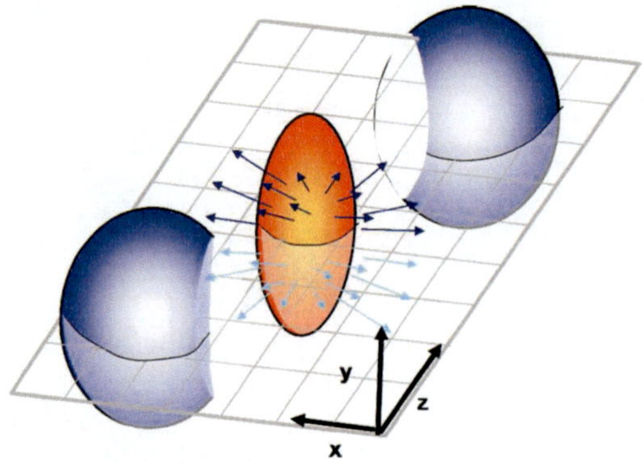

Figure 11. Schematic representation of a noncentral heavy-ion collision, characterized by an almond-shaped initial overlap zone, and a subsequent pressure-driven build-up of elliptic flow.

of the observed v_2 is on the order of the system size or QGP lifetime, $\tau_{\rm QGP} \simeq 5$ fm/c. These features have triggered the notion of a "strongly-coupled" QGP (sQGP); its initial energy density at RHIC, as implied by the above thermalization times, amounts to $\varepsilon_0 \simeq$ 10-20 GeV/fm^3, which is a factor of $\sim$10 above the estimated critical energy density for the phase transition (and a factor of $\sim$100 above that for normal nuclear matter).

In the high-p_T regime, the production mechanism of hadrons changes and becomes computable in perturbative QCD, in terms of hard parton-parton collisions upon first impact of the incoming nucleons (cf. left panel in Fig. 9). The produced high-energy partons subsequently fragment into a (rather collimated) spray of hadrons, called jet. Back-to-back jets are routinely observed in high-energy collisions of elementary particles, but are difficult to identify in the high-multiplicity environment of a heavy-ion collision. However, a jet typically contains a "leading" particle which carries most of the momentum of the parent parton (as described by an empirical "fragmentation" function). High-p_T spectra in heavy-ion collisions thus essentially determine the modification underlying the production of the leading hadron in a jet. This modification is quantified via the "nuclear modification factor",

$$R_{AA}(p_T) = \frac{\mathrm{d}N_h^{AA}/\mathrm{d}p_T}{N_{\rm coll}\,\mathrm{d}N_h^{NN}/\mathrm{d}p_T}\ , \tag{6}$$

where the numerator denotes the hadron spectrum in the nucleus-nucleus collision and the denominator represents the spectrum in an elementary nucleon-nucleon collision, weighted with the number of binary N-N collisions in the primordial stage of the A-A collision. Thus, if there is no modification of the leading hadron spectrum in the heavy-ion collision, then $R_{AA}(p_T \geq 5~\mathrm{GeV}) = 1$. RHIC data on π production in central Au-Au collisions have found a large suppression by a factor of 4-5 up to $p_T \simeq 20$ GeV. Originally, this has been attributed to radiative energy-loss, i.e., induced gluon radiation off a high-energy quark (or gluon) traversing a gluon plasma [38, 39], as computed within perturbative QCD (pQCD). In the approach of Ref. [38], the initially extracted gluon densities turned out to be consistent with those inferred from hydrodynamic calculations in the low-p_T regime. In the approach of Ref. [39], the interaction strength of the energy-loss process is quantified via the transport coefficient $\hat{q} = Q^2/\lambda$, which characterizes the (squared) momentum transfer per mean free path of the fast parton; the experimentally observed suppression requires this quantity to take on values of 5-15 GeV^2/fm, which is several times larger than expected in pQCD [40]. More recently, the importance of elastic energy loss has been realized, in particular in the context of heavy-quark propagation [32,41–44], as we will discuss in more detail below.

In the intermediate-p_T regime, RHIC experiments have observed an unexpectedly large ratio of baryons to mesons, e.g., $p/\pi \simeq 1$ or $(\Lambda + \bar{\Lambda})/(4K_s^0) \simeq 1.3$ in central Au-Au collisions. On the one hand, within the pQCD energy loss picture, the typical value of p/π is close to $\sim$0.2 as observed in elementary p-p collisions. On the other hand, within hydrodynamic calculations, a collective flow could, in principle, account for such an effect, but the applicability of hydrodynamics appears to cease at momenta of $p_T \geq 3$ GeV (as, e.g., indicated by the saturation (leveling off) of the elliptic flow, which in hydrodynamics continues to grow). Another remarkable observation in intermediate-p_T hadron spectra is a constituent-quark number scaling (CQNS) of the hadronic $v_{2,h}(p_T)$, as determined by the number, n, of constituent quarks in each hadron, h, giving rise to a single, universal function,

$$v_{2,q}(p_T/n) = v_{2,h}(p_T)/n \,, \tag{7}$$

which is interpreted as the partonic (quark) v_2 at the time of hadronization. Both CQNS and the large baryon-to-meson ratios are naturally explained in terms of quark coalescence processes of a collectively expanding partonic source at the phase transition [45–47]. The most recent experimental data [33,48] indicate that the scaling persists at a surprisingly accurate level even at low p_T, but only when applied as a function of the transverse kinetic energy, $\text{KE}_T = m_T - m_h$, of the hadrons (where $m_T = (p_T^2 + m_h^2)^{1/2}$ is the total transverse energy of the hadron, and m_h its rest mass). In Ref. [49], the quark coalescence model has been reformulated utilizing a Boltzmann transport equation where the hadron formation process is realized via the formation of mesonic resonances close to T_c. This approach overcomes the instantaneous approximation of previous models, ensuring energy conservation in the coalescence process, as well as the proper thermodynamic equilibrium limit. This, in particular, allows for an extension of the coalescence idea into the low-p_T regime, and initial calculations are consistent with KE_T scaling for $v_{2,h}$.

To summarize this section, we conclude that RHIC data have provided clear evidence for the formation of an equilibrated medium with very small viscosity (an almost "perfect liquid") and large opacity with associated energy densities well above the critical one; in addition, indications for the presence of partonic degrees of freedom have been observed. This medium has been named the strongly coupled QGP, or sQGP. However, the understanding of its microscopic properties remains an open issue at this point: What are the relevant degrees of freedom and their interactions around and above T_c? Are the 3 main phenomena described above related, and, if so, how? Is there direct evidence for deconfinement and/or chiral symmetry restoration? Results from lattice QCD, as discussed in the previous section, may already have provided several important hints, but more tight connections to RHIC data need to be established.

Toward this goal, heavy-quark observables are hoped to provide new decisive and quantitative insights: Do charm and bottom quarks participate in the flow of the medium, despite their large mass? Do they even thermalize at low p_T? Do they suffer jet quenching at high

p_T? Do they corroborate evidence for quark coalescence processes? Initial measurements of heavy-quark observables have been performed providing tantalizing evidence that the answer to these questions may indeed be largely positive: a substantial elliptic flow and suppression of single-electron spectra associated with semileptonic heavy-meson decays has been observed, i.e., electrons (and positrons) arising via decays of the type $D \to e\nu X$ and $B \to e\nu X$ (the heavy-light mesons, $D = (c\bar{q})$ and $B = (b\bar{q})$, are composed of a heavy quark (c, b) and a light antiquark ($\bar{q} = \bar{u}, \bar{d}$), and as such are the main carriers of heavy quarks in the system).

We now turn to the main subject of this article, i.e., the theoretical and phenomenological description of heavy-quark interactions in the QGP and pertinent observables at RHIC.

3. Heavy Quarks in the Quark-Gluon Plasma

The special role of heavy quarks ($Q = b, c$) as a probe of the medium created in heavy-ion collisions resides on the fact that their mass is significantly larger than the typically attained ambient temperatures or other nonperturbative scales, $m_Q \gg T_c, \Lambda_{\rm QCD}$.[4] This has several implications:

(i) The production of heavy quarks is essentially constrained to the early, primordial stages of a heavy-ion collision. Thus the knowledge of the initial heavy-quark (HQ) spectrum (from, say, p-p collisions) can serve as a well defined initial state, even for low-momentum heavy quarks. The latter feature renders heavy-quark observables a prime tool to extract transport properties of the medium.

(ii) Thermalization of heavy quarks is "delayed" relative to light quarks by a factor of $\sim m_Q/T \simeq$ 5-15. While the bulk thermalization time is of order $\sim$0.5 fm/c, the thermal relaxation of heavy quarks is expected to occur on a timescale comparable to the lifetime of the QGP at RHIC, $\tau_{\rm QGP} \simeq 5$ fm/c. Based on the thus far inferred properties of the sQGP, charm quarks could "thermalize" to a certain extent, but not fully. Therefore, their spectra should be significantly modified, but still retain memory about their interaction history – an "optimal" probe. According to this estimate, bottom quarks are expected to exhibit notably less modification.

(iii) As is well-known from electrodynamics, Bremsstrahlung off an accelerated (or decelerated) charged particle is suppressed by a large power of its the mass, $\sim (m_q/m_c)^4$ for heavy relative to light quarks. Therefore, induced gluon radiation off heavy quarks (i.e., radiative energy loss) is much suppressed relative to elastic scattering.

(iv) The typical momentum transfer from a thermal medium to a heavy quark is small compared to the HQ momentum, $p_{\rm th}^2 \sim m_Q T \gg T^2$ (where $p_{\rm th}$ is the thermal momentum of a heavy quark in nonrelativistic approximation, $p_{\rm th}^2/2m_Q \simeq 3\ T$; for a relativistically moving quark, it is parametrically even larger, $p_{\rm th} \sim m_Q v$). This renders a Brownian Motion approach (Fokker-Planck equation) a suitable and

[4] Note that the production of the heaviest known quark, the top, is out of reach at RHIC; moreover, its lifetime, $\tau_t = 1/\Gamma_t \simeq 0.1$ fm/c, is by far too short to render it a useful probe.

controlled theoretical framework for describing the diffusion of a heavy quark in the (s)QGP.

(v) The HQ mass can furthermore be utilized as a large scale in developing effective (nonperturbative) interactions, such as HQ effective theory or potential model approaches. In principle, this allows to make contact with static interaction potentials from finite-temperature lattice QCD, although a number of issues have to be resolved before reliable quantitative predictions can be made.

(vi) At low momenta, nonperturbative (resummation) effects become relatively more important with increasing mass. E.g., for bound state formation, the binding energy is known to increase with the (reduced) mass of the constituents. For the concrete problem of the charm diffusion constant, it has recently been shown that the perturbation series is badly convergent even for values of the strong coupling constant as small as $\alpha_s = 0.1$ [50].

Before first RHIC data on HQ observables became available, the expectation based on pQCD radiative energy loss was that the D-meson spectra are much less suppressed than light hadron spectra [51], with a small elliptic flow of up to $v_2^D \simeq 4\%$ [52]. At the same time, the importance of elastic collisions was emphasized in Refs. [41–43]. In particular, in Ref. [41] nonperturbative HQ resonance interactions in the QGP (motivated by lattice QCD results, cf. Fig. 8) were introduced and found to reduce the HQ thermalization times by a factor of $\sim$3 compared to elastic pQCD scattering. Predictions for the D-meson v_2 and p_T spectra in the limiting case in which the degree of HQ thermalization is similar to light quarks can be found in Ref. [53]; including the effects of coalescence with light quarks, the D-meson elliptic flow reaches up to around $v_2^D \simeq 15\%$, about a factor of $\sim$4 larger than in the pQCD energy-loss calculations. This analysis also demonstrated the important feature that the single-electron ($e^\pm$) spectra arising from the semileptonic decays, $D \to e\nu X$, closely trace the suppression (R_{AA}) and elliptic flow of the parent D meson (see also Ref. [54]). In the following, we elaborate on the underlying approaches and further developments with applications to HQ transport properties and RHIC data.

3.1. Heavy-Quark Interactions in the QGP

The scattering of charm and bottom quarks in a Quark-Gluon Plasma is dominated by interactions with the most abundant particles in the medium, i.e., gluons as well as light (u and d) and strange quarks. The basic quantity to be evaluated is therefore the scattering matrix (or cross section), which can be further utilized to compute in-medium HQ selfenergies and transport coefficients. For reasons given in the previous Section we focus on elastic $2 \to 2$ scattering processes, $Q + i \to Q + i$ with $i = g, u, d, s$. Our discussion is organized into perturbative (Sec. 3.1.1.) and nonperturbative approaches (Secs. 3.1.2. and 3.1.3.).

3.1.1. Perturbative Scattering

In QCD, the simplest possible diagrams for HQ interactions with light partons are given by leading order (LO) perturbation theory. The pertinent Feynman diagrams are very similar to the ones depicted in Fig. 3, and are summarized in Fig. 12. As discussed in Sec. 2.1., these

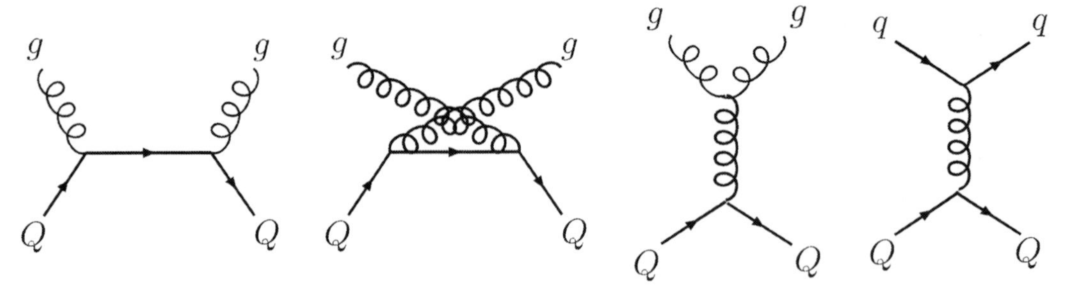

Figure 12. Feynman diagrams for leading-order perturbative HQ scattering off light partons.

processes can be expected to constitute a realistic description of HQ scattering in regimes where the strong coupling constant, α_s, is small (higher order diagrams will contribute with higher powers in α_s). In principle, this can be realized either for large HQ momenta (implying the relevant momentum transfers to be large) or at high temperatures where the interaction is screened and/or the typical momentum transfer of order $Q^2 \sim T^2$ is large. Since the color charge of gluons is larger than that of quarks (by a factor of 9/4 to order α_s), the dominant contribution to pQCD scattering arises from interactions with gluons, more precisely the t-channel gluon exchange in $Q(p_1^{(4)}) + g(p_2^{(4)}) \to Q(p_3^{(4)}) + g(p_4^{(4)})$, where $p_i^{(4)} = (E_i, \vec{p}_i)$ denotes the energy-momentum vector (or 4-momentum) of particle i (we use $p_i = |\vec{p}_i|$ for the magnitude of the 3-momentum). The corresponding differential cross section is given by

$$\frac{\mathrm{d}\sigma_{gQ}}{\mathrm{d}t} = \frac{1}{16\pi(s-M^2)^2}\overline{|\mathcal{M}|^2}\,, \tag{8}$$

where $t = (p_3^{(4)} - p_1^{(4)})^2 = 2(m_Q^2 - E_1E_2 + \vec{p}_1 \cdot \vec{p}_3)$ is the energy-momentum transfer on the heavy quark with $\vec{p}_1 \cdot \vec{p}_3 = p_1 p_3 \cos\Theta$, s is the squared center-of-mass energy $s = (p_1^{(4)} + p_2^{(4)})^2$, and Θ the scattering angle of the heavy quark. The squared scattering amplitude, averaged over initial and summed over final spin polarizations,

$$\overline{|\mathcal{M}|^2} = \pi^2\alpha_s^2\left[\frac{32(s-M^2)(s+t-M^2)}{t^2} + \ldots\right]\,, \tag{9}$$

turns out to be dominated by the t-channel gluon exchange diagram (third panel in Fig. 12), corresponding to the expression explicitly written in the brackets. The factor $1/t^2$ represents the (squared) propagator of the exchanged (free) gluon which leads to a large cross section for small t implying a small scattering angle Θ, i.e., forward scattering. However, as discussed in Sec. 2.2., the interaction is screened in the medium which, in a simplistic form, can be implemented as an in-medium gluon propagator $1/(t-\mu_D^2)$, leading to a reduction of the cross section (note that for t-channel gluon exchange, the 4-momentum transfer, t, is negative). To lowest order in perturbation theory, the gluon Debye mass is given by $\mu_D \sim gT$. The total cross section follows from integrating the expression in Eq. (8) over t; the results including all LO diagrams for charm-quark scattering off quarks and gluons are displayed in Fig. 13. The total pQCD cross sections, computed for an optimistically large $\alpha_s = 0.4$, are, in fact, quite sizable, at around $\sigma_{\rm tot} = 4(1.5)$ mb for gluons (quarks).

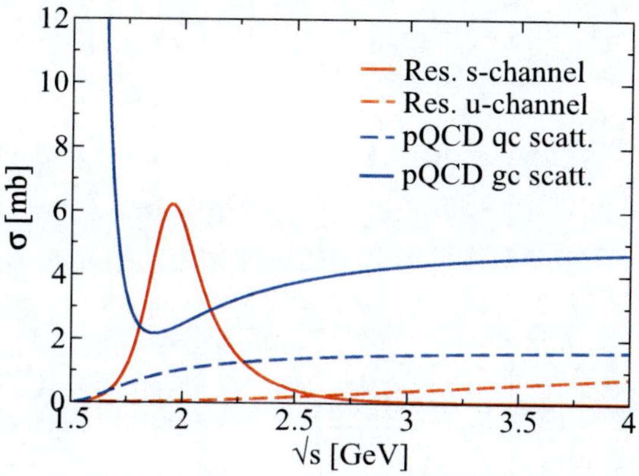

Figure 13. Total HQ scattering cross sections off light partons in pQCD (blue lines) and within the effective resonance model (red lines) [55].

In a schematic estimate these cross sections may be converted into a reaction rate by using the "pocket formula" $\Gamma_c = \sigma_{ci} n_i v_{\rm rel}$. To obtain an upper limit, one may use ideal-gas massless parton densities, $n_i = d_i \pi^2 T^3$, with quark, antiquark and gluon degeneracies of $d_{q\bar{q}} = 10.5 N_f$ with $N_f = 2.5$ and $d_g = 16$ (recall Sec. 2.2. and Fig. 5), in connection with a relative velocity $v_{\rm rel} = c$. One finds $\Gamma_c \simeq 0.4$ GeV=$(0.5{\rm fm}/c)^{-1}$, a rather large elastic scattering rate. However, as we will see below, the relevant quantity for determining the thermalization time scale is the transport cross section, which involves an extra angular weight proportional to $\sin^2 \Theta$ when integrating over the differential cross section. This renders pQCD scattering rather ineffective in isotropizing HQ distributions, due to its predominantly forward scattering angles; the resulting thermal relaxation times, $\tau_Q^{\rm therm}$, are therefore much larger than what one would naively expect from the large scattering rate estimated above.

3.1.2. Resonance Model

As was discussed in Sec. 2.2., lattice calculations suggest the existence of hadronic resonances (or bound states) in the QGP for temperatures of 1-2 T_c. In Ref. [41] the idea has been introduced that such resonances are present in the heavy-light quark sector and are operative in a significant increase of the interaction strength of heavy quarks in the QGP. The starting point of this investigation is an effective Lagrangian, *assuming* the presence of heavy-light fields, Φ, which couple to a heavy quark and light antiquark according to

$$\mathcal{L}_{\Phi Qq} = \mathcal{L}^0_\Phi + \mathcal{L}^0_Q + \mathcal{L}^0_q + \sum_m G_m Q \, \frac{1+\not{v}}{2} \, \Phi_m \, \Gamma_m \, \bar{q} + \text{h.c.} \, . \tag{10}$$

The first 3 terms on the right-hand-side represent the kinetic energy and mass terms giving rise to free particle propagation while the last term (h.c. = hermitean conjugate), roughly speaking, indicates all terms with particles and antiparticles interchanged (i.e., $Q \to \bar{Q}$, $\bar{q} \to q$, etc.). The key term is the explicitly written interaction term generating Φ-Q-$\bar{q}$ (3-point) vertices whose strength is controlled by pertinent coupling constants, G_m, which

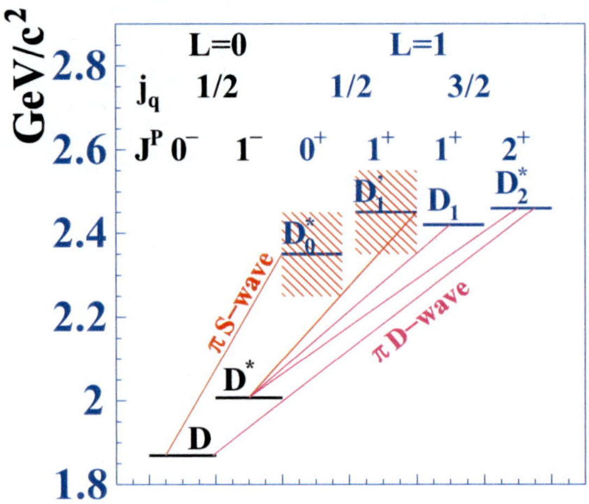

Figure 14. Level spectrum of D-mesons, $D = (c\bar{q})$, in the vacuum according to recent measurements reported in Ref. [56]. The masses of the states are shown as a function of their quantum numbers indicated in the upper portion of the figure. Note, on the one hand, the splitting (non-degeneracy) of the chiral partners D and D_0^*, as well as of D^* and D_1, by about $\Delta M = 0.4$ GeV (believed to be a consequence of spontaneous breaking of chiral symmetry in the vacuum). On the other hand, HQ symmetry, implying degeneracy of D-D^* and of D_0^*-D_1, is satisfied within $\Delta M = 0.15$ GeV.

in this approach are free parameters. The summation over m accounts for the different quantum numbers of the Φ-fields (which in turn are related to the structure of the coupling matrices, Γ_m). The couplings can be largely inferred from symmetry considerations. As alluded to toward the end of Sec. 2.1., the spontaneous breaking of chiral symmetry (SBCS) in the vacuum implies hadronic chiral partners to split in mass; in the D-meson sector, this applies to the chiral partners in the scalar ($J^P = 0^+$) - pseudoscalar ($J^P = 0^-$) multiplet, D and D_0^*, as well as to the vector ($J^P = 1^-$) and axialvector ($J^P = 1^+$) chiral multiplet, D^* and D_1. The pertinent chiral breaking is nicely borne out of recent measurements of the D-meson spectrum in vacuum [56], where the chiral splitting has been established to amount to about $\Delta M = 0.4$ GeV, cf. Fig. 14. In the QGP, however, the spontaneously broken chiral symmetry will be restored, recall, e.g., the left panel in Fig. 6. Chiral restoration is necessarily accompanied by the degeneracy of chiral partners, as is indeed observed in lQCD computations of meson spectral functions above T_c, cf. right panel of Fig. 8. We thus infer that the chiral partners in the D-meson spectrum, i.e., scalar and pseudoscalar (as well as vector and axialvector) should have the same mass and width in the QGP.

In addition, QCD possesses heavy-quark symmetries, in particular a spin symmetry which states that hadrons containing heavy quarks are degenerate if the HQ spin is flipped inside the hadron. Within the constituent quark model, one therefore expects a degeneracy of pseudoscalar and vector mesons, where the heavy and light quark are coupled in a relative S-wave (orbital angular momentum $l = 0$) and the total spin, J, of the meson is solely determined by the coupling of the two quark spins. The asserted symmetry is accurate within ~0.1-0.15 GeV in the D-meson spectrum (D-D^* and D_0^*-D_1^*, cf. Fig. 14), and

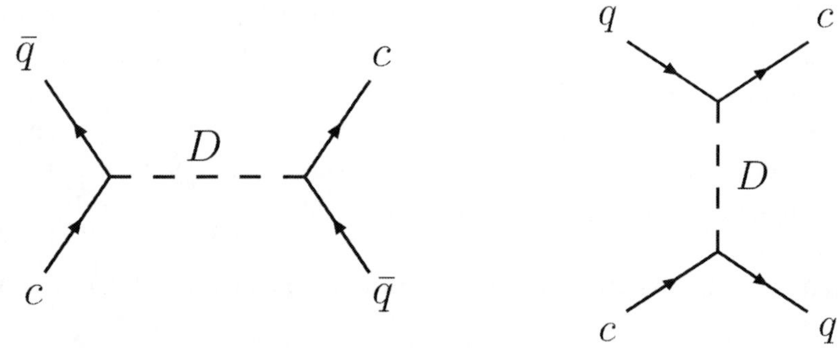

Figure 15. Feynman diagrams for nonperturbative charm-quark interactions via effective D-meson exchange in s-channel scattering off antiquarks (left panel) and u channel scattering off quarks (right panel) [41].

within ~0.05 GeV in the B-meson spectrum (as given by the $B(5280)$-$B^*(5325)$ mass difference; note that the accuracy of the HQ symmetry indeed appears to scale with the inverse HQ mass, $m_c/m_b \simeq 1/3$). Since the heavy-light resonance mass in the QGP itself is subject to uncertainties on the order of possibly up to a few hundred MeV, there is little point in accounting for the relatively small violations of HQ spin symmetry. In Ref. [41] it was therefore assumed that also the pseudoscalar-vector (as well as scalar-axialvector) states are degenerate.

With both chiral and HQ symmetry, the effective Lagrangian, Eq. (10), essentially contains 2 parameters in both the charm and bottom sector, which are the (universal) resonance mass, $m_{D,B}$, and coupling constant, $G_{D,B}$. The former has been varied over a rather broad window above the Q-$\bar{q}$ threshold. Note that bound states cannot be accessed in two-body scattering due to energy conservation, i.e., the bound state mass is by definition below the 2-particle threshold, $E_{\rm thr}$=m_Q+$m_{\bar{q}}$. Once the mass is fixed, the (energy-dependent) width for the two-body decay of the resonance, $\Phi \to Q + \bar{q}$, is determined by the coupling constant as

$$\Gamma_\Phi(M) = \frac{3G_\Phi^2}{8\pi} \frac{(M^2 - m_Q^2)^2}{M^3} \, , \Theta(M - m_Q) \tag{11}$$

where the mass of the light antiquark has been put to zero. To determine the coupling constant, some guidance can be obtained from effective quark models at finite temperature [57,58], where an in-medium D-meson width of several hundred MeV was found (see also Ref. [59] and Fig. 18 below).

The effective Lagrangian, Eq. (10), generates 2 diagrams for resonance exchange in heavy-light quark scattering, so-called s- and u-channels, as shown in Fig. 15. The pertinent total cross sections for charm-quark scattering are displayed by the red lines in Fig. 13, assuming D-meson masses and widths of m_D = 2 GeV and Γ_D = 0.4 GeV (the charm- and bottom-quark masses have been set to $m_{c,b}$ = 1.5, 4.5 GeV). In the energy regime relevant for scattering of thermal partons in the QGP, $E_{\rm cm} \simeq m_Q + E_q^{\rm th} \simeq 2.2$ GeV ($E_q^{\rm th} \simeq 3T \simeq 0.75$ GeV at $T = 0.25$ GeV), the cross section is largely dominated by the s-channel resonance formation, and is substantially larger than the LO pQCD result for Q-q scattering. This is not so compared to pQCD scattering off gluons. However, the angular distribution of the differential s-channel resonance cross section (not shown) is essentially isotropic (in the rest system of the resonance), which renders it significantly more efficient for thermalizing the HQ distributions (as we will see in Sec. 3.2. below). The interaction introduced via the Lagrangian, Eq. (10), cannot predict at what temperatures the effective resonance fields dissolve. This, however, can be overcome if the interaction underlying the resonance formation is treated on a more microscopic level. Even more importantly, a microscopic treatment could, in principle, eliminate the coupling-constant and mass parameters. First steps in this direction will be discussed in the next section.

3.1.3. Potential Scattering Based on Lattice QCD

The advances in finite-temperature lQCD to compute the in-medium free energy of a Q-$\bar{Q}$ pair as a function of its size have, in principle, opened the possibility to extract their static (chromo-electric) interaction potential, see the discussion around Eq. (4). If this problem can be well defined and solved, the next desirable step is to check the quantitative consequences of such an interaction in the light quark sector (where lQCD has also found indications for resonance formation), eventually including transport and bulk properties. In Refs. [18, 60] a relativistic correction has been suggested in terms of a velocity-velocity interaction (known as the Breit interaction in electrodynamics). In those works the focus has been on the bound state problem, by solving an underlying Schrödinger equation. In Ref. [19], a T-matrix approach to the q-$\bar{q}$ interaction has been set up, which allows for a simultaneous treatment of bound and scattering states. This framework has been applied for heavy-light quark scattering in Ref. [59]. Thus far the discussion was constrained to the color neutral (singlet) Q-$\bar{q}$ channel (i.e., a blue-antiblue, green-antigreen or red-antired color-charge combination), but a quark and an antiquark can also combine into a color-octet (a combination of red-antiblue, blue-antigreen etc.). In addition, one can extend the approach to the diquark (Q-q) channel, where color-antitriplet and sextet combinations are possible (see also Ref. [18]). Finite-temperature lattice computations of the free energy of a heavy diquark [61] suggest that the relative interaction strength in the meson and diquark systems follows the expectations of perturbative QCD (so-called Casimir scaling, which essentially reflects the color-charges of the partons), namely $V^{(1)}_{Q\bar{Q}} = 2V^{(\bar{3})}_{QQ} = -4V^{(6)}_{QQ}$; that is, the interaction in the color-triplet diquark channel is half as attractive as in the color-neutral diquark channel while it is weakly repulsive for a color-sextet diquark.

The starting point for the calculations of the heavy-light quark T-matrix is the relativistic Bethe-Salpeter equation for elastic $1+2 \to 3+4$ scattering,

$$T_m(1,2;3,4) = K_m(1,2,3,4) + \int \frac{d^4k}{(2\pi)^4} K_m(1,2;5,6)\, G^{(2)}(5,6)\, T_m(5,6;3,4)\ , \quad (12)$$

which accounts for the full 4-dimensional energy-momentum dependence of the scattering process (including "off-shell" particles for which energy and 3-momentum are independent variables); m characterizes all quantum numbers of the composite meson or diquark (color, total spin and flavor), K_m is the interaction kernel and $G^{(2)}(5,6)$ is the two-particle propagator in the intermediate state of the scattering process. The integration in the second term of Eq. (12) accounts for all possible momentum transfers in compliance with energy-momentum conservation. The labels $j = 1, \ldots, 6$ denote the quantum numbers (including 4-momentum) of the scattered single particles. Eq. (12) represents a rather involved integral equation for the full scattering T-matrix. However, the (static) potentials extracted from lQCD do not contain the rich information required for the 4-D interaction kernel K_m. It is therefore in order to adopt suitable reduction schemes [62, 63], which are well-known in nuclear physics, e.g., in the context of nucleon-nucleon scattering [64]. A reduction amounts to neglecting additional particle-antiparticle fluctuations for the intermediate 2-particle propagator and puts the latter on the energy shell, which allows to perform the energy integration (over k_0) in Eq. (12). Importantly, it furthermore enables to identify the reduced interaction kernel with a 2-body potential, V_m, and therefore to establish the connection to the potentials extracted from lattice QCD. The now 3-dimensional scattering equation, known as Lippmann-Schwinger (LS) equation, is amenable to an expansion in partial waves characterized by the angular momentum quantum number, $l = 0, 1, \ldots$, corresponding to relative S- and P-waves, etc. Assuming HQ spin symmetry, the spin quantum number does not explicitly enter, and one arrives at

$$T_{a,l}(E;p',p) = V_{a,l}(p',p) + \frac{2}{\pi}\int \mathrm{d}k\, k^2\, V_{a,l}(p',k)\, G_{Qq}(E;k)\, T_{a,l}(E;k,p)\,, \tag{13}$$

where $a = 1, \bar{3}, 6, 8$ labels the color channel. In the Thompson reduction scheme, the intermediate heavy-light quark-quark (or quark-antiquark) propagator takes the form

$$G_{Qq}(E;k) = \frac{1}{4}\frac{1 - f(\omega_k^Q) - f(\omega_k^q)}{E - (\omega_k^q + \mathrm{i}\Sigma_I^q(\omega_k^q, k)) - (\omega_k^Q + \mathrm{i}\Sigma_I^Q(\omega_k^Q, k))} \tag{14}$$

with $\omega_k^{q,Q} = (m_{q,Q}^2 + k^2)^{1/2}$ the on-shell quark energies, and $f(\omega_k^{q,Q})$ the thermal Fermi distribution functions (the numerator in Eq. (14) accounts for Pauli blocking, which, however, is essentially irrelevant at the temperatures considered here). A pictorial representation of the T-matrix scattering equation is given in the upper panel of Fig. 16. Due to the interactions with the medium, the quark propagators themselves are modified which is encoded in a single-quark selfenergy, $\Sigma_{q,Q}$, figuring into the 2-particle propagator, Eq. (14). It can be related to the heavy-light T-matrix, as well as due to interactions with gluons, as

$$\Sigma_Q(k;T) = \Sigma_{Qg} + \int \frac{\mathrm{d}^3p}{2\omega_p^q}\, f(\omega_p^q)\, T_{Qq}(E;p,k)\,, \tag{15}$$

cf. the middle panel in Fig. 16. The selfenergy can be resummed in a Dyson equation for the single-quark propagator according to

$$D_Q = D_Q^0 + D_Q^0\, \Sigma_Q\, D_Q\,, \tag{16}$$

where the free propagator is given by $D_Q^0(\omega, k) = 1/(\omega - \omega_k^0)$ (omitting any quantum number structure and denoting the free particle on-shell energy by ω_k^0). The Dyson equation is graphically displayed in the lower panel of Fig. 16. It can be solved algebraically, resulting in the full in-medium single-particle propagator

$$D_Q = \frac{1}{\omega - \omega_k^0 - \Sigma_Q}\,. \tag{17}$$

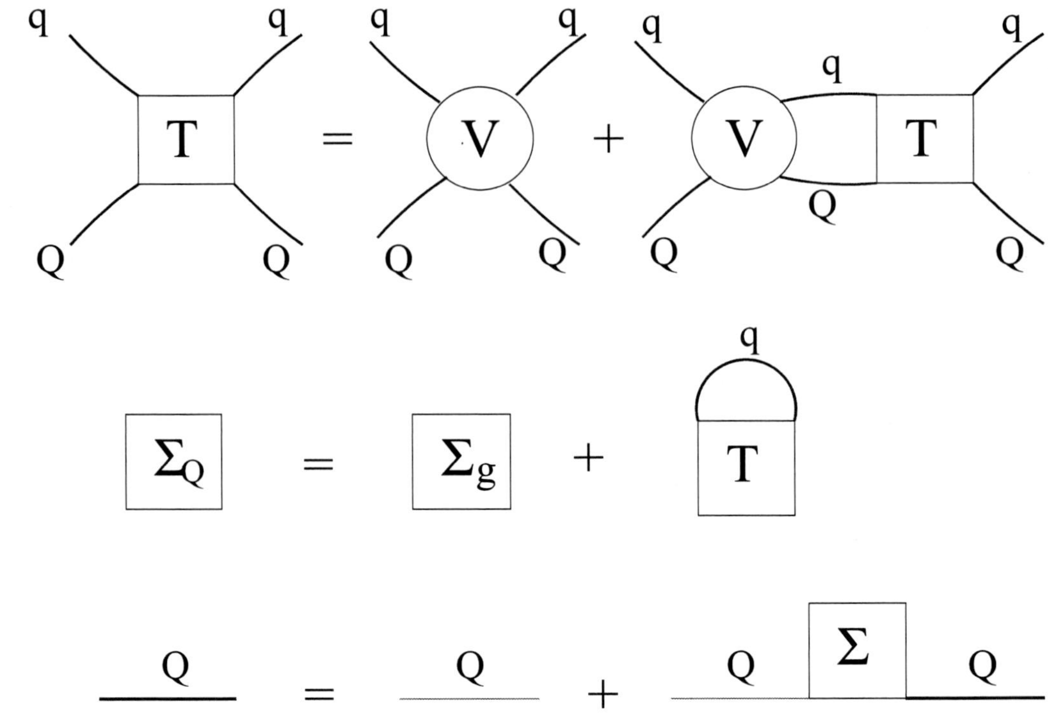

Figure 16. Diagrammatic representation of the Brueckner problem for HQ interactions in the QGP [59,65]; upper panel: T-matrix equation for HQ scattering off thermal light quarks or antiquarks; middle panel: HQ selfenergy due to interactions with gluons and quarks or antiquarks; lower panel: Dyson equation for the in-medium HQ propagator.

The convolution of the full propagators, D_q and D_Q, within the Thompson reduction scheme leads to the 2-particle propagator, Eq. (14). In general, the selfenergy is a complex-valued quantity (as is the T-matrix), with its real part affecting the in-medium quasiparticle mass while the imaginary part characterizes the attenuation (absorption) of the propagating particle (wave).

Since the HQ selfenergy depends on the heavy-light T-matrix, and the latter, in turn, depends on the selfenergy, one is facing a Brueckner-type selfconsistency problem, as illustrated in Fig. 16. In the light-quark sector, the system of Eqs. (13) and (15) has been solved by numerical iteration and moderate effects due to selfconsistency have been found. For a heavy quark, the impact of selfconsistency (which to a large extent is governed by the real parts of Σ) is weaker (since the relative corrections to the HQ mass are not as large as for light quarks). Therefore, in the calculations of the T-matrix in Ref. [59], the real and imaginary parts of the selfenergies of light and heavy quarks have been approximated by (constant) thermal mass corrections (largely being attributed to the gluon-induced selfenergy term, Σ_{Qg}, in Eq. (15) and resulting in $m_q^{\rm th} = 0.25$ GeV and $m_{c,b}^{\rm th} = 1.5, 4.5$ GeV) and quasiparticle widths of $\Gamma_{Q,q} = 0.2$ GeV. The underlying heavy-light interaction potentials, $V_{1,\bar{3},6,8}$, have been identified with the internal energy, $U_{\bar{Q}Q}$, extracted from the lattice results of the in-medium singlet free energy, $F^{(1)}_{\bar{Q}Q}$, in combination with Casimir scaling for

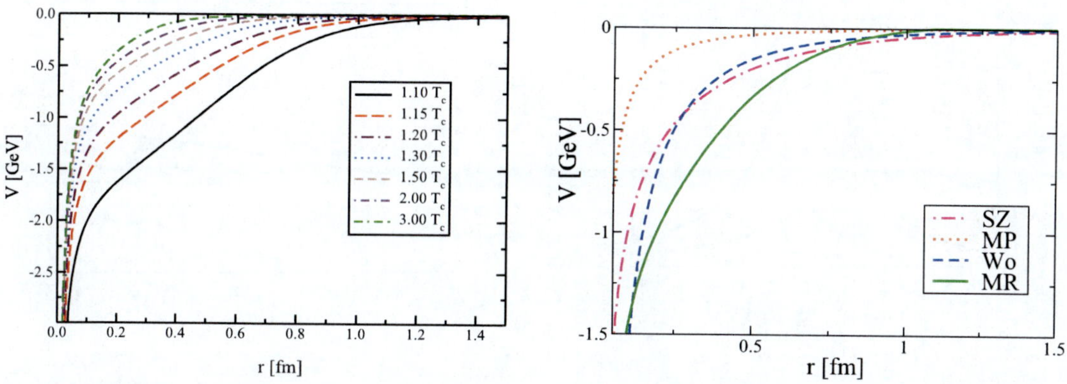

Figure 17. Left panel: subtracted color-singlet heavy-quark internal energy, $U_{Q\bar{Q}}(r)-U^{\infty}_{Q\bar{Q}}$, numerically evaluated from N_f=2 lattice QCD computations in Ref. [24], as a function of Q-$\bar{Q}$ distance for various temperatures (figure taken from Ref. [22]). Right panel: Comparison [19] of different extractions of the color-singlet Q-$\bar{Q}$ potential, Eq. (18), at T=1.5 T_c (SZ= [18], MP= [20], Wo= [23], MR= [19]).

the strengths in the non-singlet color channels. Since the long-distance limit of the internal energy, $U^{\infty}_{\bar{Q}Q} \equiv U_{\bar{Q}Q}(r \to \infty)$, does not go to zero, it needs to be subtracted to ensure the convergence of the scattering equation,

$$V_{\bar{Q}Q}(r) = U_{\bar{Q}Q}(r) - U^{\infty}_{\bar{Q}Q} \ . \tag{18}$$

Clearly, for infinite separation in a deconfined medium, the force between 2 quarks should vanish, which is consistent with the leveling off of the free (and internal) energy at large r in Fig. 7[5]. This energy contribution is naturally associated with an in-medium mass correction (in fact, in the vacuum, it can be identified with the difference between the bare charm-quark and the D-meson mass). However, a potential problem is the entropy contribution in the internal energy, $U_{\bar{Q}Q} = F_{\bar{Q}Q} + TS_{\bar{Q}Q}$, which does not vanish for $r \to \infty$; especially close to the critical temperature, the $TS^{\infty}_{\bar{Q}Q}$ term becomes uncomfortably large, and its subtraction consequently leads to a rather strong effective potential, cf. Eq. (18). This is currently the largest uncertainty in the extraction of potentials from the lQCD free energy, which maybe as large as 50%, as illustrated in Fig. 17. The use of the internal energy may be regarded as an upper estimate for the strength of the thus constructed potentials.

The numerical calculations of the in-medium heavy-light T-matrices in the scheme outlined above [59] have found that the dominant interactions are operative in the attractive color-singlet (mesons) and color-antitriplet (diquark) channels, while they are strongly suppressed in the repulsive sextet and octet channels. This can be understood from the iterative structure of the T-matrix Eq. (13), implying that higher order terms in the Born series, $T = V + VGV + VGVGV + \cdots$, are of alternating sign for repulsive potentials, while they add for attractive potentials. Moreover, from the left panel of Fig. 18 one sees that the color-singlet meson channel supports charm-light quark resonance structures in the vicinity

[5]Recall that the force is the gradient of the potential, $\vec{\mathcal{F}} = -\vec{\nabla} V(r)$

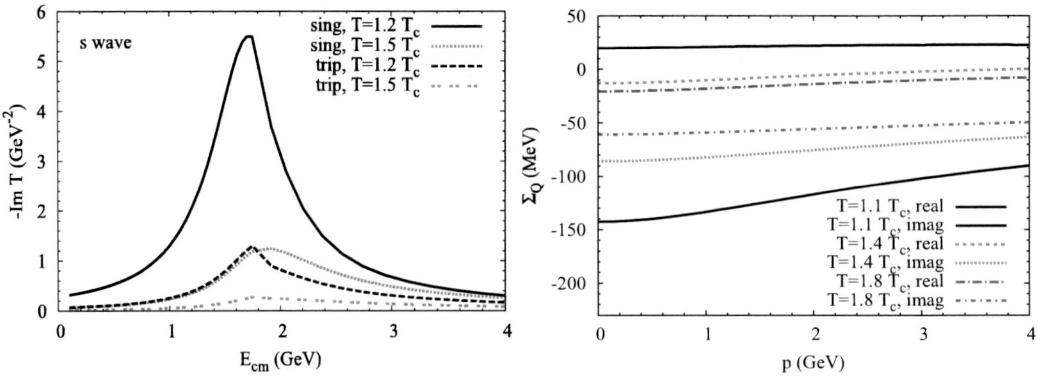

Figure 18. Left panel: in-medium T-matrix for HQ potential scattering off light quarks and antiquarks in the QGP in the color-antitriplet diquark and color-singlet meson channels, respectively, at 2 different temperatures [59]; right panel: real and imaginary parts of charm-quark selfenergies at different temperatures resulting from a sum over all T-matrix channels; the real parts are generally small, while the imaginary parts are related to the scattering rate by $\Gamma_c = -2\ \mathrm{Im}\Sigma_c$.

of the 2-body threshold up to temperatures of possibly $\sim$1.5 T_c, and up to $\sim$1.2 T_c in the antitriplet diquark channel (all numerical results related to the T-matrix approach as shown here and below are based on the potential labelled "Wo" in the right panel of Fig 17).

The interaction strength may be better quantified via the HQ selfenergies, displayed in the right panel of Fig. 18 for charm quarks. While the real parts are small at all temperatures (not exceeding 0.02 GeV), the imaginary parts are substantial, translating into scattering rates (or quasiparticle widths) of up to $\Gamma_c = -2\ \mathrm{Im}\Sigma_c \simeq$ 0.2-0.3 GeV not too far above T_c. Part of the reason for the different magnitudes in real and imaginary parts is that the real part of the T-matrix assumes both positive (repulsive) and negative (attractive) values, which compensate each other when integrated over (or from different channels). The imaginary part of the T-matrix is strictly negative (absorptive) and always adds in the selfenergy.

In the following Section we will set up a Brownian-Motion framework for HQ diffusion in the QGP and address the question of how the different interactions we have discussed above reflect themselves in the thermal relaxation for charm and bottom quarks.

3.2. Heavy-Quark Transport

The description of the motion of a heavy particle in a fluid (or heat bath) has a long history and a wide range of applications, including problems as mundane as the diffusion of a drop of ink in a glass of water. A suitable approach is given in terms of a diffusion equation for the probability density, ρ_1, of the "test particle" 1. More rigorously, one starts from the full Boltzmann equation for the phase space density, $f_1(\vec{r},\vec{p},t)$,

$$\left(\frac{\partial}{\partial t} + \frac{\vec{p}}{E}\cdot\vec{\nabla}_r - (\vec{\nabla}_r V)\cdot\vec{\nabla}_p\right) f_1(\vec{r},\vec{p},t) = I_{\mathrm{coll}}(f_1)\ , \tag{19}$$

where $\vec{\mathcal{F}} = -\vec{\nabla}_r V$ represents the force on the test particle due to an external (in-medium) potential, V, and $I_{\rm coll}(f_1)$ is the collision integral induced by scattering off particles in the heat bath. The application to heavy-quark motion has first been advocated in Ref. [66]. Neglecting the mean-field term in Eq. (19), and assuming a uniform medium, one can integrate the Boltzmann equation over the spatial coordinates to obtain an equation for the distribution function, f_Q, of the heavy quark in momentum space,

$$f_Q(\vec{p};t) \equiv \int d^3r\, f_Q(\vec{r},\vec{p},t) \ , \tag{20}$$

which is solely determined by the collision term,

$$\frac{\partial f_Q(\vec{p};t)}{\partial t} = I_{\rm coll}(f_Q) \ , \tag{21}$$

The latter can be written as an integral over all momentum transfers, k,

$$I_{\rm coll}(f_Q) = \int {\rm d}^3k[w(p+k,k)f_Q(p+k) - w(p,k)f_Q(p)] \ , \tag{22}$$

where the key ingredient is a transition rate, $w(p,k)$, for the HQ momentum to change from p to $p-k$. The two terms in the integral represent the scattering of the heavy quark into ("gain term") and out ("loss term") of the momentum state p. The transport equation (21) still constitutes a differential-integral equation for f_Q which in general is not easily solved. At this point, one can take advantage of the HQ quark mass providing a large scale, so that the momentum p of the heavy quark can be considered to be much larger than the typical momentum transfer, $k \sim T$, imparted on it from the surrounding medium. Under these conditions, the transition rate in Eq. (22) can be expanded for small k. Keeping the first two terms of this expansion, Eq. (21) is approximated by the Fokker-Planck equation,

$$\frac{\partial f_Q(p,t)}{\partial t} = \frac{\partial}{\partial p_i}\left[A_i(p) + \frac{\partial}{\partial p_j}B_{ij}(p)\right] f_Q(p,t) \ . \tag{23}$$

where the transport coefficients, A_i and B_{ij}, are given by

$$A_i(p) = \int {\rm d}^3k\, w(p,k)\, k_i, \quad B_{ij}(p) = \frac{1}{2}\int {\rm d}^3k\, w(p,k)\, k_i k_j \tag{24}$$

in terms of the transition rate w; A_i encodes the average momentum change of the heavy quark per unit time and thus describes the friction in the medium, while B_{ij} represents the average momentum broadening per unit time, i.e., the diffusion in momentum-space. For an isotropic medium (in particular a medium in thermal equilibrium), the transport coefficients can be reduced to

$$A_i(p) = \gamma(p^2)p_i, \quad B_{ij}(p) = \left[\delta_{ij} - \frac{p_i p_j}{p^2}\right] B_0(p^2) + \frac{p_i p_j}{p^2} B_1(p^2) \ , \tag{25}$$

where the friction coefficient $\gamma = \tau_{\rm therm}^{-1}$ is equivalent to a thermal relaxation time, and B_0 and B_1 are diffusion coefficients. It is very instructive to examine the limit of momentum independent coefficients (which in general is not the case and will not be assumed below), in which case the Fokker-Planck equation reduces to a particularly simple form,

$$\frac{\partial f_Q}{\partial t} = \gamma \frac{\partial}{\partial p_i}(p_i f_Q) + D \frac{\partial}{\partial p_i}\frac{\partial}{\partial p_i} f_Q \,, \tag{26}$$

which clearly illustrates the form of the diffusion term, proportional to a single diffusion constant D. The diffusion and friction constants are, in fact, related via Einstein's famous fluctuation-dissipation relation,

$$\gamma = \frac{D}{T m_Q} \,, \tag{27}$$

reducing the problem to a single transport coefficient. Note the intimate connection between the HQ transport coefficients and the temperature of the surrounding medium. This demonstrates that the Fokker-Planck equation is a consistent approximation to the Boltzmann equation in the sense that it recovers the proper equilibrium limit: both friction and diffusion terms are essential to implement the principle of detailed balance. Even in the presence of momentum dependent transport coefficients, the Einstein relation (27), remains valid in the zero-momentum limit ($p \to 0$) and provides for a valuable check whether the computed HQ diffusion constants D and γ recover the temperature of the ambient medium.

In phenomenological applications to heavy-ion collisions, the evolution of high-p_T particles is often approximated within an energy-loss treatment, which amounts to neglecting the diffusion term, i.e., $D \to 0$. This means that only momentum- (or energy-) degrading processes are taken into account, which is reflected in the Einstein equation as the $T \to 0$ limit. The lack of momentum diffusion prohibits the particles to become part of the collectively expanding medium. Nevertheless, at high momentum this approximation may be in order if the microscopic processes underlying energy loss are rare and at high momentum transfer. In this case the Fokker-Planck equation is not reliably applicable.

Let us now turn to the microscopic input to the calculation of the transport coefficients. Throughout the remainder of this paper, we employ fixed HQ masses at $m_c = 1.5$ GeV and $m_b = 4.5$ GeV. Using Fermi's Golden Rule of quantum mechanics (or quantum field theory), the transition rate w can be expressed via the quantum mechanical scattering amplitude, $\mathcal{M}$, underlying the pertinent scattering process in the medium. For elastic $Q + i \to Q + i$ scattering ($i = q, \bar{q}, g$), the rate can be written as

$$w(p,k) = \int \frac{d_i \, \mathrm{d}^3 q}{16(2\pi)^9 \omega_p \omega_q \omega_{q+k} \omega_{p-k}} f_i(q) \, \overline{|\mathcal{M}_{iQ}|^2} \, (2\pi)^4 \delta(\omega_p + \omega_q - \omega_{p-k} - \omega_{q+k}) \,. \tag{28}$$

In perturbative QCD, the amplitude is explicitly given by Eq. (9) representing the Feynman diagrams depicted in Fig. 12. Likewise, the T-matrix discussed in the previous Section can be directly related to the $\mathcal{M}$ amplitude, see Ref. [59]. In Eq. (28), d_i denotes the spin-color degeneracy of the parton, $f_i(q)$ is its phase-space distribution in the medium, while p and q ($p - k$ and $q + k$) are the initial (final) momenta of the heavy quark and the parton, respectively. All in- and outgoing particles are on their mass shell, i.e., $\omega_p = \sqrt{m^2 + p^2}$ with their respective masses, $m = m_{Q,i}$; the δ-function enforces energy conservation in the process.

In Fig. 19, we compare the temperature dependence of the inverse friction coefficient (thermal relaxation time) for charm quarks at zero momentum for elastic scattering in pQCD

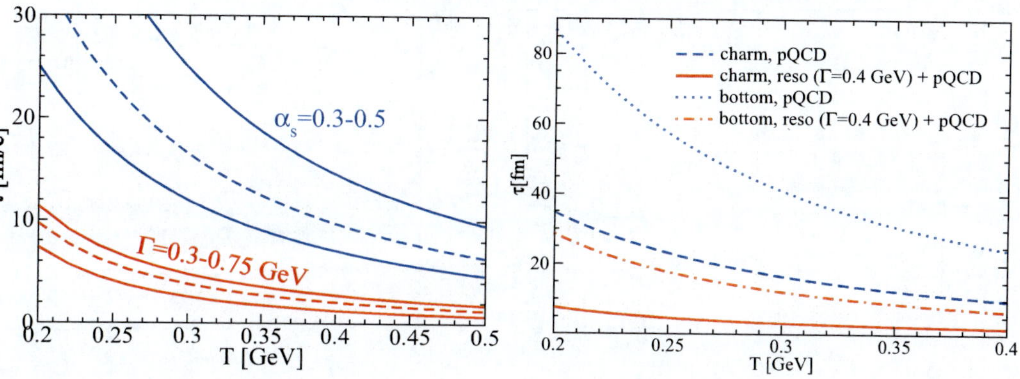

Figure 19. Thermal relaxation times of heavy quarks at zero 3-momentum as a function of temperature in the QGP [41]; left panel: charm quarks in the resonance model (with a D-meson width Γ_D=0.3-0.8 GeV; lower band) compared to LO-pQCD (with a strong coupling constant α_s=0.3-0.5; upper band) [74]. Right panel: comparison of resonance+pQCD interactions (red lines) and pQCD only (blue lines) for charm and bottom quarks.

(as discussed in Sec. 3.1.1.) with the effective resonance model (Sec. 3.1.2.). The latter leads to substantially lower thermalization times than pQCD scattering, by around a factor of $\sim$3 at temperatures $T \simeq$ 1-2 T_c. In contrast to pQCD, the values of $\tau_{c,\mathrm{therm}} \simeq$ 2-10 fm/c within the resonance model are comparable to the expected QGP lifetime at RHIC, $\tau_{\mathrm{QGP}} \simeq 5$ fm/c; thus, if resonances are operative, significant modifications of charm spectra at RHIC are anticipated due to (the approach to) thermalization. The uncertainty band covered by varying the effective coupling constant, G, over a wide range is comparatively moderate. We recall that the magnitude of the underlying total cross sections for pQCD and the resonance model (cf. left panel of Fig. 13) are not largely different; an important effect thus arises due to the angular dependence of the differential cross section (or scattering amplitudes), which is forward dominated in pQCD (corresponding to a small 3- or 4-momentum transfer, t or k) while isotropic in the rest frame of a D-meson resonance implying larger momentum transfers, k. Since the expression for A_i (and thus for γ), Eqs. (24) and (25), directly involves k, large-angle scattering is more efficient in thermalizing the c-quark distributions. Charm-quark relaxation in the resonance model appears to become less efficient at high temperatures. This is not due to the disappearance of the resonances (as will be the case in the lattice-QCD based potential approach), but due to a mismatch between the excitation energy of the resonances and the increasing average thermal energy of the partons in the heat bath. E.g., at T = 0.5 GeV, the latter amounts to $\omega_{\mathrm{thermal}} = 3\,T = 1.5$ GeV, which is well above the optimal energy for forming a D-meson resonance in collisions with a zero-momentum charm quark. Thus, with increasing temperatures the efficiency of the resonances in c-quark scattering is diminished since the partons in the surrounding medium become too energetic.

The effect of B-meson resonances on bottom quarks is relatively similar to the charm sector, i.e., a factor of $\sim$3 reduction in the thermal relaxation time compared to pQCD, cf. right panel of Fig. 19. However, the magnitude of $\tau_{b,\mathrm{therm}}$ stays above typical QGP lifetimes at RHIC, especially below 2 T_c where almost all of the QGP evolution is expected

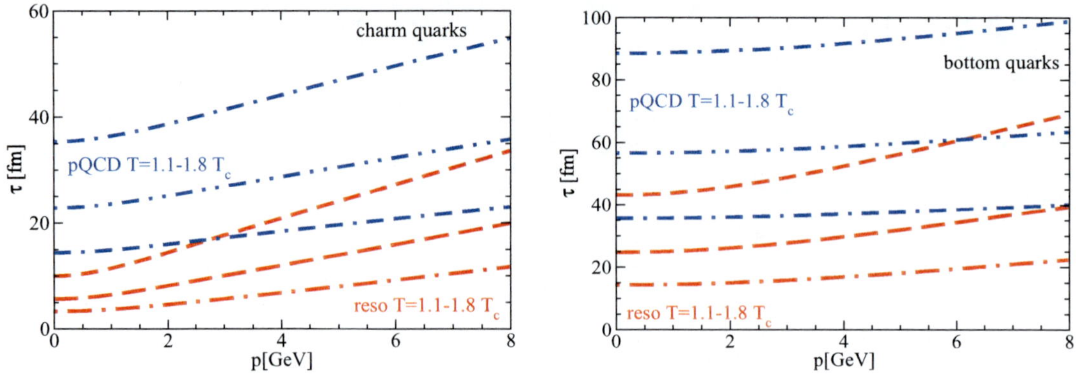

Figure 20. Thermal relaxation times of charm (left panel) and bottom quarks (right panel) in the QGP as a function of 3-momentum at 3 different temperatures, using either pQCD (α_s=0.4; upper 3 lines at p=0 for charm, and lines 1, 2 and 4 from above at p=0 for bottom) or resonance scattering (Γ_Φ=0.4 GeV; lower 3 lines at p=0 for charm, and lines 1, 2 and 4 from below at p=0 for bottom) [41]. The temperatures are T=1.1 T_c, 1.4 T_c and 1.8 T_c from top to bottom within each set of interaction.

to occur (based, e.g., on hydrodynamic simulations). The momentum dependence of the relaxation times for LO-pQCD and resonance interactions is illustrated in Fig. 20. The latter show a more pronounced increase of the relaxation time with increasing p, since their interaction strength is concentrated at low energies. It has been found that even at high momenta, the main interaction is still via resonance formation with a "comoving" parton from the heat bath (rather than from tails of the resonance in interactions with partons of typical thermal energies).

Next, we examine the transport coefficients as computed from the T-matrix approach using lQCD-based potentials (as elaborated in in Sec. 3.1.3.). In Fig. 21 we display the temperature dependence of the thermal relaxation time for charm quarks. Close to T_c, the strength of the T-matrix based interactions (including all color channels) is very comparable to the effective resonance model. It turns out that the color-singlet meson channel and the color-antitriplet diquark channel contribute to the friction coefficient by about equal parts; the somewhat smaller T-matrix in the diquark channel is compensated by the 3-fold color degeneracy in the intermediate scattering states. The contribution of the repulsive sextet and octet channels is essentially negligible. Contrary to both pQCD and the resonance model, the lQCD-based T-matrix approach leads to an increase of the c-quark relaxation time with increasing temperature, despite the substantial increase of the parton densities with approximately T^3. The increase in scattering partners is overcompensated by the loss of interaction strength as determined by weakening of the lQCD-based potentials, which is largely attributed to color screening. As a consequence, thermalization due to LO-pQCD scattering becomes more efficient than the nonperturbative T-matrix for temperatures above $T \simeq 1.8\,T_c$. This feature is very much in line with the generally expected behavior that the QGP becomes a more weakly coupled gas at sufficiently large temperatures. It is also encoded in lattice QCD computations of the bulk matter properties, where the noninter-

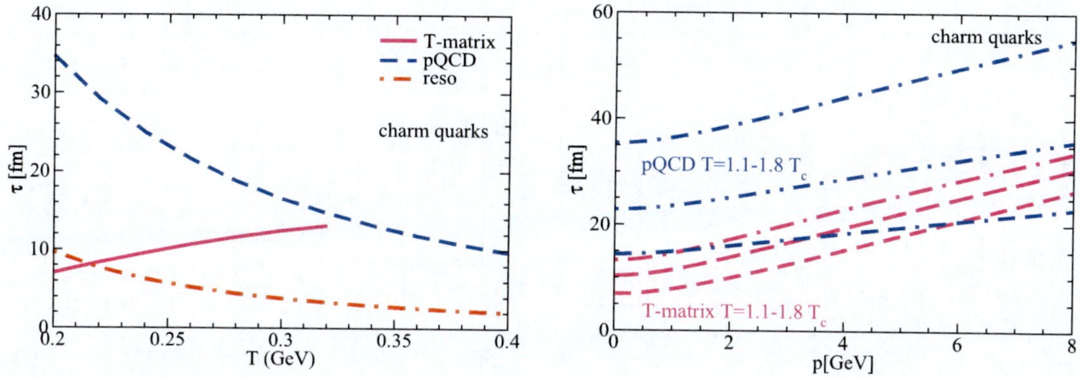

Figure 21. Thermal relaxation times for charm quarks in the QGP as computed from the heavy-light quark T-matrices utilizing potentials estimated from lattice QCD, including all 4 color combinations in $c\bar{q}$ and cq channels [59, 65]. The temperature dependence at zero 3-momentum (middle line in the left panel) and the 3-momentum dependence at 3 temperatures (lower 3 lines at p=0 in the right panel) of $\tau_{c,\mathrm{therm}}$ are compared to LO-pQCD scattering (upper blue curves at p=0). Note that in the right panel, the curves are for temperatures T=1.1 T_c, 1.4 T_c, 1.8 T_c from bottom top (top to bottom) for the T-matrix (pQCD) interactions.

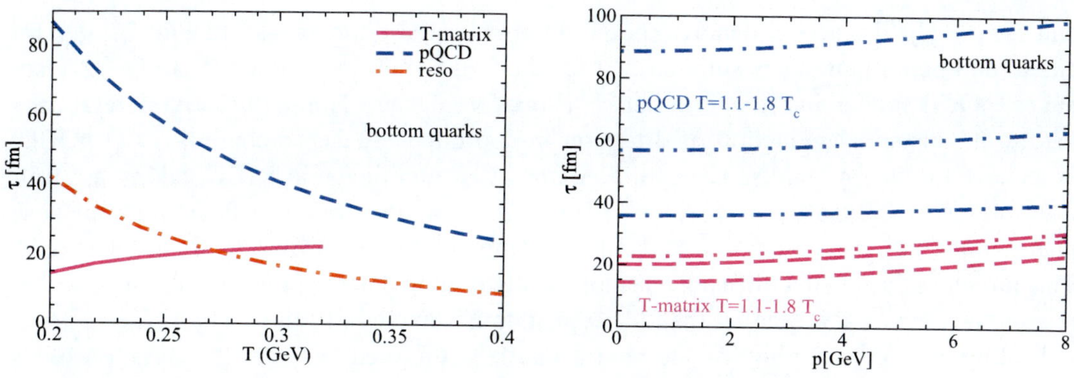

Figure 22. Same as Fig. 21 but for bottom quarks [59, 65].

acting Stefan-Boltzmann limit is recovered within 20% starting at temperatures of roughly $2\,T_c \simeq 0.35$-0.4 GeV, cf. Fig. 5. The 3-momentum dependence of the T-matrix based relaxation times, shown in the right panel of Fig. 21, reconfirms the loss of interaction strength at high momenta as found in the resonance model. The transport properties of the bottom quarks as evaluated in the T-matrix approach are summarized in Fig. 22; one finds very similar features as in the charm sector, quantitatively differing by a factor of $\sim m_b/m_c = 3$.

The momentum-space diffusion (or friction) coefficient can be converted into a spatial diffusion constant, defined in the standard way via the variance of the time (t) evolution of the particle's position,

$$\langle x^2 \rangle - \langle x \rangle^2 = 2D_x t\,. \tag{29}$$

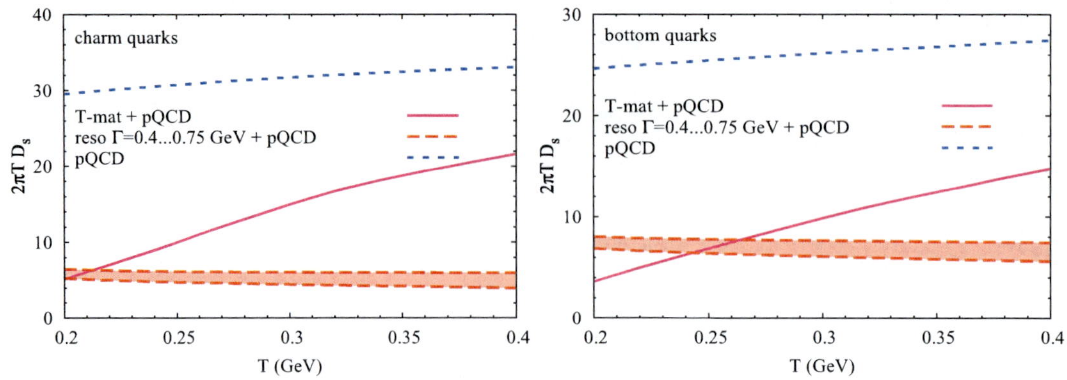

Figure 23. Spatial diffusion constant, D_s, for charm (left panel) and bottom (right panel) quarks in units of the thermal wave length, $1/(2\pi T)$, as a function of temperature. Compared are results from elastic pQCD scattering, resonance+pQCD model and T-matrix+pQCD model.

with

$$D_x = \frac{dT}{m_Q \gamma} \tag{30}$$

and $D_s = D_x/d$ where d denotes the number of spatial dimensions. In Fig. 23, D_s for charm and bottom quarks is summarized for the 3 different interactions (LO-pQCD, resonance+pQCD model and nonperturbative T-matrix+pQCD). Nonperturbative interactions lead to a substantial reduction of diffusion in coordinate space compared to LO pQCD, especially for charm quarks. E.g., for a central Au-Au collision at RHIC, taking as an illustration $T \simeq 0.2$ GeV, $t = \tau_{\rm QGP} \simeq 5$ fm/c, $D_s = 6/2\pi T$ and $d = 2$ (in the transverse plane), one finds $\Delta x = \sqrt{2D_x t} \simeq 4.3$ fm/c compared to ~ 8.6 fm/c for LO pQCD only. This indicates that spatial diffusion is significantly inhibited in the presence of nonperturbative interactions, and is smaller than the typical transverse size of the fireball, $R \simeq 8$ fm/c, indicating a strong coupling of the charm quark to the medium over the duration of its evolution.

3.3. Heavy-Quark Observables at RHIC

In this Section we elaborate on how the above developed Brownian Motion approach can be implemented into a description of HQ observables in ultrarelativistic heavy-ion collisions (URHICs) [42, 44, 68, 69][6]. This requires the following ingredients: (i) a realistic evolution of the expanding QGP fireball; (ii) Langevin simulations of the heavy quarks in the fireball background with realistic input spectra; (iii) hadronization of the HQ spectra into D- and B-meson spectra at the end of the QGP fireball evolution, and (iv) semileptonic decays of the D- and B-mesons to compare to experimental single-electron ($e^{\pm}$) spectra. We will focus on Au-Au ($\sqrt{s} = 200$ AGeV) collisions at RHIC where first measurements of the

[6]For implementations into parton transport models, see, e.g., Refs. [70, 71].

nuclear modification factor, $R_{AA}(p_T)$, and elliptic flow, $v_2(p_T)$, have become available over the last ~3-4 years.

3.3.1. Langevin Simulations

The Fokker-Planck equation for the time evolution of the phase-space distribution of a heavy particle moving through a fluid can be solved stochastically utilizing a Langevin process. The change in position and momentum of the heavy quark over a discrete but small time interval, δt, are evaluated in the rest frame of the medium (QGP) according to

$$\delta\vec{x} = \frac{\vec{p}}{\omega_p}\,\delta t\,, \quad \delta\vec{p} = -A(t,\vec{p}+\delta\vec{p})\,\vec{p}\,\delta t + \delta\vec{W}(t,\vec{p}+\delta\vec{p})\,, \tag{31}$$

where ω_p denotes the on-shell HQ energy and $\vec{p}/\omega_p$ is its relativistic velocity. The drag and diffusion terms of the Fokker-Planck equation determine the change of momentum, $\delta\vec{p}$. The momentum diffusion is realized by a random force $\delta\vec{W}$ which is assumed to be distributed according to Gaussian noise [67],

$$P(\delta\vec{W}) \propto \exp\left[-\frac{\hat{B}_{jk}\delta W^j \delta W^k}{4\delta t}\right]\,, \tag{32}$$

and controlled by the momentum diffusion coefficients B_{ij} of Eq. (25). The Gaussian form of this force is inherently consistent with the underlying Fokker-Planck equation which was derived from the Boltzmann equation in the limit of many small momentum transfers (central limit theorem).

A realistic application to URHICs hinges on a proper description of the evolving medium. As discussed in Sec. 2.3., ideal hydrodynamic simulations, especially for the QGP phase, reproduce the observed collective expansion properties in central and semicentral Au-Au collisions at RHIC very well, and have been employed for HQ Langevin simulations in connection with LO-pQCD transport coefficients in Ref. [42]. Alternatively, the basic features of hydrodynamic evolutions (collective flow and expansion timescales) may be parametrized using expanding fireball models. In Ref. [44], an earlier developed fireball model for central Au-Au collisions [72] has been extended to account for the azimuthally asymmetric (elliptic) expansion dynamics in close reminiscence to the hydrodynamic simulations of Ref. [73]. Let us briefly outline the main components of such a description. Starting point is the time dependent volume expansion of a fire cylinder, $V_{\rm FB} = z(t)\pi a(t)b(t)$ where $a(t)$ and $b(t)$ characterize the elliptical expansion in the transverse plane and $z(t)$ the longitudinal size (typically covering $\Delta y = 1.8$ units in rapidity, corresponding to the width of a thermal distribution). As in ideal hydrodynamics, the evolution is assumed to be isentropic, i.e., to proceed at a fixed total entropy, S, which is matched to the number of observed hadrons at the empirically inferred chemical freezeout, cf. Fig. 2. The time dependence of the entropy density, $s(t) = S/V_{FB}(t) = s(T)$, then determines the temperature evolution, $T(t)$ of the medium using $s_{\rm QGP}(T) = d_{\rm eff}\frac{4\pi^2}{90}T^3$ in the QGP (with $d_{\rm eff} \simeq 40$ to account for deviations from the ideal gas, cf. Fig. 5) and a (numerical) hadron resonance gas equation of state, $s_{\rm HG}(T)$, in the hadronic phase. At T_c, which at RHIC energies is assumed to coincide with chemical freezeout at $T_{\rm chem} = 180$ MeV, the hadron

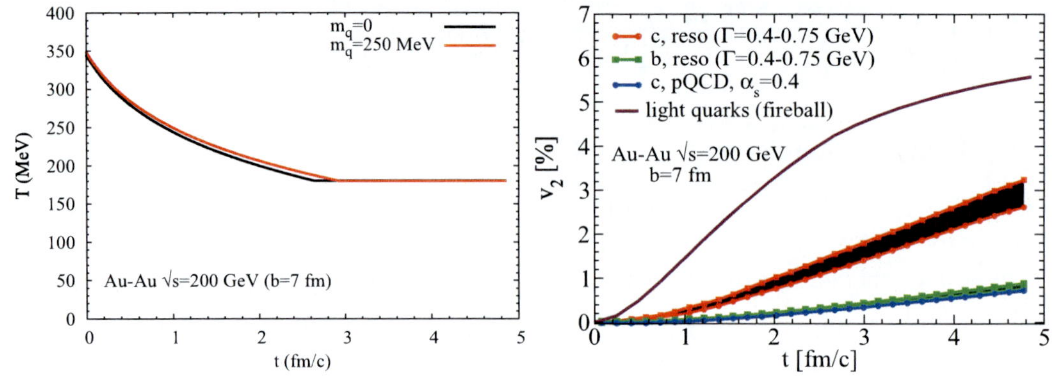

Figure 24. Thermal fireball expansion for semicentral Au-Au collisions at RHIC. Left panel: temperature evolution for either a massless gas with $d_{\rm eff}$=42 or a massive quasi-particle gas with $d_{\rm QGP}$=48 and thermal parton masses of m_i=0.25 GeV; right panel: time evolution of the inclusive elliptic flow, v_2, of the bulk medium (upper solid curve) and for heavy quarks as following from relativistic Langevin calculations: charm quarks with LO-pQCD interactions only (α_s=0.4, lowest line), as well as charm (upper band) and bottom quarks (lower band) for the resonance model (Γ_D=0.4 GeV) + pQCD.

gas is connected to the QGP phase in a standard mixed-phase construction. Assuming a formation time of the thermal medium of $\tau_0 = 1/3$ fm/c (translating into a longitudinal size of $z_0 = \tau_0 \Delta y = 0.6$ fm) after the initial overlap of the colliding nuclei, the initial temperature for semicentral (central) Au-Au($\sqrt{s} = 200$ AGeV) collisions amounts to $T_0 = 0.34(0.37)$ GeV. The subsequent cooling curve and elliptic flow of the bulk medium are displayed in Fig. 24. The last ingredient needed for the HQ Langevin simulations are the initial charm and bottom-quark spectra. They have been constructed to reproduce available experimental information on D-meson and $e^{\pm}$ spectra in elementary p-p and d-Au collisions at RHIC [74], where no significant medium formation (and thus modification of their production spectra) is expected.

In Fig. 25 we compare the results for HQ p_T spectra and elliptic flow of Langevin simulations in the above described fireball expansion using either LO-pQCD scattering with $\alpha_s = 0.4$ (which may be considered as an upper estimate) or a combination of the $Q + \bar{q} \to \Phi$ resonance interaction with LO-pQCD [44]. This combination is motivated by the fact that in the resonance model the interaction of a heavy quark is restricted to (light) antiquarks from the medium, while LO-pQCD is dominated by interactions with thermal gluons (the contribution from antiquarks is small, at the ~10-15% level of the total pQCD part). One finds that both the suppression at intermediate p_T and the elliptic flow of charm quarks are augmented by a factor 3-5 over LO-pQCD interactions, quite reminiscent to what has been found at the level of the transport coefficients. The uncertainty due to variations in the effective resonance parameters is moderate, as large as ±30%. It is remarkable that the Langevin simulations naturally provide for a leveling off of the elliptic flow as a characteristic signature of the transition for a quasi-thermal regime at low p_T to a kinetic regime for $p_T \geq 2$ GeV, very reminiscent to the empirical KE$_T$ scaling shown in Fig. 10. Even the

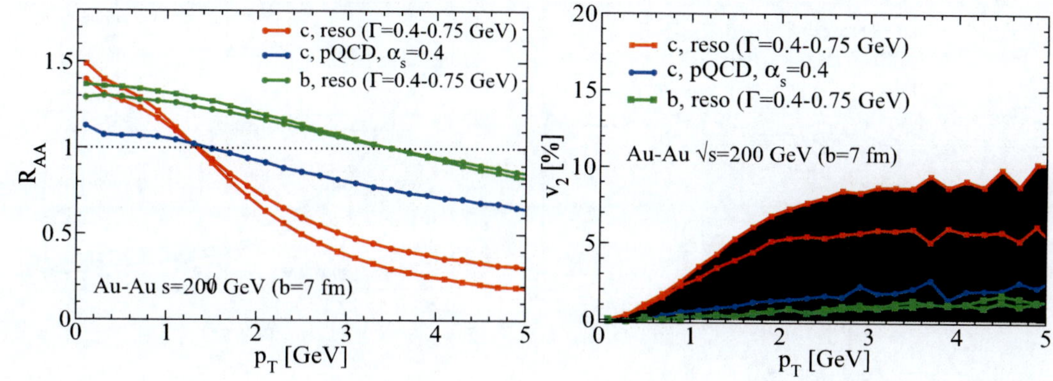

Figure 25. Nuclear modification factor (left panel) and elliptic flow (right panel) of heavy quarks as a function of their transverse momentum in semicentral (b=7 fm) Au-Au collisions [44].

quantitative plateau value of 7-8% is recovered, indicating that with the resonance+pQCD model the c-quarks are largely participating in the collective expansion of the medium. On the other hand, bottom quarks strongly deviate from the universal behavior, which is of course due to their much larger mass, $m_b = 4.5$ GeV.

The results of the HQ Langevin calculations in a hydrodynamic simulation for semicentral Au-Au collisions are summarized in Fig. 26 [42]. In these calculations, the pQCD HQ scattering amplitudes have been augmented by a full perturbative in-medium gluon exchange propagator with a fixed Debye mass of μ_D=1.5 T, while the strong coupling constant α_s has been varied to produce a rather large range of diffusion constants. The hydrodynamic evolution is performed for Au-Au collisions at impact parameter b=6.5 fm

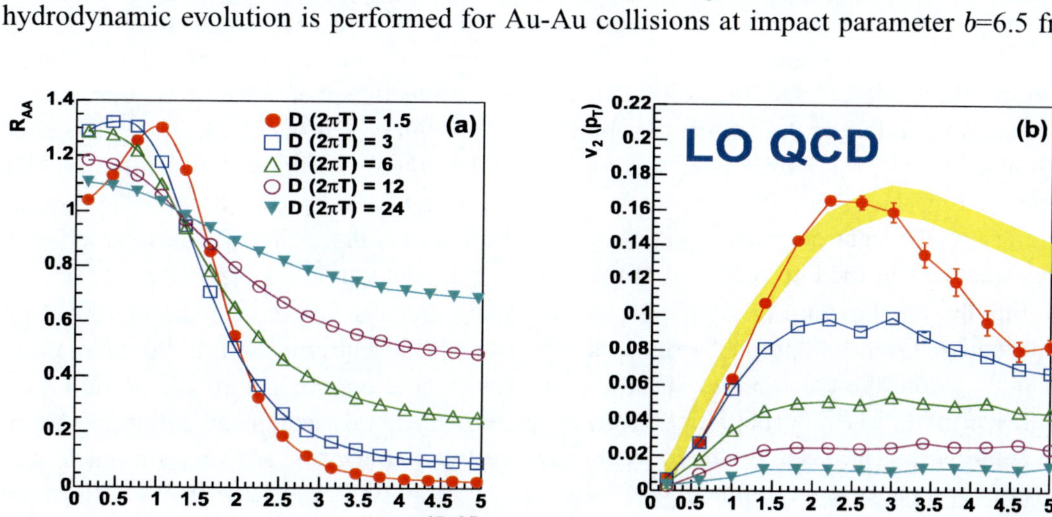

Figure 26. Nuclear modification factor and elliptic flow of charm quarks as a function of their transverse momentum in semicentral (b=6.5 fm) Au-Au collisions using a hydrodynamic evolution of the bulk medium at RHIC [42].

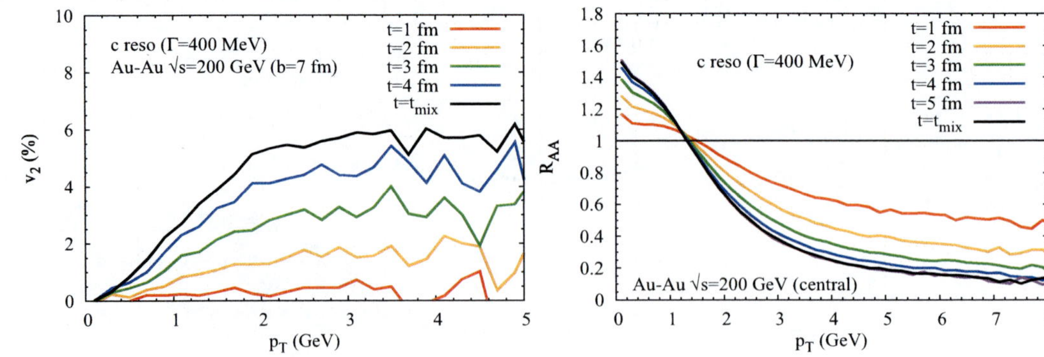

Figure 27. Time evolution of the nuclear modification factor (left panel, central collisions) and elliptic flow (right panel, semicentral collisions) of charm quarks in the resonance+pQCD model in Au-Au collisions at RHIC [44, 55].

with a thermalization time of τ_0=1 fm/c (corresponding to an initial temperature of $T_0 = 0.265$ GeV) and a critical temperature of T_c=0.165 GeV. The basic trends of the HQ spectra and elliptic flow are consistent with the fireball simulations of Ref. [44], indicating a strong correlation between large v_2 and small R_{AA} (strong suppression). Comparing more quantitatively the simulations with a "realistic" pQCD HQ diffusion constant of $D_s = 24/2\pi T$ to LO-pQCD in the fireball evolution (corresponding to $D_s \simeq 30/2\pi T$), reasonable agreement is found, with $R_{AA}(p_T$=5 GeV)$\simeq$0.7 and $v_2(p_T$=5 GeV)$\simeq$1.5-2 %, especially when accounting for the lower T_0 in the hydro evolution (implying less suppression) and the smaller impact parameter (implying less v_2).

It is very instructive to investigate the time evolution of the suppression and elliptic flow, displayed in Fig. 27 for the resonance+pQCD model in the fireball evolution. It turns out that the suppression in R_{AA} is actually largely built up in the early stages of the medium expansion, where the latter is the hottest and densest, i.e., characterized by a large (local) opacity. On the other hand, the elliptic flow, being a collective phenomenon, requires about ~4 fm/c to build up in the ambient fireball matter, implying that the charm-quark v_2 starts building up only 1-1.5 fm/c after thermalization, and then rises rather gradually, see left panel of Fig. 24. Thus, the time evolution of R_{AA} and v_2 is not much correlated, quite contrary to the tight correlation suggested by the final results. This will have important consequences for the interpretation of the experimental data in Sec. 3.3.2..

Finally, we show in Fig. 28 the fireball simulation results for heavy quarks in the non-perturbative T-matrix approach augmented by LO-pQCD scattering. Here, the latter only includes gluonic interactions to strictly avoid any double counting with the perturbative (Born) term of the T-matrix (which involves scattering off both quarks and antiquarks from the heat bath). The results, which do not involve free parameters, are very similar to the effective resonance+pQCD model.

3.3.2. Heavy-Meson and Single-Electron Spectra

To make contact with experiment, the quark spectra as computed in the previous section require further processing. First, the quarks need to be converted into hadrons (D and B

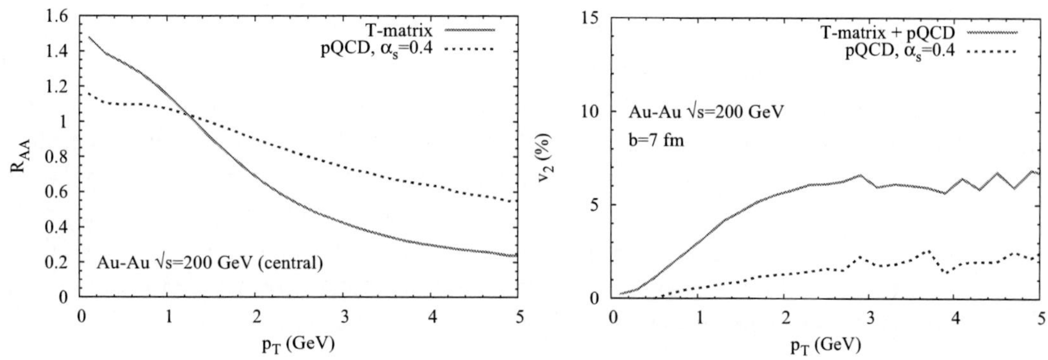

Figure 28. Nuclear modification factor (left panel) and elliptic flow (right panel) of charm quarks in central/semicentral Au-Au collisions at RHIC as computed within HQ Langevin simulations in a thermal fireball background with nonperturbative potential scattering off thermal quarks and antiquarks plus LO-pQCD scattering off gluons [59, 65].

mesons, their excited states, and possibly baryons containing heavy quarks). Second, since the current RHIC data are primarily for single electrons, the pertinent semileptonic decays, $D, B \to e\nu_e X$, need to be evaluated.

The hadronization of quarks produced in energetic collisions of elementary particles (e.g., e^+e^- annihilation or hadronic collisions) is a notoriously difficult problem that has so far evaded a strict treatment within QCD and thus requires phenomenological input. A commonly employed empirical procedure to describe hadronization of quarks produced at large transverse momentum is to define a fragmentation function, $D_{h/i}(z)$, which represents a probability distribution that a parton, i, of momentum p_i hadronizes into a hadron, h, carrying a momentum fraction $z = p_h/p_i$ of the parent parton (with $0 < z \leq 1$, reflecting the fact that color-neutralization in the fragmentation process requires the production of extra "soft" partons, which, in general, do not end up in h). At large enough p_T, the parton production occurs at a very short time scale, $\tau_{\rm prod} \simeq 1/p_T$, and thus hadronization, characterized by a typical hadronic scale, $\tau_{\rm had} \simeq 1/\Lambda_{\rm QCD}$, becomes independent of the production process (this is roughly the essence of the "factorization theorem" of QCD [75]). Therefore, the distribution $D(z)$ is supposed to be universal, i.e., can be determined (or fit) in, e.g., $e^+e^- \to \mathit{hadrons}$ and then be applied to hadronic collisions. For light quarks and gluons, $D(z)$ is typically a rather broad distribution centered around $z \simeq 0.5$, while for heavy quarks it is increasingly peaked toward $z = 1$, sometimes even approximated by a Dirac δ-function, $D(z) = \delta(z-1)$ (so called δ-function fragmentation). Toward lower p_T, other hadronization processes are expected to come into play. In hadronic collisions, a possibility is that a produced quark recombines with another quark or antiquark from its environment, e.g., valence quarks of the colliding hadrons [76]. There is ample empirical evidence for the presence (and even dominance) of the recombination mechanism in both p-p and π-p collisions, in terms of flavor asymmetries of hadrons (including charmed hadrons [77–79]) produced at forward/backward rapidities, where recombination with valence quarks is favored). E.g., for charm production in $\pi^- N$ collisions, the D^-/D^+ ratio is enhanced at large y, indicating the presence of $\bar{c}d \to D^-$ recombination with a d-quark from the $\pi^- = d\bar{u}$ (but not $c\bar{d} \to D^+$).

As discussed in Sec. 2.3., the quark recombination (or coalescence) model has received renewed interest in the context of RHIC data, by providing a successful explanation of 2 phenomena observed in intermediate-p_T hadron spectra, namely the constituent quark-number scaling of the elliptic flow and the large baryon-to-meson ratios. It is therefore natural to also apply it to the hadronization of heavy quarks [53, 80], where it appears to be even more suited since the HQ mass provides a large scale relative to which corrections to the coalescence model are relatively suppressed even at low momentum. We here follow the approach of Ref. [53] where the p_T spectrum of a D-meson is given in terms of the light and charm quark or antiquark phase space distributions, $f_{\bar{q},c}$, as

$$\frac{\mathrm{d}N_D^{\mathrm{coal}}}{\mathrm{d}y\,\mathrm{d}^2p_T} = g_D \int \frac{p \cdot \mathrm{d}\sigma}{(2\pi)^3} \int \mathrm{d}^3q \; f_D(q,x) \; f_{\bar{q}}(\vec{p}_{\bar{q}}, \vec{r}_{\bar{q}}) \; f_c(\vec{p}_c, \vec{r}_c) \; , \tag{33}$$

where $\vec{p} = \vec{p}_{\bar{q}} + \vec{p}_c$ denotes the momentum of the D meson, g_D a combinatorial factor (ensuring color-neutrality and spin-isospin averaging), $f_D(q,x)$ the Wigner function of the D-meson which is usually assumed to be a double Gaussian in relative momentum, $\vec{q} = \vec{p}_c - \vec{p}_{\bar{q}}$ and size, $\vec{r} = \vec{r}_c - \vec{r}_{\bar{q}}$, and $\mathrm{d}\sigma$ represents an integration over the hadronization volume. The charm-quark distribution function is directly taken from the output of the Langevin simulations discussed in the previous section, while the light-quark distributions are taken as determined from the successful application of the coalescence model of Ref. [45] to light hadron observables at RHIC. The coalescence mechanism, however, does not exhaust all charm quarks for hadronization, especially at high p_T where the light-quark phase-space density becomes very small; in Ref. [44], the "left-over" charm quarks, $\hat{N}_c$, have been hadronized with δ-function fragmentation (as was done when constructing the input spectrum in connection with p-p and d-Au data). The total D-meson spectrum thus takes the form

$$\frac{\mathrm{d}N_D^{\mathrm{tot}}}{\mathrm{d}y\,\mathrm{d}^2p_T} = \frac{\mathrm{d}N_D^{\mathrm{coal}}}{\mathrm{d}y\,\mathrm{d}^2p_T} + \frac{\mathrm{d}\hat{N}_c^{\mathrm{frac}}}{\mathrm{d}y\,\mathrm{d}^2p_T} \, . \tag{34}$$

In the approach of Ref. [44] the formation of baryons containing charm quarks (most notably $\Lambda_c = udc$) has been estimated to be rather small, with a Λ_c/D ratio significantly smaller than 1, and is therefore neglected. The same procedure as for charm quarks is also applied for bottom-quark hadronization. In principle, the QGP and mixed phase is followed by an interacting hadronic phase (cf. Fig. 9), where D and B mesons are subject to further reinteractions. In Ref. [81] D-meson reaction rates, Γ_D^{HG}, have been estimated in a hot pion gas. Even at temperatures close to $T_c \simeq 0.18$ GeV, $\Gamma_D^{\mathrm{HG}} \leq 0.05$ GeV, which is significantly smaller than in the QGP at all considered temperatures, cf. Fig. 18, and are therefore neglected in the calculations of Ref. [44]. Finally, the D- and B-meson spectra are decayed with their (weighted average) semileptonic decay branching (~10% for D mesons), assuming a dominance of 3-body decays (e.g., $D \to Ke\nu$). Before discussing the pertinent single-$e^\pm$ spectra in more detail, two important features should be recalled: (i) the shape and magnitude of the decay electrons closely follows those from the parent D mesons [53, 54]; (ii) the electron spectra are a combination of charm and bottom decays which experimentally cannot be separated. Since bottom-quark spectra (and thus their decay electrons) are predicted to be much less modified than charm spectra, a reliable interpretation of the $e^\pm$ spectra mandates a realistic partitioning of the 2 contributions. Following the strategy of

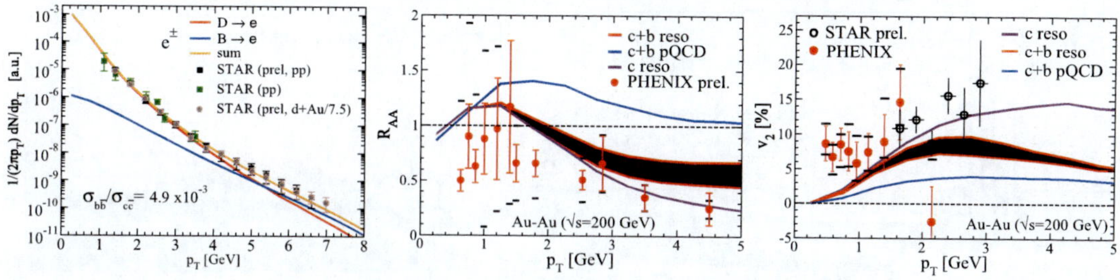

Figure 29. Left panel: single-electron spectra from heavy-flavor decays in *p*-*p* and *d*-Au collisions; the empirically inferred decomposition [44, 74] is compared to STAR data [83, 84]. Middle and right panel: $e^{\pm}$ elliptic flow and nuclear modification factor in semicentral Au-Au collisions (*b*=7fm/*c*) [44, 74] compared to first RHIC data [86–88].

Ref. [44], the input charm and electron spectra are constructed as follows: one first reproduces available D-meson spectra in *d*-Au collisions [82], calculates the pertinent electron decays and then adjusts the bottom contribution to reproduce the $e^{\pm}$ spectra in *p*-*p* and *d*-Au reactions. As a result of this procedure, the bottom contribution to the $e^{\pm}$ spectra in the elementary system exceeds the charm contribution at momenta $p_T \simeq 5$-5.5 GeV, see left panel of Fig. 29; this is consistent with the (rather large) margin predicted by perturbative QCD [85].

The first comparison of the $e^{\pm}$ spectra obtained within the Langevin resonance+pQCD+coalescence model [44, 74] to RHIC data available at the time is shown in the middle and right panel of Fig. 29. While no quantitative conclusions could be drawn, a calculation with pQCD elastic scattering alone was disfavored. One also notices that the bottom contribution leads to a significant reduction of the v_2^e at $p_T \geq 3$ GeV, as well as a reduced suppression in R_{AA}^e, where the c and b contributions become comparable for $p_T \geq 4.5$ GeV in semicentral collisions.

The resonance+pQCD+coalescence model is compared to improved $e^{\pm}$ data [89–91] in the upper panels of Fig. 30. For central Au-Au collisions, the $e^{\pm}$ suppression appears to be underpredicted starting at $p_T \simeq 4$-5 GeV (with the b contribution exceeding the charm at $p_T \geq 3.7$ GeV), indicating the presence of additional suppression mechanisms at higher momenta. The lower panels in Fig. 30 show the results of calculations without quark coalescence, i.e., all c and b quarks are hadronized with δ-function fragmentation. The shape of the $R_{AA}^e(p_T)$, as well as the magnitude of v_2^e, are not properly reproduced. This conclusion has been consolidated by another improvement of the experimental results, displayed in the left panel of Fig. 31 together with theoretical predictions within the resonance+pQCD+coalescence model [44], the hydrodynamic Langevin simulations with pQCD-inspired transport coefficients [42], as well as the radiative energy-loss approach with a transport coefficient $\hat{q}$=14 GeV^2/fm [52]. Both Langevin simulations point at a HQ diffusion coefficient of around $D_s \simeq 5/(2\pi T)$, cf. Fig. 23. The comparison of the 2 Langevin approaches reiterates the importance of the coalescence contribution: the latter is absent in the hydro calculations of Ref. [42] which cannot simultaneously describe the measured elliptic flow and suppression with a single value of the diffusion coefficient. Quark coalescence, on the other hand, introduces an "anti-correlation" of v_2 and R_{AA} into the spectra which increases the v_2 but decreases the suppression (larger R_{AA}), which clearly

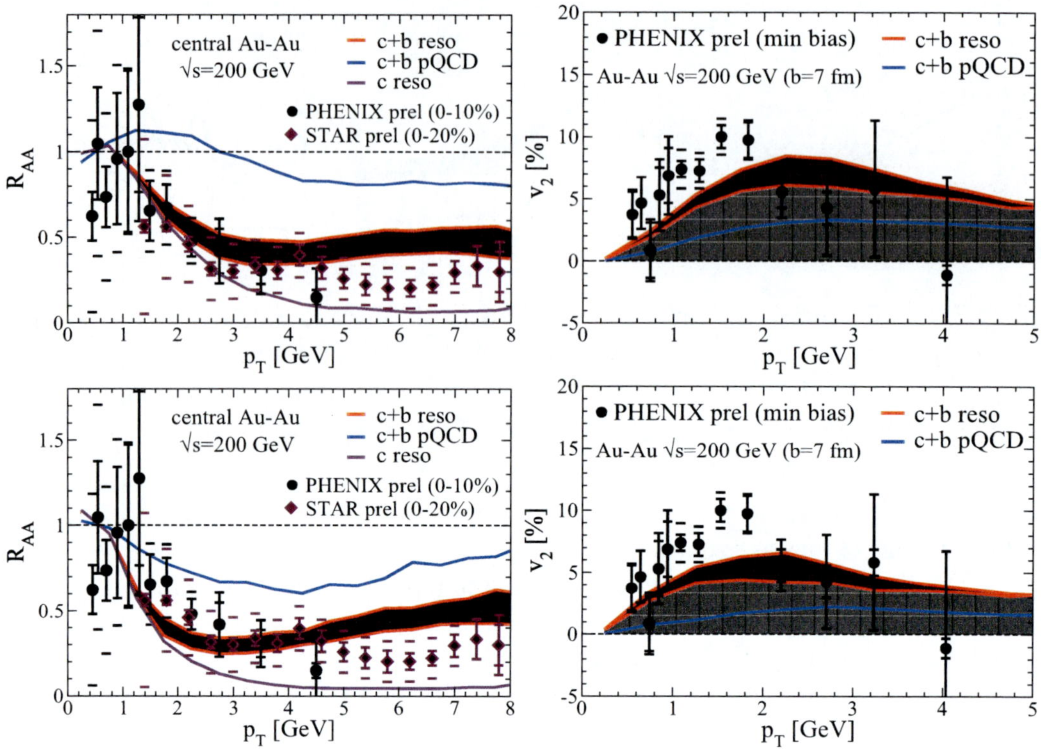

Figure 30. Elliptic flow and nuclear modification factor of electrons from heavy-flavor decays in Au-Au collisions at RHIC as computed within HQ Langevin simulations in a thermal fireball background employing the resonance+pQCD model for HQ interactions in the QGP [44, 55]. In the lower panels, quark coalescence processes at hadronization are switched off. The data are from Refs. [89–91].

improves on a consistent description of the data. In the radiative energy loss approach [52], the suppression is approximately reproduced but the azimuthal asymmetry is too small, especially at low p_T. As discussed in connection with the Fokker-Planck Eq. (26), energy-loss calculations do not account for momentum diffusion; a non-zero v_2 is therefore solely due to the geometric path length difference across the long and the short axes of the almond-shaped transverse fireball area (a shorter path length inducing less suppression). The lack of v_2 thus corroborates the interpretation that the charm (and maybe bottom) quarks become part of the collectively expanding medium, while the large transport coefficient supports the strongly coupled nature of the medium, even without diffusion and coalescence.

Finally, the right panel of Fig. 31 shows the predictions of fireball Langevin simulations employing the nonperturbative T-matrix+pQCD+coalescence approach for HQ interactions [59]. Given that the calculations are free of any adjustable parameter, the agreement with PHENIX data is surprisingly good. One should keep in mind, however, that the inherent uncertainties in this approach, e.g., in the extraction and definition of an in-medium two-body potential from the lattice-QCD free energy, are still appreciable, estimated at around $\pm 30\%$ at the level of $e^{\pm}$ observables [59]. A conceptually attractive feature of

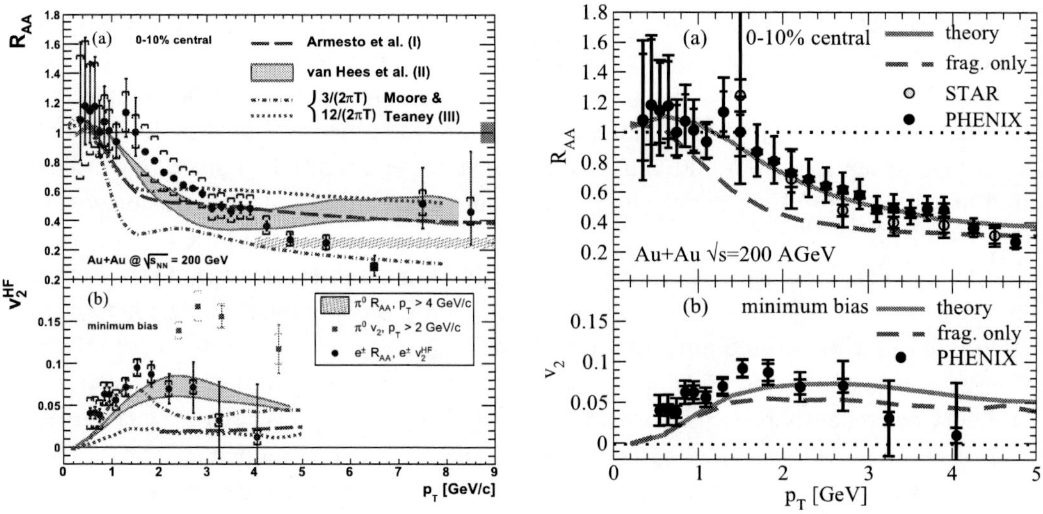

Figure 31. Elliptic flow and nuclear modification factor of electrons from heavy-flavor decays in Au-Au collisions at RHIC. Left panel: PHENIX data [92] compared to theoretical predictions based on the Langevin simulations in the resonance+pQCD+coalescence model (bands) [44] or with upscaled pQCD interactions (dash-dotted and dotted lines) [42], as well as with radiative energy-loss calculations (dashed lines) [52]. Right panel: PHENIX [92] and STAR [93] are compared to Langevin simulations employing HQ T-matrix interactions plus coalescence [59].

the T-matrix+coalescence approach is that it directly connects two thus far disconnected phenomena observed at RHIC, namely the strongly coupled QGP and quark coalescence: the very same interaction that induces the strong coupling of the heavy quarks to the QGP leads to the formation of "pre-hadronic" resonance structures close to T_c. The latter are naturally identified with the meson (M) and diquark (dq) states building up hadrons in a $q + \bar{q} \to M$ and $q + q \to dq, dq + q \to B$ processes (B: baryon), as suggested by the universal constituent-quark number scaling (CQNS), see also Ref. [49].

3.4. SQGP at RHIC?

Let us now try to elaborate on the possible broader impact of the current status of the HQ observables and their interpretation. It is gratifying to see that the available electron data thus far confirm the strongly coupled nature of the QGP produced at RHIC; relativistic Langevin simulations have quantified this notion in terms of extracted transport coefficients which are a factor of 3-5 stronger than expectations based on elastic perturbative QCD interactions. At least in the low-momentum regime, this should be a reliable statement since the heavy-quark mass warrants the main underlying assumptions, namely: (i) the applicability of the Brownian motion approach, and (ii) the dominance of elastic interactions. Clearly, an important next step is to augment these calculations by a controlled implementation of radiative energy-loss mechanisms, see, e.g., Ref. [94] for a first estimate. As for the microscopic understanding of the relevant interactions underlying the HQ rescattering, it should be noted that large couplings necessarily require resummations of some sort, especially for the problem at hand, i.e., HQ diffusion, for which the perturbation series converges especially poorly [50]. An example of such a (partial) resummation is given by the T-matrix approach [19, 59] discussed above, which is particularly valuable when it can be combined with model-independent input from lattice QCD. Here the objective must be to reduce the uncertainties in the definition and extraction of a suitable potential. This task can be facilitated by computing "Euclidean" (imaginary time) correlation functions for heavy-light mesons and check them against direct lattice computations which can be carried out with good accuracy. This kind of constraints are currently pursued in the heavy quarkonium sector [20–22], i.e., for Q-$\bar{Q}$ correlation functions, indicating that potential models are a viable framework to describe in-medium HQ interactions.

An alternative nonperturbative approach to describe the medium of a strongly-coupled gauge theory has recently been put forward by exploiting connections between string theory and Conformal Field Theory (CFT), the so-called AdS/CFT correspondence. The key point here is a conjectured duality between the weak-coupling limit of a certain string theory (defined in Anti-de-Sitter (AdS) space) and the strong-coupling limit of a supersymmetric gauge theory ("conformal" indicates that the theory does not carry any intrinsic scale, such as Λ_{QCD} in QCD; this difference may, in fact, be the weakest link in the identification of the CFT plasma with the QGP; it implies, e.g., the absence of a critical temperature in CFT). A remarkable result of such a correspondence is the derivation of a universal value for the ratio of shear viscosity to entropy density pertaining to a large class of strongly coupled quantum field theories, $\eta/s = 1/(4\pi)$, which was furthermore conjectured to be an absolute lower bound for any quantum liquid [95]. Within the same framework, the HQ diffusion has been computed, with the result $D_s \simeq 1/(2\pi T)$ [96, 97]. The assumption that the QGP is indeed in a strongly coupled regime (sQGP) can then be utilized to establish a relation between η/s and the heavy-quark diffusion constant, and thus obtain a quantitative estimate of η/s. Based on AdS/CFT, and exploiting the proportionality $\eta/s \propto D_s$ (as, e.g., borne out of kinetic theory) one has

$$\frac{\eta}{s} \approx \frac{1}{4\pi} D_s(2\pi T) = \frac{1}{2}\, T\, D_s \qquad \text{(AdS/CFT)} \tag{35}$$

using the empirically inferred $D_s(2\pi T)$=4-6, one arrives at η/s=(4-6)/(4π)=(1-1.5)/π. This may be compared to a rather recent lattice QCD computation displayed in the left panel of Fig. 32 [98], which also contains the result of a perturbative calculation. A caveat here is that the lattice computations are for a pure gluon plasma (GP).

An alternative estimate for η/s might be obtained in the weak-coupling regime. Starting point is a kinetic theory estimate of η for an ultrarelativistic gas [101, 102]

$$\eta \approx \frac{4}{15} n \,\langle p\rangle\, \lambda_{\rm tr} , \tag{36}$$

where n is the particle density, $\lambda_{\rm tr}$ the transport mean free path over which a particle's momentum is degraded by an average momentum $\langle p\rangle$. Assuming the latter to be of order of the average thermal energy of a massless parton, one has $n\langle p\rangle \approx \varepsilon$ (the energy density of the gas). Further using $Ts = \varepsilon + P = \frac{4}{3}\varepsilon$ and $\lambda_{\rm tr} = \tau_{\rm tr}$, one obtains

$$\frac{\eta}{s} \approx \frac{1}{5}\, T\, \tau_{\rm tr} . \tag{37}$$

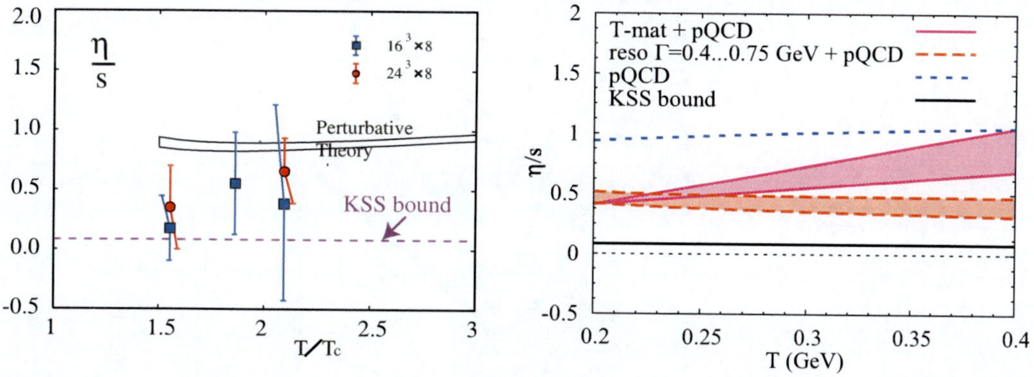

Figure 32. The ratio of shear viscosity to entropy density, η/s. Left panel: lattice QCD computations in a gluon plasma [98] compared to results inferred from perturbation theory [99, 100]. Right panel: schematic estimates utilizing the HQ diffusion constant employing either LO pQCD elastic scattering (α_s=0.4) in the weakly interacting limit (38), or the lattice-QCD potential based T matrix approach (augmented by pQCD scattering off gluons) in the strong-coupling limit (35).

Finally, accounting for the delayed thermal relaxation time of a heavy quark via $\tau_Q \approx (T/m_Q)\tau_{\rm tr}$, and using the expression, Eq. (30), for the HQ diffusion constant, one arrives at the following rough estimate for a weakly coupled (perturbative) QGP (wQGP),

$$\frac{\eta}{s} \approx \frac{1}{5}\, T\, D_s \qquad (\text{wQGP})\ . \tag{38}$$

Note the significantly smaller coefficient in this estimate compared to the one in expression (35), which reflects the expected underestimation of the shear viscosity if a gas estimate is applied in a liquid-like regime (as emphasized in Ref. [101]). As a rough application we may use the LO pQCD results for the HQ diffusion coefficient. With $D(2\pi T) \simeq 40$ for a gluon plasma (GP) at T=0.4 GeV (see Fig. 23, with a ~25% increase for removing the contributions from thermal quarks and antiquarks), one finds $\eta/s \simeq 1.25$, which is surprisingly close to the perturbative estimate constructed in Ref. [98] which is based on a next-to-leading logarithm of the shear viscosity [100] and a hard-thermal-loop calculation of the entropy density [99] (both of which represent pQCD calculations beyond the LO estimate of the HQ diffusion constant in Fig. 23).

In the right panel of Fig. 23 we attempt a schematic estimate of η/s in the Quark-Gluon Plasma based on the 3 basic calculations of the HQ diffusion coefficient discussed throughout this article. For the LO-pQCD calculation, we adopt the estimate (38) for a weakly-coupled gas, while for the resonance+pQCD model we use the strongly-coupled estimate (35). The most realistic estimate is presumably represented by the T-matrix+pQCD calculation, for which we constructed a pertinent band in η/s as follows: at the low temperature end (T=0.2 GeV) we adopt the the strong-coupling limit (35), while at the high-T end (T=0.4 GeV) we define a range whose lower limit utilizes the weak-coupling estimate and whose upper limit is simply given by the LO-pQCD only result; these 2 limits are then linearly interpolated with the T=0.2 GeV point (the latter carries, of course, significant uncertainty itself which is not accounted for in the plot).

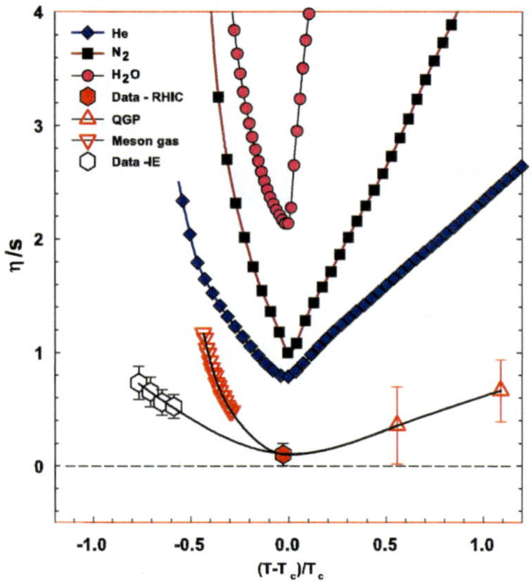

Figure 33. Compilation of the ratio of shear viscosity to entropy density for various substances [103]: atomic He, molecular N_2 and H_2O [104] (upper 3 symbols), pure-glue lattice QCD (upward triangles above T_c) [98], pion gas (downward triangles below T_c) [105] and empirical estimates from heavy-ion data (hexagons) [103].

A remarkable feature of the lattice-QCD potential based T-matrix approach is that the interaction strength *decreases* with increasing temperature - in other words, the most strongly coupled regime appears to be close to the critical temperature. It turns out that the occurrence of a maximal interaction strength at a phase transition is a rather generic phenomenon which is present in a large variety of substances: at their critical pressure, helium, nitrogen and even water have a very pronounced minimum in η/s at the critical temperature, as is nicely demonstrated in Ref. [104] and compiled in Fig. 33 taken from Ref. [103]. This plot also contains calculations for η/s in the hadronic phase, using free π-π interactions in a pion gas [105] (similar results are obtained for a π-K gas with empirical (vacuum) scattering phase shifts [107]). The decrease of η/s in a hot meson gas with increasing T corroborates the presence of a minimum around the critical temperature. The finite-temperature QCD phase transition at μ_q=0 is presumably a cross-over, but the minimum structure of η/s close to T_c is likely to persist, in analogy to the atomic and molecular system above the critical pressure [104]. If such a minimum is indeed intimately linked to the critical temperature, the AdS/CFT correspondence may be of limited applicability to establish rigorous connections between the sQGP (and RHIC phenomenology) and CFTs, since, as mentioned above, the latter do not possess an intrinsic scale.

The question of how a gluon plasma (GP) compares to a quark-gluon plasma is not merely a theoretical one, but also of practical relevance. In heavy-ion collisions at collider energies (RHIC and LHC), the (very) early phases of the reaction are presumably dominated by the (virtual) gluon fields in the incoming nuclei. A lot of progress has been made in recent years in the determination of these gluon distributions and their early evolution in URHICs [106]. The modifications of the HQ spectra due to these strong "color" fields should certainly be addressed in future work. In the subsequent evolution, an early formed GP is estimated to chemically equilibrate into a QGP rather rapidly, and closer to T_c quark coalescence models are suggestive for the dominance of quark degrees of freedom - this is where the T-matrix approach with lQCD-based potentials is predominantly operative (recall that it hinges on the presence of quarks and antiquarks in the medium). The underlying interactions could therefore provide a unified and quantitative framework for HQ diffusion, quark coalescence and the sQGP in the vicinity of T_c, with, in principle, no adjustable parameters (other than input from lQCD). Its further development should aim at a better determination of potentials extracted from lQCD, include constraints from lattice correlation functions and applications to quarkonia, and implement contributions due to gluon radiation processes.

4. Conclusion

The study of elementary particle matter has been an extremely active research field over the last 20-30 years, and it may not have reached its peak time yet. We are beginning to understand better what the key features of media are whose forces are directly governed by gauge theories. Here the strong nuclear force between quarks and gluons (as described by Quantum Chromodynamics) occupies a special role due to its large interaction strength and the self-couplings of its field quanta. On the one hand, QCD gives rise to novel nonperturbative phenomena in the vacuum, most notably the confinement of quarks into hadrons and the chiral symmetry breaking (generating the major part of the visible mass in the universe), whose underlying mechanisms are, however, not yet understood. On the other hand, the strong force generates a very rich phase structure of its different matter states (upon varying temperature and baryon density), which are even less understood. Pertinent phase transitions are, in fact, closely connected with deconfinement and chiral symmetry restoration. First principle numerical calculations of discretized (lattice) QCD at finite temperature have clearly established a transition (or a rapid cross over) from hadronic matter into a deconfined Quark-Gluon Plasma with restored chiral symmetry at a temperature of $T_c \simeq 0.175$ GeV. This state of matter is believed to have prevailed in the early Universe in the first few microseconds after the Big Bang. A particularly exciting aspect of this research field is that such kind of matter can be reproduced, at least for a short moment, in present-day laboratory experiments, by accelerating and colliding heavy atomic nuclei. However, to analyze these reactions, and to extract possible evidence for QGP formation out of the debris of hundreds to thousands of produced hadrons, is a formidable task that requires a broad approach, combining information from lattice QCD, effective models of QCD in a well-defined applicability range, and their implementation into heavy-ion phenomenology. Large progress has been made at the Relativistic Heavy-Ion Collider (RHIC), where clear evidence for the formation of thermalized QCD matter well above the critical energy density has been deduced. The apparently small viscosity of the medium is inconsistent with a weakly interacting gas of quarks and gluons, but is possibly related to recent results from lattice QCD which suggest the presence of resonant correlations for temperatures up to $\sim 2\ T_c$. We have argued that heavy quarks can serve as controlled probe of the transport properties of the strongly coupled QGP (sQGP). The modifications of heavy-quark spectra in Au-Au collisions at RHIC can be evaluated in a Brownian motion framework which allows to establish quantitative connections between the heavy-quark diffusion coefficients and observables, such as the suppression of their spectra and especially their elliptic flow. Theoretical analyses have confirmed that perturbative interactions are too weak to account for the measured heavy-quark observables (i.e., their electron decay spectra), while an effective resonance model seems to furnish the required nonperturbative interaction strength. An appealing framework to calculate heavy-quark interactions (and transport properties) in the medium is to extract interaction potentials from lattice QCD and iterate them in a nonperturbative T-matrix equation. This approach is, in principle, free of adjustable parameters, but currently subject to significant uncertainties, primarily in the definition of the potentials and their applicability to light quarks and at high momentum. However, promising results have been obtained in that the T-matrix builds up resonance-like structures close to the phase transition, which could be instrumental in explaining the observed elliptic flow. In addition, the resonance correlations naturally explain the importance of quark coalescence processes for the hadronization of the QGP (as indicated by the measured light hadron spectra). Clearly, a lot more work is required to elaborate these connections more rigorously and quantitatively, to implement additional components toward a more complete description (e.g., energy loss via gluon radiation or the effects of strong color fields), to scrutinize the results in comparison to improved lattice QCD computations, and to confront calculations with high precision RHIC (and LHC) data. The latter are expected to emerge in the coming years and will surely hold new surprises, further pushing the frontier of our knowledge of the Quark-Gluon Plasma.

Acknowledgment

We are indebted to our colleagues Vincenzo Greco, Che-Ming Ko and Massimo Mannarelli for many fruitful discussions and collaboration. This work is supported in part by a U.S. National Science Foundation CAREER award under grant no. PHY-0449489.

References

[1] M. Gell-Mann, *Phys. Lett.* **8**, 214 (1964).

[2] D. J. Gross and F. Wilczek, *Phys. Rev. Lett.* **30**, 1343 (1973).

[3] H. D. Politzer, *Phys. Rev. Lett.* **30**, 1346 (1973).

[4] H. Fritzsch, M. Gell-Mann and H. Leutwyler, *Phys. Lett. B* **47**, 365 (1973).

[5] M. Schmelling, *Proc. of 28. International Conference on High-Energy Physics (ICHEP 96), Warsaw, Poland, 25-31 Jul 1996, World Scientific*, Z. Ajduk and A.K. Wroblewski (eds.), *ICHEP* **96**, 91 (1996); hep-ex/9701002.

[6] M. Cheng *et al.*, *Phys. Rev. D* **74**, 054507 (2006).

[7] Y. Aoki, Z. Fodor, S. D. Katz and K. K. Szabo, *Phys. Lett. B* **643**, 46 (2006).

[8] R. Rapp, T. Schäfer, E. V. Shuryak and M. Velkovsky, *Phys. Rev. Lett.* **81**, 53 (1998).

[9] M. G. Alford, K. Rajagopal and F. Wilczek, *Phys. Lett. B* **422**, 247 (1998).

[10] P. Braun-Munzinger, K. Redlich and J. Stachel, in Hwa, R.C. (ed.) et al.: *Quark gluon plasma 3*, 491 (2007); arXiv:nucl-th/0304013.

[11] M. Cheng *et al.*, arXiv:0710.0354 [hep-lat].

[12] O. Kaczmarek and F. Zantow, *Phys. Rev. D* **71**, 114510 (2005).

[13] N. Brambilla *et al.* [Quarkonium Working Group], arXiv:hep-ph/0412158.

[14] T. Schäfer and E. V. Shuryak, *Rev. Mod. Phys.* **70**, 323 (1998).

[15] C. E. DeTar and R. Gupta [HotQCD Collaboration], *PoS (LAT2007)* **179** (2007); and arXiv:0710.1655 [hep-lat].

[16] P. Petreczky and K. Petrov, *Phys. Rev. D* **70**, 054503 (2004).

[17] S. Digal, P. Petreczky and H. Satz, *Phys. Lett. B* **514**, 57 (2001).

[18] E. V. Shuryak and I. Zahed, *Phys. Rev. D* **70**, 054507 (2004).

[19] M. Mannarelli and R. Rapp, *Phys. Rev. C* **72**, 064905 (2005).

[20] A. Mocsy and P. Petreczky, *Phys. Rev. D* **73**, 074007 (2006).

[21] W. M. Alberico, A. Beraudo, A. De Pace and A. Molinari, *Phys. Rev. D* **72**, 114011 (2005).

[22] D. Cabrera and R. Rapp, *Phys. Rev. D* **76**, 114506 (2007).

[23] C. Y. Wong, *Phys. Rev. C* **72**, (2005) 034906.

[24] O. Kaczmarek and F. Zantow, arXiv:hep-lat/0506019.

[25] M. Asakawa, T. Hatsuda and Y. Nakahara, *Prog. Part. Nucl. Phys.* **46**, 459 (2001).

[26] G. Aarts, C. Allton, M. B. Oktay, M. Peardon and J. I. Skullerud, *Phys. Rev. D* **76**, 094513 (2007);
M. B. Oktay, M. J. Peardon, J. I. Skullerud, G. Aarts and C. R. Allton, *PoS LAT2007*, 227 (2007).

[27] M. Asakawa, T. Hatsuda and Y. Nakahara, *Nucl. Phys. A* **715**, 863 (2003).

[28] F. Karsch and E. Laermann, in Hwa, R.C. (ed.) et al.: *Quark gluon plasma 3*, 1 (2007); arXiv:hep-lat/0305025.

[29] I. Arsene *et al.* [BRAHMS Collaboration], *Nucl. Phys. A* **757**, 1 (2005);
B.B. Back *et al.* [PHOBOS Collaboration], *Nucl. Phys. A* **757**, 28 (2005);
J. Adams *et al.* [STAR Collaboration], *Nucl. Phys. A* **757**, 102 (2005);
K. Adcox *et al.* [PHENIX Collaboration], *Nucl. Phys. A* **757**, 184 (2005).

[30] U.W. Heinz and M. Jacob, arXiv:nucl-th/0002042.

[31] P. Huovinen, P. F. Kolb, U. W. Heinz, P. V. Ruuskanen and S. A. Voloshin, *Phys. Lett. B* **503**, 58 (2001).

[32] S. Wicks, W. Horowitz, M. Djordjevic and M. Gyulassy, *Nucl. Phys. A* **784**, 426 (2007).

[33] A. Adare *et al.* [PHENIX Collaboration], *Phys. Rev. Lett.* **98**, 162301 (2007).

[34] D. Teaney, J. Lauret and E. V. Shuryak, *Phys. Rev. Lett.* **86**, 4783 (2001).

[35] T. Hirano and K. Tsuda, *Phys. Rev. C* **66**, 054905 (2002).

[36] P. F. Kolb and U. W. Heinz, in Hwa, R.C. (ed.) et al.: *Quark gluon plasma* **3**, 634 (2007); arXiv:nucl-th/0305084.

[37] C. Nonaka and S. A. Bass, *Phys. Rev. C* **75**, 014902 (2007).

[38] M. Gyulassy, I. Vitev, X. N. Wang and B. W. Zhang, in Hwa, R.C. (ed.) et al.: *Quark gluon plasma 3*, 69 (2007); arXiv:nucl-th/0302077.

[39] N. Armesto, C. A. Salgado and U. A. Wiedemann, *Phys. Rev. D* **69**, 114003 (2004).

[40] R. Baier, *Nucl. Phys. A* **715**, 209 (2003).

[41] H. van Hees and R. Rapp, *Phys. Rev. C* **71**, 034907 (2005).

[42] G. D. Moore and D. Teaney, *Phys. Rev. C* **71**, 064904 (2005).

[43] M. G. Mustafa, *Phys. Rev. C* **72**, 014905 (2005).

[44] H. van Hees, V. Greco and R. Rapp, *Phys. Rev. C* **73**, 034913 (2006).

[45] V. Greco, C. M. Ko and P. Levai, *Phys. Rev. Lett.* **90**, 202302 (2003).

[46] R. J. Fries, B. Müller, C. Nonaka and S. A. Bass, *Phys. Rev. Lett.* **90**, 202303 (2003).

[47] R. C. Hwa and C. B. Yang, *Phys. Rev. C* **67**, 034902 (2003).

[48] B. I. Abelev *et al.* [STAR Collaboration], *Phys. Rev. C* **75**, 054906 (2007).

[49] L. Ravagli and R. Rapp, *Phys. Lett. B* **655**, 126 (2007).

[50] S. Caron-Huot and G. D. Moore, arXiv:0708.4232 [hep-ph].

[51] M. Djordjevic, M. Gyulassy and S. Wicks, *Phys. Rev. Lett.* **94**, 112301 (2005).

[52] N. Armesto, M. Cacciari, A. Dainese, C. A. Salgado and U. A. Wiedemann, *Phys. Lett. B* **637**, 362 (2006).

[53] V. Greco, C. M. Ko and R. Rapp, *Phys. Lett. B* **595**, 202 (2004).

[54] X. Dong, S. Esumi, P. Sörensen, N. Xu and Z. Xu, *Phys. Lett. B* **597**, 328 (2004).

[55] R. Rapp and H. van Hees, *J. Phys. G* **32**, S351 (2006).

[56] K. Abe *et al.* [Belle Collaboration], *Phys. Rev. D* **69**, 112002 (2004).

[57] D. Blaschke, G. Burau, T. Barnes, Yu. Kalinovsky and E. Swanson, *Heavy Ion Phys.* **18**, 49 (2003).

[58] D. Blaschke, G. Burau, Yu. L. Kalinovsky and V. L. Yudichev, *Prog. Theor. Phys. Suppl.* **149**, 182 (2003).

[59] H. van Hees, M. Mannarelli, V. Greco and R. Rapp, arXiv:0709.2884 [hep-ph].

[60] G. E. Brown, C. H. Lee, M. Rho and E. Shuryak, *Nucl. Phys. A* **740**, 171 (2004).

[61] M. Döring, K. Hübner, O. Kaczmarek, F. Karsch, *Phys. Rev. D* **75**, 054504 (2007).

[62] R. Blankenbecler, R. Sugar, *Phys. Rev.* **142**, 1051 (1966).

[63] R.H. Thompson, *Phys. Rev. D* **1**, 110 (1970).

[64] R. Machleidt, *Adv. Nucl. Phys.* **19**, 189 (1989).

[65] H. van Hees, M. Mannarelli, V. Greco, R. Rapp, in preparation.

[66] B. Svetitsky, *Phys. Rev. D* **37**, 2484 (1988).

[67] J. Dunkel and P. Hänggi, *Phys. Rev. E* **71**, 016124 (2005).

[68] P. B. Gossiaux, V. Guiho and J. Aichelin, *J. Phys. G* **32**, S359 (2006).

[69] W. Liu and C. M. Ko, arXiv:nucl-th/0603004.

[70] D. Molnar, *J. Phys. G* **31**, S421 (2005).

[71] B. Zhang, L. W. Chen and C. M. Ko, *Phys. Rev. C* **72**, 024906 (2005).

[72] R. Rapp, *Phys. Rev. C* **63**, 054907 (2001).

[73] P. F. Kolb, J. Sollfrank and U. W. Heinz, *Phys. Rev. C* **62**, 054909 (2000).

[74] R. Rapp, V. Greco and H. van Hees, *Nucl. Phys. A* **774**, 685 (2006).

[75] J.C. Collins, D.E. Soper, G. Sterman, *Nucl. Phys.* **B263** (1986) 37.

[76] K.P. Das and R.C. Hwa, *Phys. Lett.* **68B**, 459 (1977).

[77] V.G. Kartvelishvili, A.K. Likhoded and S.R. Slabospitsky, *Sov. J. Nucl. Phys.* **33** (1981) 434.

[78] E. Braaten, Y. Jia and T. Mehen, *Phys. Rev. Lett.* **89**, 122001 (2002).

[79] R. Rapp and E. V. Shuryak, *Phys. Rev. D* **67**, 074036 (2003).

[80] Z. w. Lin and D. Molnar, *Phys. Rev. C* **68**, 044901 (2003).

[81] C. Fuchs, B. V. Martemyanov, A. Faessler and M. I. Krivoruchenko, *Phys. Rev. C* **73**, 035204 (2006).

[82] J. Adams *et al.* [STAR Collaboration], *Phys. Rev. Lett.* **94**, 062301 (2005).

[83] A. Tai *et al.* [STAR Collaboration], *J. Phys. G* **30**, S809 (2004).

[84] A.A.P. Suaide *et al.* [STAR Collaboration] *J. Phys. G* **30**, S1179 (2004).

[85] M. Cacciari, P. Nason and R. Vogt, *Phys. Rev. Lett.* **95**, 122001 (2005).

[86] S.S. Adler *et al.* [PHENIX Collaboration], *Phys. Rev. C* **72**, 024901 (2005).

[87] F. Laue *et al.* [STAR Collaboration], *J. Phys. G* **31**, S27 (2005).

[88] B. Jacak [PHENIX Collaboration], *J.* Phys. Conf. Ser. **50**, 22 (2006).

[89] S. S. Adler *et al.* [PHENIX Collaboration], *Phys. Rev. Lett.* **96**, 032301 (2006)

[90] S. A. Butsyk, *Nucl. Phys. A* **774**, 669 (2006)

[91] J. Bielcik [STAR Collaboration], *Nucl. Phys. A* **774**, 697 (2006)

[92] A. Adare *et al.* [PHENIX Collaboration], *Phys. Rev. Lett.* **98**, 172301 (2007).

[93] B. I. Abelev *et al.* [STAR Collaboration], *Phys. Rev. Lett.* **98**, 192301 (2007).

[94] I. Vitev, A. Adil and H. van Hees, *J. Phys. G* **34**, S769 (2007).

[95] P. Kovtun, D. T. Son and A. O. Starinets, *Phys. Rev. Lett.* **94**, 111601 (2005).

[96] C. P. Herzog, A. Karch, P. Kovtun, C. Kozcaz and L. G. Yaffe, *JHEP* **0607**, 013 (2006).

[97] J. Casalderrey-Solana and D. Teaney, *Phys. Rev. D* **74**, 085012 (2006).

[98] A. Nakamura and S. Sakai, *Phys. Rev. Lett.* **94**, 072305 (2005).

[99] J. P. Blaizot, E. Iancu and A. Rebhan, *Phys. Rev. Lett.* **83**, 2906 (1999).

[100] P. Arnold, G. D. Moore and L. G. Yaffe, *JHEP* **0305**, 051 (2003).

[101] P. Danielewicz and M. Gyulassy, *Phys. Rev. D* **31**, 53 (1985).

[102] W. Israel and J.N. Vandalas, *Lett. Nuovo Cim.* **19**, 887 (1970).

[103] R. A. Lacey and A. Taranenko, *PoS C* **FRNC2006**, 021 (2006).

[104] L. P. Csernai, J. I. Kapusta and L. D. McLerran, *Phys. Rev. Lett.* **97**, 152303 (2006).

[105] J. W. Chen and E. Nakano, *Phys. Lett. B* **647**, 371 (2007).

[106] L. McLerran, *Nucl. Phys. A* **752**, 355 (2005).

[107] M. Prakash, M. Prakash, R. Venugopalan and G. Welke, *Phys. Rept.* **227**, 321 (1993).

In: The Physics of Quarks: New Research ISBN 978-1-60456-802-8
Editors: N.L. Watson and T.M. Grant , pp. 139-162

Chapter 7

CAN THE QUARK MODEL BE RELATIVISTIC ENOUGH TO INCLUDE THE PARTON MODEL?

***Y.S. Kim*[1*] *and Marilyn E. Noz*[2†]**
[1]Department of Physics, University of Maryland,
College Park, MD 20742, U.S.A.
[2]Department of Radiology, New York University,
New York, NY, 10016, U.S.A.

Abstract

Since quarks are regarded as the most fundamental particles which constitute hadrons that we observe in the real world, there are many theories about how many of them are needed and what quantum numbers they carry. Another important question is what keeps them inside the hadron, which is known to have space-time extension. Since they are relativistic objects, how would the hadron appear to observers in different Lorentz frames? The hadron moving with speed close to that of light appears as a collection of Feynman's partons. In other words, the same object looks differently to observers in two different frames, as Einstein's energy-momentum relation takes different forms for those observers. In order to explain this, it is necessary to construct a quantum bound-state picture valid in all Lorentz frames. It is noted that Paul A. M. Dirac studied this problem of constructing relativistic quantum mechanics beginning in 1927. It is noted further that he published major papers in this field in 1945, 1949, 1953, and in 1963. By combining these works by Dirac, it is possible to construct a Lorentz-covariant theory which can explain hadronic phenomena in the static and high-speed limits, as well as in between. It is shown also that this Lorentz-covariant bound-state picture can explain what we observe in high-energy laboratories, including the parton distribution function and the behavior of the proton form factor.

1. Introduction

The hydrogen atom played the pivotal role in the development of quantum mechanics. Its discrete energy levels led to the concept of a localization condition for the probability distribution, and thus to the bound-state picture of quantum mechanics. Likewise, the quark

*E-mail address: yskim@physics.umd.edu
†E-mail address: marilyne.noz@gmail.com

model is still playing the central role in high-energy physics [1]. In this model, hadrons are bound states of more fundamental particles called"quarks" with their own internal quantum numbers, such as isospins, unitary spins, and then flavors.

Thus, the symmetry of combining these quantum numbers has been and still is an important branch of physics. Unlike the proton and electron in the hydrogen atom, quarks have never been observed as free particles. They are always confined inside the hadron. Thus the only way of determining their properties is through observing the symmetry properties of hadrons.

Then there comes the question of the binding forces between them, and the dynamics governing those forces. If the hadrons are assumed to be quantum bound states, there are localized probability distributions whose boundary conditions generate discrete mass spectra. This aspect of quantum mechanics is well known. On the other hand, it is not yet completely clear how the localized probability distribution would look to observers in different Lorentz frames. Protons coming from high-energy accelerators are quantum bound states seen in a Lorentz frame moving very fast with respect to their rest frame. This is the question on which we would like to concentrate in this review paper.

There are then three steps. First, we have to assemble the physical principles needed to construct this scheme. We shall need space-time transformation laws of special relativity and uncertainty principles of quantum mechanics applicable to position and momentum variables. Since we are interested in constructing a Lorentz-covariant theory, we need the time-energy uncertainty relation. However, this time-energy relation does not allow excited states, and has to be treated differently. This is the first hurdle we have to overcome.

The second step is to construct a mathematical formalism which will accommodate all the physical conditions presented in the first step. As always, harmonic oscillators serve as test models for all new theories. We shall construct a formalism based on harmonic oscillators, whose wave functions satisfy Lorentz-covariant boundary conditions, orthogonality conditions, the difference between position-momentum and time-energy uncertainty relations. This covariant oscillator formalism will satisfy all physical laws of quantum mechanics and special relativity.

The third step is to see whether the theory tells the story of the real world. For this purpose, we discuss in detail the proton form factors and Feynman's parton picture [2, 3]. Indeed, it has been the most outstanding issue in high energy physics whether the quark model and parton model are two different limiting cases of one Lorentz-covariant entity. We examine this issue in detail.

This review paper is largely based on the papers published by the present authors. But we are not the first ones to approach the difficult problem of constructing a Lorentz-covariant picture of quantum bound states. Indeed, this problem was recognized earlier by Dirac, Wigner, and Feynman. We shall present a review of their valiant efforts in this direction. These great physicists constructed big lakes. We are connecting these lakes to construct a canal leading to an understanding of relativistic bound states applicable to the quark model and the parton model.

Einstein was against the Copenhagen interpretation of quantum mechanics. Why was he so against it? The present form of quantum mechanics is regarded as unsatisfactory because of its probabilistic interpretation. At the same time, it is unsatisfactory because it does not appear to be Lorentz-covariant. We still do not know how the hydrogen atom appears to a moving observer.

While relativity was Einstein's main domain of interest, why did he not complain about the lack of Lorentz covariance? It is possible that Einstein was too modest to mention relativity, and instead concentrated his complaint against its probabilistic interpretation. It is also possible that Einstein did not want to send his most valuable physics asset to a battle ground. We cannot find a definite answer to this question, but it is gratifying to note that the present authors are not the first ones to question whether the Copenhagen school of thought is consistent with the concept of relativity [4].

Paul A. M. Dirac was never completely happy with the Copenhagen interpretation of quantum mechanics, but he thought it was a necessary temporary step. In that case, he thought we should examine whether quantum mechanics is consistent with special relativity.

As for combining quantum mechanics with special relativity, there was a giant step forward with the construction of the present form of quantum field theory. It leads to a Lorentz covariant S-matrix which enables us to calculate scattering amplitudes using Feynman diagrams. However, we cannot solve bound-state problems or localized probability distributions using Feynman diagrams [5].

Dirac was never happy with the present form of field theory [6], particularly with infinite quantities in its renormalization processes. Furthermore, field theory never addresses the issue of localized probability. Indeed, Dirac concentrated his efforts on determining whether localized probability distribution is consistent with Lorentz covariance.

In 1927 [7], Dirac noted that there is a time-energy uncertainty relation without time-like excitations. He pointed out that this space-time asymmetry causes a difficulty in combining quantum mechanics with special relativity.

In 1945 [8], Dirac constructed four-dimensional harmonic oscillator wave functions including the time variable. His oscillator wave functions took a normalizable Gaussian form, but he did not attempt to give a physical interpretation to this mathematical device.

In 1949 [9], Dirac emphasized that the task of building a relativistic quantum mechanics is equivalent to constructing a representation of the Poincaré group. He then pointed out difficulties in constructing such a representation. He also introduced the light-cone coordinate system.

In 1963 [10], Dirac used two coupled oscillators to construct a representation of the $O(3,2)$ deSitter group which later became the basic mathematics for two-photon coherent states known as squeezed states of light [11].

In this paper, we combine all of these works by Dirac to make the present form of uncertainty relations consistent with special relativity. Once this task is complete, we can start examining whether the probability interpretation is ultimately valid for quantum mechanics.

We then use this Lorentz-covariant model to understand the relativistic aspect of the quark model. The most outstanding problem is whether the quark model and the parton model can be combined into one Lorentz-covariant model, as in the case of Einstein's energy-momentum relation $E = \sqrt{m^2 + p^2}$ valid for all values of (p/m).

We know that the quark model is valid for small values of (p/m). We consider Feynman's parton picture for the limit of large (p/m). We then consider the proton proton form factor for (p/m) between these two limiting cases.

In Sec. 2., we review four of Dirac's major papers by giving graphical illustrations. Section 3. is devoted to combining all four of Dirac's papers into one Lorentz covariant model for quantum bound states. In Sec. 4., we discuss the parton model and the proton form factor to show that the model is consistent with what we observe in high-energy laboratories.

2. Dirac's Attempts to Make Quantum Mechanics Lorentz Covariant

Paul A. M. Dirac made it his lifelong effort to formulate quantum mechanics so that it would be consistent with special relativity. In this section, we review four of his major papers on this subject. In each of these papers, Dirac points out fundamental difficulties in this problem.

In 1927 [7], Dirac notes that there is an uncertainty relation between the time and energy variables which manifests itself in emission of photons from atoms. He notes further that there are no excitations along the time or energy axis, unlike Heisenberg's uncertainty relation which allows quantum excitations. Thus, there is a serious difficulty in combining these relations in the Lorentz- covariant world.

In 1945 [8], Dirac considers the four-dimensional harmonic oscillator and attempts to construct a representation of the Lorentz group using the oscillator wave functions. However, he ends up with the wave functions which do not appear to be Lorentz-covariant.

In 1949 [9], Dirac considers three forms of relativistic dynamics which can be constructed from the ten generators of the Poincaré group. He then imposes subsidiary conditions necessitated by the existing form of quantum mechanics. In so doing, he ends up with inconsistencies in all three of the cases he considers.

In 1963 [10], he constructed a representation of the $O(3,2)$ deSitter group using coupled harmonic oscillators. Using step-up and step-down operators, he constructs a beautiful algebra, but he makes no attempts to exploit the physical contents of his algebra.

In spite of the shortcomings mentioned above, it is indeed remarkable that Dirac worked so tirelessly on this important subject. We are interested in combining all of his works to achieve his goal of making quantum mechanics consistent with special relativity. Let us review the contents of these papers in detail, by transforming Dirac's formulas into geometrical figures.

2.1. Dirac's C-Number Time-Energy Uncertainty Relation

Even before Heisenberg formulated his uncertainty principle in 1927, Dirac studied the uncertainty relation applicable to the time and energy variables [7, 12]. This time-energy uncertainty relation was known before 1927 from the transition time and line broadening in atomic spectroscopy. As soon as Heisenberg formulated his uncertainty relation, Dirac considered whether the two uncertainty relations could be combined to form a Lorentz covariant uncertainty relation [7].

He noted one major difficulty. There are excitations along the space-like longitudinal direction starting from the position-momentum uncertainty, while there are no excitations along the time-like direction. The time variable is a c-number. How then can this space-time asymmetry be made consistent with Lorentz covariance, where the space and time coordinates are mixed up for moving observers.

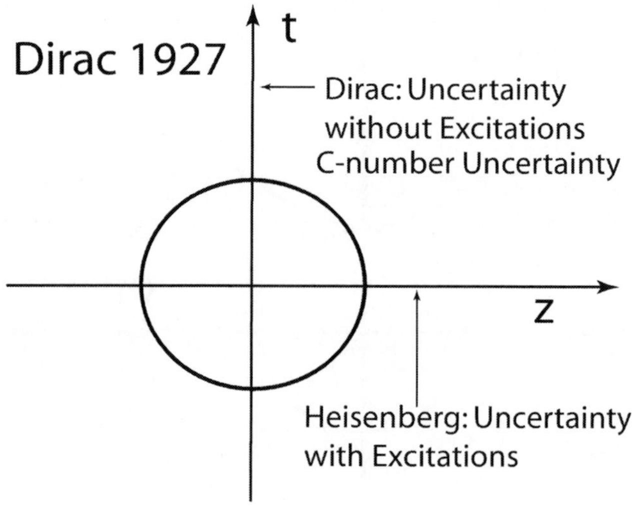

Figure 1. Space-time picture of quantum mechanics. There are quantum excitations along the space-like longitudinal direction, but there are no excitations along the time-like direction. The time-energy relation is a c-number uncertainty relation.

Heisenberg's uncertainty relation is applicable to space separation variables. For instance, the Bohr radius measures the difference between the proton and electron. Dirac never addressed the question of the separation in time variable or the time interval even in his later papers.

As for the space-time asymmetry, Dirac came back to this question in his 1949 paper [9] where he discusses the "instant form" of relativistic dynamics. He talks indirectly about the possibility of freezing three of the six parameters of the Lorentz group, and thus working only with the remaining free parameters.

This idea was presented earlier by Wigner [13, 14] who observed that the internal space-time symmetries of particles are dictated by his little groups with three independent parameters.

2.2. Dirac's Four-Dimensional Oscillators

During World War II, Dirac was looking into the possibility of constructing representations of the Lorentz group using harmonic oscillator wave functions [8]. The Lorentz group is the language of special relativity, and the present form of quantum mechanics starts with harmonic oscillators. Therefore, he was interested in making quantum mechanics Lorentz-covariant by constructing representations of the Lorentz group using harmonic oscillators.

In his 1945 paper [8], Dirac considers the Gaussian form

$$\exp\left\{-\frac{1}{2}\left(x^2 + y^2 + z^2 + t^2\right)\right\}. \tag{1}$$

We note that this Gaussian form is in the $(x,\ y,\ z,\ t)$ coordinate variables. Thus, if we consider a Lorentz boost along the z direction, we can drop the x and y variables, and write

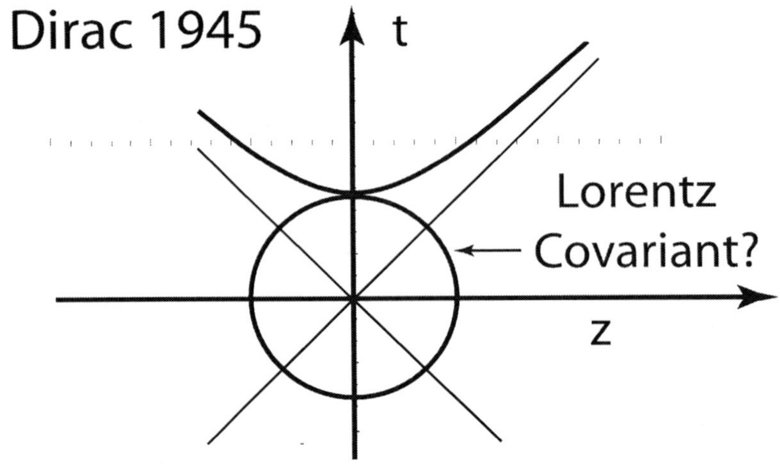

Figure 2. Dirac's four-dimensional oscillators localized in a closed space-time region. This is not a Lorentz-invariant concept. How about Lorentz covariance?

the above equation as

$$\exp\left\{-\frac{1}{2}\left(z^2 + t^2\right)\right\}. \tag{2}$$

This is a strange expression for those who believe in Lorentz invariance where $(z^2 - t^2)$ is an invariant quantity.

On the other hand, this expression is consistent with his earlier papers on the time-energy uncertainty relation [7]. In those papers, Dirac observed that there is a time-energy uncertainty relation, while there are no excitations along the time axis.

Let us look at Fig. 1 carefully. This figure is a pictorial representation of Dirac's Eq.(2), with localization in both space and time coordinates. Then Dirac's fundamental question would be how to make this figure covariant? This question is illustrated in Fig. 2. This is where Dirac stops. However, this is not the end of the Dirac story.

2.3. Dirac's Light-Cone Coordinate System

In 1949, the Reviews of Modern Physics published a special issue to celebrate Einstein's 70th birthday. This issue contains Dirac paper entitled "Forms of Relativistic Dynamics" [9]. In this paper, he introduced his light-cone coordinate system, in which a Lorentz boost becomes a squeeze transformation, where one axis expands while the other contracts in such a way that their product remains invariant.

When the system is boosted along the z direction, the transformation takes the form

$$\begin{pmatrix} z' \\ t' \end{pmatrix} = \begin{pmatrix} \cosh(\eta/2) & \sinh(\eta/2) \\ \sinh(\eta/2) & \cosh(\eta/2) \end{pmatrix} \begin{pmatrix} z \\ t \end{pmatrix}. \tag{3}$$

This is not a rotation, and people still feel strange about this form of transformation. In 1949 [9], Dirac introduced his light-cone variables defined as [9]

$$u = (z + t)/\sqrt{2}, \qquad v = (z - t)/\sqrt{2}, \tag{4}$$

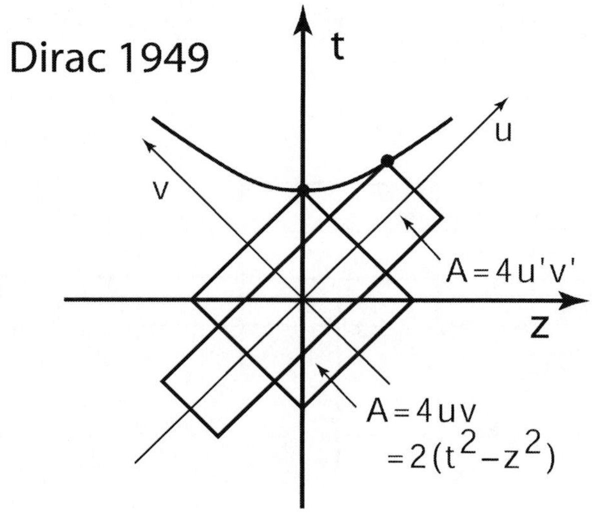

Figure 3. Lorentz boost in the light-cone coordinate system. The boost traces a point along the hyperbola. The boost also squeezes the square into a rectangle.

the boost transformation of Eq.(3) takes the form

$$u' = e^{\eta/2} u, \qquad v' = e^{-\eta/2} v. \tag{5}$$

The u variable becomes expanded while the v variable becomes contracted, as is illustrated in Fig. 3. Their product

$$uv = \frac{1}{2}(z+t)(z-t) = \frac{1}{2}\left(z^2 - t^2\right) \tag{6}$$

remains invariant. Indeed, in Dirac's picture, the Lorentz boost is a squeeze transformation.

2.4. Dirac's Coupled Oscillators

In 1963 [10], Dirac published a paper on symmetries of coupled harmonic oscillators. Starting from step-up and step-down operators for the two oscillators, he was able to construct a representation of the deSitter group $O(3,2)$. Since this group contains two $O(3,1)$ Lorentz groups, we can extract Lorentz-covariance properties from his mathematics. It is even possible to extend this symmetry group to $O(3,3)$ to include damping effects of the oscillators.

In the present paper, we avoid group theory and use a set of two-by-two matrices to exploit the physical contents of Dirac's 1963 paper. Let us start with the Hamiltonian for this system of two oscillators, which takes the form

$$H = \frac{1}{2}\left\{\frac{1}{m_1} p_1^2 + \frac{1}{m_2} p_2^2 + A x_1^2 + B x_2^2 + C x_1 x_2\right\}. \tag{7}$$

It is possible to diagonalize by a single rotation the quadratic form of x_1 and x_2. However, the momentum variables undergo the same rotation. Therefore, the uncoupling of the potential energy by rotation alone will lead to a coupling of the two kinetic energy terms.

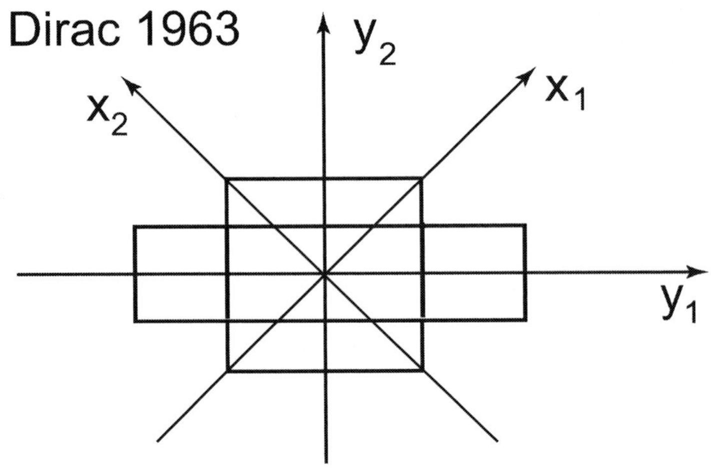

Figure 4. System of two coupled oscillators in the normal coordinate system. If the coupling becomes stronger, one coordinate variable becomes contracted while the other becomes expanded. However, the product of the two coordinates remains constant. This is an area-preserving transformation in this graph, just like Lorentz-boost in the light-cone coordinate system as described in Fig. 3.

In order to avoid this complication, we have to bring the kinetic energy portion into a rotationally invariant form. For this purpose, we will need the transformation

$$\begin{pmatrix} p_1' \\ p_2' \end{pmatrix} = \begin{pmatrix} (m_2/m_1)^{1/4} & 0 \\ 0 & (m_1/m_2)^{1/4} \end{pmatrix} \begin{pmatrix} p_1 \\ p_2 \end{pmatrix}. \tag{8}$$

This transformation will change the kinetic energy portion to

$$\frac{1}{2m}\left\{p_1'^2 + p_2'^2\right\} \tag{9}$$

with $m = (m_1 m_2)^{1/2}$. This scale transformation does not leave the x_1 and x_2 variables invariant. If we insist on canonical transformations [15], the transformation becomes

$$\begin{pmatrix} x_1' \\ x_2' \end{pmatrix} = \begin{pmatrix} (m_1/m_2)^{1/4} & 0 \\ 0 & (m_2/m_1)^{1/4} \end{pmatrix} \begin{pmatrix} x_1 \\ x_2 \end{pmatrix}. \tag{10}$$

The scale transformations on the position variables are inversely proportional to those of their conjugate momentum variables. This is based on the Hamiltonian formalism where the position and momentum variables are independent variables.

On the other hand, in the Lagrangian formalism, where the momentum is proportional to the velocity which is the time derivative of the position coordinate, we have to apply the same scale transformation for both momentum and position variables [16]. In this case, the scale transformation takes the form

$$\begin{pmatrix} x_1' \\ x_2' \end{pmatrix} = \begin{pmatrix} (m_2/m_1)^{1/4} & 0 \\ 0 & (m_1/m_2)^{1/4} \end{pmatrix} \begin{pmatrix} x_1 \\ x_2 \end{pmatrix}. \tag{11}$$

With Eq.(8) for the momentum variables, this expression does not constitute a canonical transformation.

The canonical transformation leads to a unitary transformation in quantum mechanics. The issue of non-canonical transformation is not yet completely settled in quantum mechanics and is still an open question [15]. In either case, the Hamiltonian will take the form

$$H = \frac{1}{2m}\left\{p_1^2 + p_2^2\right\} + \frac{1}{2}\left\{Ax_1^2 + Bx_2^2 + Cx_1x_2\right\}, \tag{12}$$

Here, we have deleted for simplicity the primes on the x and p variables.

We are now ready to decouple this Hamiltonian by making the coordinate rotation:

$$\begin{pmatrix} y_1 \\ y_2 \end{pmatrix} = \begin{pmatrix} \cos\alpha & -\sin\alpha \\ \sin\alpha & \cos\alpha \end{pmatrix} \begin{pmatrix} x_1 \\ x_2 \end{pmatrix}. \tag{13}$$

Under this rotation, the kinetic energy portion of the Hamiltonian in Eq.(12) remains invariant. Thus we can achieve the decoupling by diagonalizing the potential energy. Indeed, the system becomes diagonal if the angle α becomes

$$\tan 2\alpha = \frac{C}{B - A}. \tag{14}$$

This diagonalization procedure is well known. What is new in this note is the introduction of the new parameters K and η defined as

$$K = \sqrt{AB - C^2/4},$$
$$\exp(\eta) = \frac{A + B + \sqrt{(A-B)^2 + C^2}}{\sqrt{4AB - C^2}}. \tag{15}$$

In terms of this new set of variables, the Hamiltonian can be written as

$$H = \frac{1}{2m}\left\{p_1^2 + p_2^2\right\} + \frac{K}{2}\left\{e^{2\eta}y_1^2 + e^{-2\eta}y_2^2\right\}, \tag{16}$$

with

$$y_1 = x_1\cos\alpha - x_2\sin\alpha,$$
$$y_2 = x_1\sin\alpha + x_2\cos\alpha. \tag{17}$$

This completes the diagonalization process. The normal frequencies are

$$\omega_1 = e^{\eta}\omega, \qquad \omega_2 = e^{-\eta}\omega, \tag{18}$$

with

$$\omega = \sqrt{\frac{K}{m}}. \tag{19}$$

This relatively new set of parameters has been discussed in in connection with Feynman's rest of the universe [17].

Let us go back to Eq.(12) and Eq.(14). If $\alpha = 0$, C becomes zero and the oscillators become decoupled. If $\alpha = 45^o$, then $A = B$, which means that the system consists of two identical oscillators coupled together by the C term. In this case,

$$\exp(\eta) = \sqrt{\frac{2A + C}{2A - C}}, \quad or \quad \eta = \frac{1}{2}\ln\left(\frac{2A + C}{2A - C}\right). \tag{20}$$

Thus η measures the strength of the coupling.

The mathematics becomes very simple for $\alpha = 45^o$, and this simple case can be applied to many physical problems, including the present problem of combining quantum mechanics with relativity. Indeed the y_1, y_2 variables become

$$y_1 = \frac{x_1 - x_2}{\sqrt{2}}, \qquad y_2 = \frac{x_1 + x_2}{\sqrt{2}}. \tag{21}$$

If y_1 and y_2 are measured in units of $(mK)^{1/4}$, the ground-state wave function of this oscillator system is

$$\psi_\eta(x_1, x_2) = \frac{1}{\sqrt{\pi}} \exp\left\{-\frac{1}{2}(e^\eta y_1^2 + e^{-\eta} y_2^2)\right\}, \tag{22}$$

The wave function is separable in the y_1 and y_2 variables, when $\eta = 0$, remains separable while they are expanded and contracted by $e^{\eta/2}$ and $e^{-\eta/2}$ respectively as illustrated in Fig. 4.

On the other hand, for the variables x_1 and x_2, the wave function takes the form

$$\psi_\eta(x_1, x_2) = \frac{1}{\sqrt{\pi}} \exp\left\{-\frac{1}{4}\left[e^\eta (x_1 - x_2)^2 + e^{-\eta}(x_1 + x_2)^2\right]\right\}. \tag{23}$$

In his 1963 paper [10], Dirac strictly worked with step-up and step-down operators. He made no attempt to use a normal coordinate system. It is indeed gratifying to translate his algebraic formulas into a geometry. Let us now compare Fig. 4 with Fig. 3. They are the same! Indeed, the geometry of Lorentz boost in Dirac's light-cone coordinate system is identical to that of the coupled oscillators. The coupling constant is translated into the boost parameter as given in Eq.(20).

3. Lorentz-Covariant Picture of Quantum Bound States

If we combine Fig. 1 and Fig. 3, then we end up with Fig. 5. In mathematical formula, this transformation changes the Gaussian form of Eq.(2) into

$$\psi_\eta(z, t) = \left(\frac{1}{\pi}\right)^{1/2} \exp\left\{-\frac{1}{4}\left[e^{-\eta}(z + t)^2 + e^\eta (z - t)^2\right]\right\}. \tag{24}$$

This formula together with Fig. 5 is known to describe all essential high-energy features observed in high-energy laboratories [3, 18, 19].

Indeed, this elliptic deformation explains one of the most controversial issues in high-energy physics. Hadrons are known to be bound states of quarks. Its bound-state quantum mechanics is assumed to be the same as that of the hydrogen atom. The question is how the hadron would look to an observer on a train. If the train moves with a speed close to that of light, the hadron appears like a collection of partons, according to Feynman [3]. Feynman's partons have properties quite different from those of the quarks. For instance, they interact incoherently with external signals. The elliptic deformation property described in Fig. 5 explains that the quark and parton models are two different manifestations of the same covariant entity.

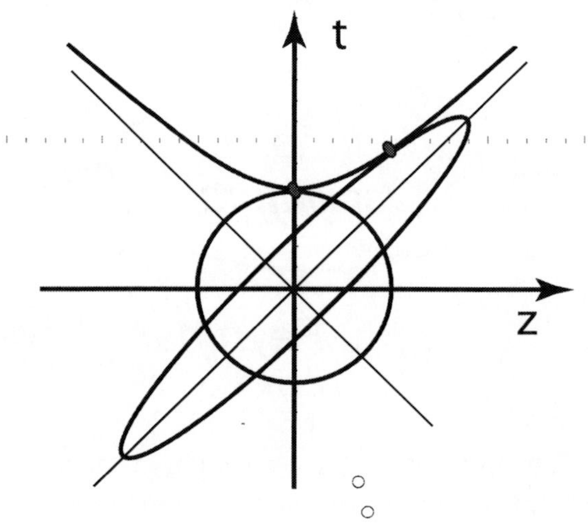

Figure 5. Effect of the Lorentz boost on the space-time wave function. The circular space-time distribution in the rest frame becomes Lorentz-squeezed to become an elliptic distribution.

Quantum field theory has been quite successful in terms of Feynman diagrams based on the S-matrix formalism, but is useful only for physical processes where a set of free particles becomes another set of free particles after interaction. Quantum field theory does not address the question of localized probability distributions and their covariance under Lorentz transformations. In order to address this question, Feynman *et al.* suggested harmonic oscillators to tackle the problem [5]. Their idea is indicated in Fig. 6.

In this report, we are concerned with the quantum bound system, and we have examined the four-papers of Dirac on the question of making the uncertainty relations consistent with special relativity. Indeed, Dirac discussed this fundamental problem with mathematical devices which are both elegant and transparent.

Dirac of course noted that the time variable plays the essential role in the Lorentz-covariant world. On the other hand, he did not take into consideration the concept of time separation. When we talk about the hydrogen atom, we are concerned with the distance between the proton and electron. To a moving observer, there is also a time-separation between the two particles.

Instead of the hydrogen atom, we use these days the hadron consisting of two quarks bound together with an attractive force, and consider their space-time positions x_a and x_b,and use the variables [5]

$$X = \frac{x_a + x_b}{2}, \qquad x = \frac{x_a - x_b}{2\sqrt{2}}. \tag{25}$$

The four-vector X specifies where the hadron is located in space and time, while the variable x measures the space-time separation between the quarks. Let us call their time components T and t. These variables actively participate in Lorentz transformations. The existence of the T variable is known, but the Copenhagen school was not able to see the existence of this t variable.

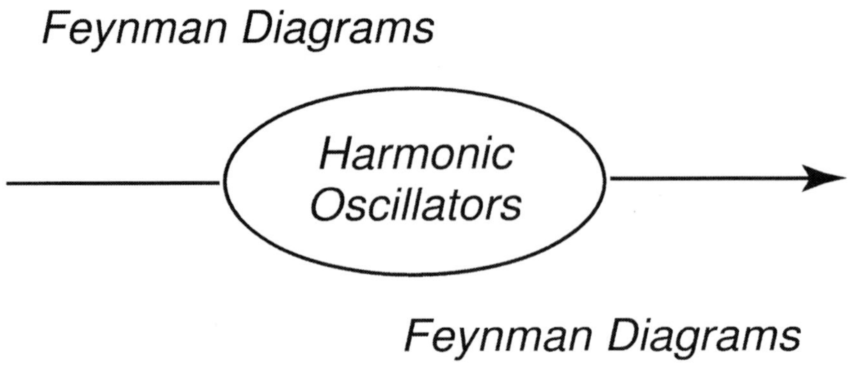

Figure 6. Feynman's roadmap for combining quantum mechanics with special relativity. Feynman diagrams work for running waves, and they provide a satisfactory resolution for scattering states in Einstein's world. For standing waves trapped inside an extended hadron, Feynman suggested harmonic oscillators as the first step.

Paul A. M. Dirac was concerned with the time variable throughout his four papers discussed in this report. However, he did not make a distinction between the T and t variables. The T variable ranges from $-\infty$ to $+\infty$, and is constantly increasing. On the other hand, the t variable is the time interval, and remains unchanged in a given Lorentz frame.

Indeed, when Feynman *et al.* wrote down the Lorentz-invariant differential equation [5]

$$\frac{1}{2}\left\{x_\mu^2 - \frac{\partial^2}{\partial x_\mu^2}\right\}\psi(x) = \lambda\psi(x), \tag{26}$$

x_μ was for the space-time separation between the quarks.

This four-dimensional differential equation has more than 200 forms of solutions depending on boundary conditions. However, there is only one set of solutions to which we can give a physical interpretation. Indeed, the Gaussian form of Eq.(1) is a solution of the above differential equation. If we boost the system along the z direction, we can separate away the x and y components in the Gaussian form and write the wave function in the form of Eq.(2).

It is then possible to construct a representation of the Poincaré group from the solutions of the above differential equation [14]. If the system is boosted, the wave function becomes the Gaussian form given in Eq.(24), which becomes Eq.(2) if η becomes zero. This wave function is also a solution of the Lorentz-invariant differential equation of Eq.(26). The transition from Eq.(2) to Eq.(24) is illustrated in Fig. 5.

4. Lorentz-Covariant Quark Model

Early successes in the quark model includes the calculation of the ratio of the neutron and proton magnetic moments [20], and the hadronic mass spectra [5, 21]. These are based on hadrons at rest. We are interested in this paper how the hadrons in the quark model appear to observers in different Lorentz frames.

The idea that the proton or neutron has a space-time extension had been developed long before Gell-Mann's proposal for the quark model [1]. Yukawa [22] developed this idea as early as 1953, and his idea was followed up by Markov, Ginzburg, and Man'ko [23, 24].

Since Einstein formulated special relativity for point particles, it has been and still is a challenge to formulate a theory for particles with space-time extensions. The most naive idea would be to study rigid spherical objects, and there were many papers on this subjects. But we do not know where that story stands these days. We can however replace these extended rigid bodies by extended wave packets or standing waves, thus by localized probability entities. Then what are the constituents within those localized waves? The quark model gives the natural answer to this question.

The first experimental discovery of the non-zero size of the proton was made by Hofstadter and McAllister [25], who used electron-proton scattering to measure the charge distribution inside the proton. If the proton were a point particle, the scattering amplitude would just be a Rutherford formula. However, Hofstadter and MacAllister found a tangible departure from this formula which can only be explained by a spread-out charge distribution inside the proton.

In this section, we are interested in how well the bound-state picture developed in Sec. 4. works in explaining relativistic phenomena of hadrons, specifically the proton. The Lorentz-covariant model of Sec. 3. is of course based on Dirac's four papers discussed in Sec. 2..

First, we show that the static quark model and Feynman's parton picture are two limiting cases of one Lorentz-covariant entity. In the quark model, the hadron appears like quantum bound states with discrete energy spectra. In the parton model, the rapidly moving hadron appears like a collection of infinite number of free independent particles. Can these be explained with one theory? This is what we would like to address in subsection 4.1..

As for the case between these two limits, we discuss the hadronic form factor which occupies an important place in every theoretical model in strong interactions since Hofstadter's discovery in 1955 [25]. The key question is the proton form factor decreases as $1/(\text{momentum transfer})^4$ as the momentum transfer becomes large. This is called the dipole cut-off in the literature. We shall see in subsection 4.2. that the covariant model of Sec. 3. gives this dipole cut-off for spinless quarks.

4.1. Feynman's Parton Picture

In a hydrogen atom or a hadron consisting of two quarks, there is a spacial separation between two constituent elements. In the case of the hydrogen atom we call it the Bohr radius. If the atom or hadron is at rest, the time-separation variable does not play any visible role in quantum mechanics. However, if the system is boosted to the Lorentz frame which moves with a speed close to that of light, this time-separation variable becomes as important as the space separation of the Bohr radius. Thus, the time-separation variable plays a visible role in high-energy physics which studies fast-moving bound states. Let us study this problem in more detail.

It is a widely accepted view that hadrons are quantum bound states of quarks having a localized probability distribution. As in all bound-state cases, this localization condition is responsible for the existence of discrete mass spectra. The most convincing evidence for this bound-state picture is the hadronic mass spectra [5, 14].

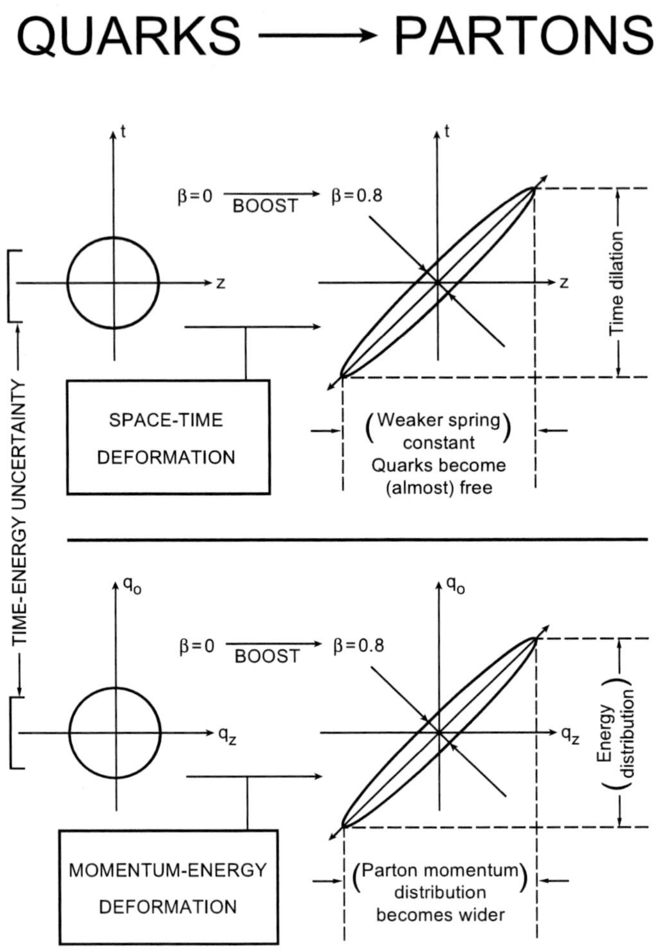

Figure 7. Lorentz-squeezed space-time and momentum-energy wave functions. As the hadron's speed approaches that of light, both wave functions become concentrated along their respective positive light-cone axes. These light-cone concentrations lead to Feynman's parton picture.

However, this picture of bound states is applicable only to observers in the Lorentz frame in which the hadron is at rest. How would the hadrons appear to observers in other Lorentz frames?

In 1969, Feynman observed that a fast-moving hadron can be regarded as a collection of many "partons" whose properties appear to be quite different from those of the quarks [3, 14]. For example, the number of quarks inside a static proton is three, while the number of partons in a rapidly moving proton appears to be infinite. The question then is how the proton looking like a bound state of quarks to one observer can appear different to an observer in a different Lorentz frame? Feynman made the following systematic observations.

a. The picture is valid only for hadrons moving with velocity close to that of light.

b. The interaction time between the quarks becomes dilated, and partons behave as free independent particles.

c. The momentum distribution of partons becomes widespread as the hadron moves fast.

d. The number of partons seems to be infinite or much larger than that of quarks.

Because the hadron is believed to be a bound state of two or three quarks, each of the above phenomena appears as a paradox, particularly b) and c) together. How can a free particle have a wide-spread momentum distribution?

In order to resolve this paradox, let us construct the momentum-energy wave function corresponding to Eq.(24). If the quarks have the four-momenta p_a and p_b, we can construct two independent four-momentum variables [5]

$$P = p_a + p_b, \qquad q = \sqrt{2}(p_a - p_b). \tag{27}$$

The four-momentum P is the total four-momentum and is thus the hadronic four-momentum. q measures the four-momentum separation between the quarks. Their light-cone variables are

$$q_u = (q_0 - q_z)/\sqrt{2}, \qquad q_v = (q_0 + q_z)/\sqrt{2}. \tag{28}$$

The resulting momentum-energy wave function is

$$\phi_\eta(q_z, q_0) = \left(\frac{1}{\pi}\right)^{1/2} \exp\left\{-\frac{1}{2}\left[e^{-2\eta}q_u^2 + e^{2\eta}q_v^2\right]\right\}. \tag{29}$$

Because we are using here the harmonic oscillator, the mathematical form of the above momentum-energy wave function is identical to that of the space-time wave function of Eq.(24). The Lorentz squeeze properties of these wave functions are also the same. This aspect of the squeeze has been exhaustively discussed in the literature [14, 18, 26], and they are illustrated again in Fig. 7 of the present paper. The hadronic structure function calculated from this formalism is in a reasonable agreement with the experimental data [27].

When the hadron is at rest with $\eta = 0$, both wave functions behave like those for the static bound state of quarks. As η increases, the wave functions become continuously squeezed until they become concentrated along their respective positive light-cone axes. Let us look at the z-axis projection of the space-time wave function. Indeed, the width of the quark distribution increases as the hadronic speed approaches that of the speed of light. The position of each quark appears widespread to the observer in the laboratory frame, and the quarks appear like free particles.

The momentum-energy wave function is just like the space-time wave function. The longitudinal momentum distribution becomes wide-spread as the hadronic speed approaches the velocity of light. This is in contradiction with our expectation from non-relativistic quantum mechanics that the width of the momentum distribution is inversely proportional to that of the position wave function. Our expectation is that if the quarks are free, they must have their sharply defined momenta, not a wide-spread distribution.

However, according to our Lorentz-squeezed space-time and momentum-energy wave functions, the space-time width and the momentum-energy width increase in the same direction as the hadron is boosted. This is of course an effect of Lorentz covariance. This indeed leads to the resolution of one of the the quark-parton puzzles [14, 18, 26].

Another puzzling problem in the parton picture is that partons appear as incoherent particles, while quarks are coherent when the hadron is at rest. Does this mean that the coherence is destroyed by the Lorentz boost? The answer is NO, and here is the resolution to this puzzle.

When the hadron is boosted, the hadronic matter becomes squeezed and becomes concentrated in the elliptic region along the positive light-cone axis. The length of the major axis becomes expanded by e^{η}, and the minor axis is contracted by e^{η}.

This means that the interaction time of the quarks among themselves becomes dilated. Because the wave function becomes wide-spread, the distance between one end of the harmonic oscillator well and the other end increases. This effect, first noted by Feynman [2, 3], is universally observed in high-energy hadronic experiments. The period of oscillation increases like e^{η}.

On the other hand, the external signal, since it is moving in the direction opposite to the direction of the hadron, travels along the negative light-cone axis.

If the hadron contracts along the negative light-cone axis, the interaction time decreases by $e^{-\eta}$. The ratio of the interaction time to the oscillator period becomes $e^{-2\eta}$. The energy of each proton coming out of the Fermilab accelerator is $900GeV$. This leads the ratio to 10^{-6}. This is indeed a small number. The external signal is not able to sense the interaction of the quarks among themselves inside the hadron.

Indeed, Feynman's parton picture is one concrete physical example where the decoherence effect is observed. As for the entropy, the time-separation variable belongs to the rest of the universe. Because we are not able to observe this variable, the entropy increases as the hadron is boosted to exhibit the parton effect. The decoherence is thus accompanied by an entropy increase.

Let us go back to the coupled-oscillator system. The light-cone variables in Eq.(24) correspond to the normal coordinates in the coupled-oscillator system given in Eq.(16). According to Feynman's parton picture, the decoherence mechanism is determined by the ratio of widths of the wave function along the two normal coordinates.

This decoherence mechanism observed in Feynman's parton picture is quite different from other decoherences discussed in the literature. It is widely understood that the word decoherence is the loss of coherence within a system. On the other hand, Feynman's decoherence discussed in this section comes from the way the external signal interacts with the internal constituents.

4.2. Nucleon Form Factors

Let us first see what effect the charge distribution has on the scattering amplitude, using nonrelativistic scattering in the Born approximation. If we scatter electrons from a fixed charge distribution whose density is $e\rho(r)$, the scattering amplitude is

$$f(\theta) = -\left(\frac{e^2 m}{2\pi}\right)\int d^3x d^3x' \frac{\rho(r')}{R} \exp\left(-i\mathbf{Q}\cdot\mathbf{x}\right), \tag{30}$$

where $r = |\mathbf{x}|$, $R = |\mathbf{r} - \mathbf{r}'|$, and $\mathbf{Q} = \mathbf{K_f} - \mathbf{K_i}$, which is the momentum transfer. This amplitude can be reduced to

$$f(\theta) = \frac{2me^2}{Q^2} F(Q^2). \tag{31}$$

$F(Q^2)$ is the Fourier transform of the density function which can be written as

$$F\left(Q^2\right) = \int d^3x \rho(r) \exp\left(-i\mathbf{Q}\cdot\mathbf{x}\right). \tag{32}$$

The above quantity is called the form factor. It describes the charge distribution in terms of the momentum transfer. The charge density is normalized:

$$\int \rho(r) d^3x = 1. \tag{33}$$

Then F(0) = 1 from Eq.(32). If the density function is a delta function corresponding to a point charge, $F\left(Q^2\right) = 1$ for all values of Q^2, then the scattering amplitude of Eq.(30) becomes the Rutherford formula for Coulomb scattering. The deviations from Rutherford scattering for increasing values of Q^2 give a measure of the charge distribution. This was precisely what Hofstadter's experiment on the scattering of electrons from a proton target found. [25].

As the energy of incoming electrons becomes higher, we have to take into account the recoil effect of target protons, and formulate the problem relativistically. It is generally agreed that electrons and their electromagnetic interaction can be described by quantum electrodynamics, in which the method of perturbation theory using Feynman diagrams is often employed for practical calculations [28, 29]. In this perturbation approach, the scattering amplitude is expanded in power series of the fine structure constant α. Therefore, in lowest order in α, we can describe the scattering of an electron by a proton using the diagram given in Fig.(8). The corresponding matrix element is given in many textbooks on elementary particle physics [30]. It is proportional to

$$\bar{U}\left(P_f\right)\Gamma_\mu\left(P_f, P_i\right)\left(\frac{1}{Q^2}\right)\bar{U}\left(K_f\right)\gamma^\mu U\left(K_i\right), \tag{34}$$

where P_i, P_f, K_i and K_f are the initial and final four-momenta of the proton and electron respectively. $U\left(P_i\right)$ is the Dirac spinor for the initial proton. Q^2 is the (four-momentum transfer)^4 and is

$$Q^2 = \left(P_f - P_i\right)^2 = \left(K_f - K_i\right)^2. \tag{35}$$

The $(1/Q^2)$ factor in Eq.(34) comes from the virtual photon being exchanged between the electron and the proton. It is customary to use the letter t for Q^2 in form factor studies, and this t should not be confused with the time separation variable. In the metric we use, this quantity is positive for physical values of the four-momenta for the particles involved in the scattering process.

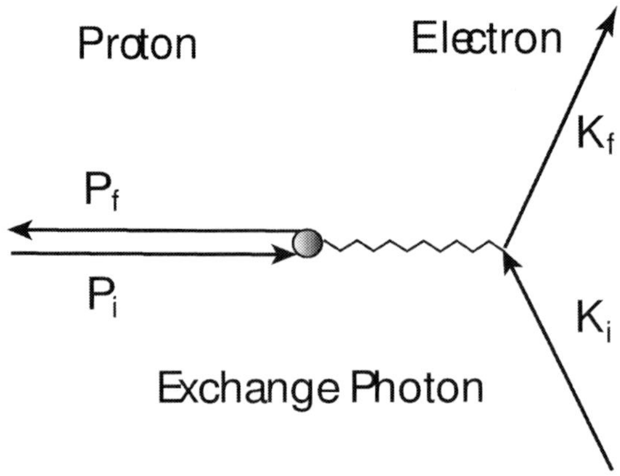

Figure 8. Breit frame for electron-nucleon scattering. The momentum of the outgoing nucleon is equal in magnitude but opposite in sign to that of the incoming nucleon.

In order to make a relativistic calculation of the form factor, let us go back to the definition of the form factor given in Eq.(32). The density function depends only on the target particle, and is proportional to $\psi(x)^{\dagger}\psi(x)$, where $\psi(x)$ is the wave function for quarks inside the proton. This expression is a special case of the more general form

$$\rho(x) = \psi_f^{\dagger}(x)\psi_i(x), \tag{36}$$

where ψ_i and ψ_f are the initial and final wave function of the target atom. Indeed, the form factor of Eq.(32) can be written as

$$F\left(Q^2\right) = \left(\psi_f(x), e^{-i\mathbf{Q}\cdot\mathbf{r}}\psi_i(x)\right). \tag{37}$$

Starting from this expression, we can make the required Lorentz generalization using the relativistic wave functions for hadrons.

In order to see the details of the transition to relativistic physics, we should be able to replace each quantity in the expression of Eq.(32) by its relativistic counterpart. Let us go to the Lorentz frame in which the momenta of the incoming and outgoing nucleons have equal magnitude but opposite signs.

$$\mathbf{p_i} + \mathbf{p_f} = 0. \tag{38}$$

This kinematical condition is illustrated in Fig. 8.

The Lorentz frame in which the above condition holds is usually called the Breit frame. We assume without loss of generality that the proton comes along the z direction before the collision and goes out along the negative z direction after the scattering process, as illustrated in Fig. 8. In this frame, the four vector $Q = (K_f - K_i) = (P_i - P_f)$ has no time-like component. Thus the exponential factor $\mathbf{Q}\cdot\mathbf{r}$ can be replaced by the Lorentz-invariant form $Q \cdot x$. As for the wave functions for the protons, we can use the covariant harmonic oscillator wave functions discussed in this paper assuming that the nucleons are in the ground state. Then the only difference between the nonrelativistic and relativistic cases is that the integral in the evaluation of Eq.(32) is four-dimensional, including that for the time-like direction. This integral in the time-separation variable does not interfere with the exponential factor which does not depend on the time-separation variable.

Let us now write down the integral:

$$g\left(Q^2\right) = \int d^4x \psi^{\dagger}_{-\beta}(x)\psi_\beta(x) \exp\left(-iQ \cdot x\right). \tag{39}$$

where β is the velocity parameter for the incoming proton, and the wave function ψ_β takes the form:

$$\psi_\beta(x) = \frac{1}{\pi} \exp\left\{-\frac{x^2+y^2}{2}\right\}$$
$$\times \exp\left\{-\frac{1}{4}\left[\frac{1-\beta}{1+\beta}(z+t)^2 + \frac{1+\beta}{1-\beta}(z-t)^2\right]\right\}. \tag{40}$$

After the above decomposition of the wave functions, we can perform the integrations in the x and y variables trivially. After dropping these trivial factors, we can write the product of the two wave functions as

$$\psi^{\dagger}_{-\beta}(x)\psi_\beta(x) = \frac{1}{\pi} \exp\left\{-\left(\frac{1+\beta^2}{1-\beta^2}\right)\left(t^2+z^2\right)\right\}. \tag{41}$$

Thus the z and t variables have been separated. Since the exponential factor in Eq.(32) does not depend on t, the t integral in Eq.(39) can also be trivially performed, and the integral of Eq.(39) can be written as

$$g\left(Q^2\right) = \sqrt{\frac{1}{\pi}\left(\frac{1-\beta^2}{1+\beta^2}\right)} \int dz e^{-2iPz} \exp\left\{-\frac{1+\beta^2}{1-\beta^2}z^2\right\}, \tag{42}$$

where P is the z component of the momentum of the incoming nucleon. The (momentum transfer)^2 variable Q^2 is $4P^2$. Indeed, the distribution of the hadronic material along the longitudinal direction became contracted [31].

We note that β can be written as

$$\beta^2 = \frac{Q^2}{Q^2 + 4M^2}, \tag{43}$$

where M is the proton mass. This equation tells $\beta = 0$ when $Q^2 = 0$, while it becomes one as Q^2 becomes infinity.

The evaluation of the above integral for $g\left(Q^2\right)$ in Eq.(42) leads to

$$g\left(Q^2\right) = \left(\frac{2M^2}{Q^2 + 2M^2}\right) \exp\left\{\frac{-Q^2}{2(Q^2+2M^2)}\right\}. \tag{44}$$

For $Q^2 = 0$, the above expression becomes 1. It decreases as

$$g\left(Q^2\right) \sim \frac{1}{Q^2} \tag{45}$$

for large values of Q^2.

We have so far carried out the calculation for an oscillator bound state of two quarks. The proton consists of three quarks. As shown in the paper of Feynman *et al.* [5], the problem becomes a product of two oscillator modes. Thus, the generalization of the above calculation to the three-quark system is straightforward, and the result is that the form factor $G\left(Q^2\right)$ takes the form

$$G\left(Q^2\right) = \left(\frac{2M^2}{Q^2 + 2M^2}\right)^2 \exp\left\{\frac{-Q^2}{(Q^2 + 2M^2)}\right\}. \tag{46}$$

which is 1 at $Q^2 = 0$, and decreases as

$$G\left(Q^2\right) \sim \left[\frac{1}{Q^2}\right]^2 \tag{47}$$

for large values of Q^2. Indeed, this function satisfies the requirement of the "dipole-cut-off" behavior for of the form factors, which has been observed in high-energy laboratories. This calculation was carried first by Fujimura *et al.* in 1970 [32].

Let us re-examine the above calculation. If we replace β by zero in Eq.(42) and ignore the elliptic deformation of the wave functions, $g\left(Q^2\right)$ will become

$$g\left(Q^2\right) = e^{-Q^2/4}, \tag{48}$$

which will leads to an exponential cut-off of the form factor. This is not what we observe in laboratories.

In order to gain a deeper understanding of the above-mentioned correlation, let us study the case using the momentum-energy wave functions:

$$\phi_\beta(q) = \left(\frac{1}{\pi}\right)^2 \int d^4 e^{-iq\cdot x}\psi_\beta(x). \tag{49}$$

As before, we can ignore the transverse components. Then $g\left(Q^2\right)$ can be written as [14]

$$\int dq_0 dq_z \phi^*_\beta\left(q_0, q_z - P\right)\phi_\beta\left(q_0, q_z + P\right). \tag{50}$$

We have sketched the above overlap integral in Fig. 9. When $Q^2 = 0$ or $P = 0$, the two wave functions overlap completely in the $q_z q_0$ plane. As P increases, the wave functions become separated. However, they maintain a small overlapping region due to the elliptic or squeeze deformation seen in Fig. 5. In the non-relativistic case, where the deformation is not taken into account, there is no overlapping region as seen in Fig. 9. This is precisely why the relativistic calculation gives a slower decrease in Q^2 than in the nonrelativistic case.

We have so far been interested only in the space-time behavior of the hadronic wave function. We should not forget the fact that quarks are spin-1/2 particles. The effect of this spin manifests itself prominently in the baryonic mass spectra. Since we are concerned here with the relativistic effects, we have to construct a relativistic spin wave function for the quarks. This quark wave function should give the hadronic spin wave function. In the case of nucleons, the quark spins should be combined in a manner to generate the form factor of Eq.(41).

The naive approach to this problem is to use free Dirac spinors for the quarks. However, it was shown by Lipes [33] that the use of free-particle Dirac spinors leads to a wrong form factor behavior. Since quarks in a hadron are not free particles, Lipes's result does not alarm us. The difficult problem is to find a suitable mechanism in which quark spins are coupled to orbital motion in a relativistic manner. This is a nontrivial research problem, and further study is needed along this direction [34].

In addition, there are recent experimental results which indicate departure from the dipole behavior of Eq.(47) [35]. In addition, there have been other theoretical attempts to calculate the proton form factor. Yes, whenever a new theoretical model appears, there appears a new attempt to calculate the form factor. In the past, there were many attempts to calculate this quantity in the framework of quantum field theory, without much success.

In 1960, Frazer and Fulco calculated the form factor using the technique of dispersion relations [36]. In so doing they had to assume the existence of the so-called ρ meson, which was later found experimentally, and which subsequently played a pivotal role in the development of the quark model.

Even these days, the form factor calculation occupies a very important place in recent theoretical models, such as QCD lattice theory [37] and the Faddeev equation [38]. However, it is still noteworthy that Dirac's form of Lorentz-covariant bound states leads to the essential dipole cut-off behavior of the proton form factor.

Conclusion

The hydrogen atom played a pivotal role in the development of quantum mechanics. Quantum mechanics had to be formulated to explain its discrete energy spectra, radiation decay rates, as well as the scattering of electrons by the proton.

The quark model still plays the central role in present-day high-energy physics. The model can explain the hadronc mass spectra, hadronic structure, and hadronic decay rates. In addition we are dealing with hadrons which appear as quantum bound states when they are at rest, how would they appear to observes in different frames, particularly in the frame moving with a velocity close to that of light. It was Feynman who proposed the parton model to describe those high-energy hadrons.

In this model, we have presented a Lorentz-covariant model which gives both the quark model and the parton model as two limiting cases. In addition we discussed the form factor as the case between these limits.

In constructing the Lorentz-covariant model, we noted Paul A. M. Dirac made life-long efforts to make quantum mechanics consistent with special relativity. We have chosen four of his papers and combined them to construct a consistent theory. It was like building a canal. The easiest way to build the canal is to link up the existing lakes. Dirac indeed dug four big lakes. It is a gratifying experience to link them up.

Dirac constructed those lakes in order to study whether the Copenhagen school of quantum mechanics can be made consistent with Einstein's Lorentz-covariant world.

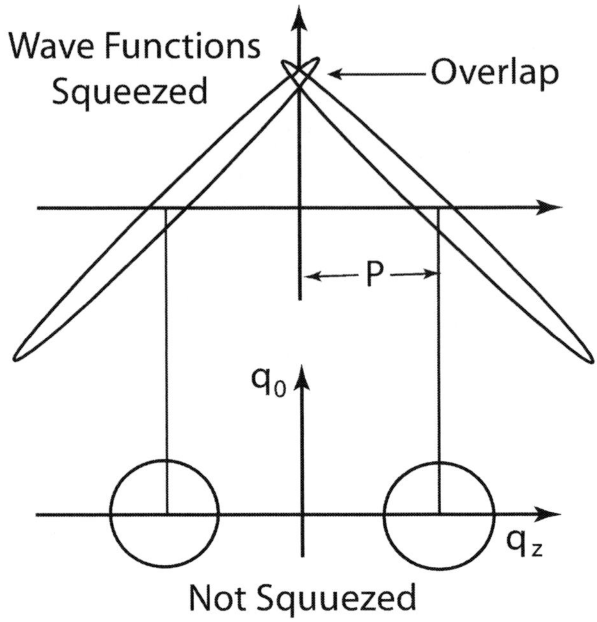

Figure 9. Lorentz-Dirac deformation of the momentum-energy wave functions in the form factor calculation. As the momentum transfer increases, the two wave functions become separated. In the relativistic case, the wave functions maintain an overlapping region. Wave functions become completely separated in the nonrelativistic calculation. This lack of overlapping region leads to an unacceptable behavior of the form factor.

After studying Dirac's papers, we arrived at the conclusion that the Copenhagen school completely forgot to take into account the question of simultaneity and time separation [4]. The question then is whether the localized probability distribution can be made consistent with Einstein's Lorentz covariance. We have addressed this question in this paper.

Acknowledgment

We would like to thank Stephen Wallace for telling us about both experimental and theoretical aspects of form factor studies.

References

[1] Gell-Mann, M. *Phys. Lett* 1964, 13, 598-602.

[2] Feynman, R. P. *Phys. Rev. Lett.* 1969, 23 1415-1417.

[3] Feynman, R. P. The Behavior of Hadron Collisions at Extreme Energies; in High Energy Collisions, *Proceedings of the Third International Conference*, Stony Brook, NY; Yang, C. N.; *et al.*; Eds.; Gordon and Breach: New York, NY 1969, pp 237-249.

[4] Kim, Y. S.; Noz, M. E. The Question of Simultaneity in Relativity and Quantum Mechanics; in QUANTUM THEORY: Reconsideration of Foundations - 3; Adenier, G; Khrennikov, A.; Nieuwenhuizen, T. M. Eds.; *AIP Conference Proceedings; American Institute of Physics: College Park,* MD,2006; 810, pp. 168-178.

[5] Feynman, R. P.; Kislinger, M.; Ravndal, F. *Phys. Rev. D* 1971, 3, 2706-2732.

[6] Dirac, P. A. M. *Physics Today* 1970, 23, No. 4, 29-31.

[7] Dirac, P. A. M. *Proc. Roy. Soc. (London)* 1927, A114, 243-265.

[8] Dirac, P. A. M. *Proc. Roy. Soc. (London)* 1945, A183, 284-295.

[9] Dirac, P. A. M. *Rev. Mod. Phys.* 1949, 21, 392-399.

[10] Dirac, P. A. M. *J. Math. Phys.* 1963, 4, 901-909.

[11] Kim, Y. S.; Noz, M. E. *Phase Space Picture of Quantum Mechanics;* World Scientific Publishing Company: Singapore, 1991.

[12] Wigner, E. P. *On the Time-Energy Uncertainty Relation; in Aspects of Quantum Theory;* Salam, A.; Wigner, E. P. Eds.; Cambridge University Press: London, England, 1972; pp 237-247.

[13] Wigner, E. *Ann. Math.* 1939, 40, 149-204.

[14] Kim, Y. S.; Noz, M. E. *Theory and Applications of the Poincaré Group*; D. Reidel Publishing Company: Dordrecht, The Netherlands, 1986.

[15] Han, D.; Kim,Y. S.; Noz, M. E. *J. Math. Phys.* 1995, 36, 3940-3954.

[16] Aravind, P. K. *Am. J. Phys.* 1989, 57,309-311.

[17] Han, D.; Kim, Y. S.; Noz, M. E. *Am. J. Phys.* 1999, 67,61-66.

[18] Kim, Y. S.; Noz, M. E. *Phys. Rev. D* 1977, 15, 335-338.

[19] Kim, Y. S.; Noz, M. E. *J. Opt. B: Quantum and Semiclass. Opt.* 2005,7 S458-S467.

[20] Beg, M. A. B; Lee, B. W.; Pais, A. *Phys. Rev. Lett.* 1964, 13 514-517

[21] Greenberg, O. W.; Resnikoff, M. *Phys. Rev.* 1967, 1844-1851

[22] Yukawa, H. *Phys. Rev.* 1953, 91 415-416.

[23] Markov, M. *Suppl. Nuovo Cimento* 1956, 3, 760-772.

[24] Ginzburg, V. L.; Man'ko, V. I. *Nucl. Phys.* 1965, 74 577-588

[25] Hofstadter, R; McAllister, R. W. *Phys. Rev.* 1955, 98 217-218.

[26] Kim, Y. S. *Phys. Rev. Lett.* 1989, 63, 348-351.

[27] Hussar, P. E. *Phys. Rev. D* 1981, 23 2781-2783.

[28] Schweber, S. S. A*n Introduction to Relativistic Quantum Field Theory*;Row-Peterson: Elmsford, NY, 1961.

[29] Itzykson, C.; Zuber, J. B. *Quantum Field Theory;* McGraw-Hill: New York,NY, 1980.

[30] Frazer, W. *Elementary Particle Physics;* Prentice Hall: Englewood Cliffs, NJ, 1966.

[31] Licht, A. L.; Pagnamenta,A. *Phys. Rev. D* 1970, 2, 1150-1156; 1156-1160.

[32] Fujimura, K; Kobayashi, T; Namiki, M. *Prog. Theor. Phys.* 1970, 43 73-79.

[33] Lipes, R. *Phys. Rev. D.* 1972, 5 2948-2863.

[34] Henriques, A. B.; Keller, B. H.; Moorhouse, R. G. 1975 *Ann. Phys. (NY)* 93 125-151.

[35] Punjabi, V; Perdrisat, C. F.; et al. *Phys.Rev C* 2005, 71 055202-27; Erratum-ibid. C 2005 71 069902.

[36] Frazer, W.; Fulco, J. *Phys. Rev. Lett.* 1960, 2 365-368.

[37] Matevosyan, H. H.; Thomas, A. W.; Miller, G. A. *Phys.Rev.* 2005, C72 065204-5.

[38] Alkofer, R; Holl, A; Kloker, M; Karssnigg A.; Roberts, C. D. *Few-Body Systems* 2005, 37 1-31.

In: The Physics of Quarks: New Research ISBN 978-1-60456-802-8
Editors: N.L. Watson and T.M. Grant , pp. 163-196

Chapter 8

RESUMMATIONS IN QCD HARD-SCATTERING AT LARGE AND SMALL x

***Nikolaos Kidonakis*[1], *Agustín Sabio Vera*[2] *and Philip Stephens*[1]**
[1]Kennesaw State University, 1000 Chastain Rd. # 1202,
Kennesaw, GA 30144, USA
[2]Physics Department, Theory Division, CERN,
CH-1211 Geneva 23, Switzerland

Abstract

We discuss different resummations of large logarithms that arise in hard-scattering cross sections of quarks and gluons in regions of large and small x. The large-x logarithms are typically dominant near threshold for the production of a specified final state. These soft and collinear gluon corrections produce large enhancements of the cross section for many processes, notably top quark and Higgs production, and typically the higher-order corrections reduce the factorization and renormalization scale dependence of the cross section. The small-x logarithms are dominant in the regime where the momentum transfer of the hard sub-process is much smaller than the total collision energy. These logarithms are important to describe multijet final states in deep inelastic scattering and hadron colliders, and in the study of parton distribution functions. The resummations at small and large x are linked by the eikonal approximation and are dominated by soft gluon anomalous dimensions. We will review their role in both contexts and provide some explicit calculations at one and two loops.

1. Introduction

Particle physics in high-energy hadron colliders depends crucially on our ability to calculate cross sections to an ever increasing theoretical accuracy, which is achieved by the incorporation of higher-order corrections. Hard-scattering cross sections in perturbative QCD obey factorization theorems [1] that play a key role in the calculation of these corrections. Typically, the cross section for a process involving the collision of two hadrons (proton-antiproton at the Fermilab Tevatron or proton-proton at the CERN LHC) into a specified final state can be described as a convolution of non-perturbative parton distribution functions that describe the parton content of the hadron, and a partonic cross section that can be calculated order-by-order in perturbation theory. The short-distance partonic cross section involves the scattering of quarks and gluons. The partonic processes are of the form

$$f_1(p_1) + f_2(p_2) \to F(p) + X\,, \tag{1}$$

where f_1 and f_2 represent partons (quarks or gluons), F represents an observed system in the final state, such as a top quark or a jet or a Higgs boson, and X represents any additional final-state particles. The factorization is described schematically by

$$\sigma_{h_1 h_2 \to F} = \sum_f \int dx_1 dx_2\, \phi_{f_1/h_1}(x_1, \mu_F)\, \phi_{f_2/h_2}(x_2, \mu_F)\, \hat{\sigma}_{f_1 f_2 \to F}(s, t, u, \mu_F, \mu_R)\,, \tag{2}$$

where $\sigma_{h_1 h_2 \to F}$ is the physical cross section (total or differential) for the production of final state F in the scattering of hadrons h_1 and h_2, ϕ_{f_i/h_i} is the distribution function for parton f_i with momentum fraction x_i of hadron h_i, and $\hat{\sigma}_{f_1 f_2 \to F}$ is the partonic cross section. The collinear singularities are factorized in a process-independent manner and absorbed into the parton distribution functions which are dependent on the factorization scale μ_F. The physical cross section is in principle independent of the factorization scale μ_F and the renormalization scale μ_R, but in practice there is a strong dependence because we truncate the infinite perturbative series at finite order (typically next-to-leading-order (NLO) or next-to-next-to-leading-order (NNLO) in the strong coupling α_s). The parton-level cross section explicitly involves the standard kinematical invariants, $s = (p_1 + p_2)^2$, $t = (p_1 - p)^2$, $u = (p_2 - p)^2$, formed from the 4-momenta of the particles in the hard scattering.

Near threshold, i.e. when the energy of the incoming partons is just sufficient to produce a final state without additional radiation, the production cross section receives significant corrections from large-x logarithms [2–5]. These logarithms arise from incomplete cancellations between virtual terms and terms that describe soft-gluon emission. Since near threshold any additional radiation has to be soft, the large-x logarithms are especially important in that kinematical region. Large-x resummation depends critically on the color structure of the process [4, 6–8] as well as the kinematics [4, 9].

Small-x logarithms arise when the perturbative scales characterizing the hard-subprocess are much smaller than the total collision energy. In this case resummation of logarithms of the form $\ln(1/x)$ becomes important. When the transverse scales of the outgoing scattered particles are similar and large this resummation can be described by the Balitsky-Fadin-Kuraev-Lipatov (BFKL) evolution equation [10–14]. This equation is a linear integral equation which leads to an exponential rise of the cross section. The slope of this rise can be interpreted as a perturbative construction of the QCD Pomeron. This Pomeron is considered the mediator of many QCD diffractive processes, such as diffractive vector meson production.

The conditions by which the BFKL evolution should be valid are satisfied by jet production with large rapidity gaps. Phenomenological studies of this process with the summation of the terms $\alpha_S^n \ln^n(1/x)$ (leading-order kernel) are not very predictive since the value of the coupling is a free parameter and the Regge energy scale, a sort of factorization scale at high energies, can only be fixed at higher orders. Inclusion of the next-to-leading order corrections, $\alpha_S^{n+1} \ln^n(1/x)$ [15, 16], brings the predictions in closer agreement with data.

In the next section we discuss large-x resummation and finite-order expansions of the resummed cross section through next-to-next-to-next-to-leading order (NNNLO). In Section 3 we present some applications of large-x resummation to various hard-scattering processes, namely top-antitop pair production, single top quark production, W-boson production at large transverse momentum, and Higgs boson production via $b\bar{b} \to H$. In Section 4 we present typical one-loop and two-loop calculations in the eikonal approximation that are needed in resummations for processes with massive quarks, such as heavy quark pair production. Section 5 discusses small-x resummation and applications of BFKL. We conclude in Section 6.

2. Large-x Resummations

Large-x resummations depend crucially on the kinematics and color structure of the process under study. In single-particle-inclusive (1PI) kinematics we identify one particle F with momentum p. In pair-invariant-mass (PIM) kinematics we identify a pair of particles (such as a heavy quark-antiquark pair) with invariant mass squared Q^2.

In general, the partonic cross section $\hat{\sigma}$ includes soft corrections in the form of plus distributions $\mathcal{D}_l(x_{th})$ with respect to a kinematical variable x_{th} that measures distance from threshold, with $l \leq 2n-1$ at n-th order in α_s beyond the leading order. In 1PI kinematics, x_{th} is usually denoted as s_4 (or s_2) and is defined by $s_4 = s + t + u - \sum m^2$, where the sum is over the squared masses of all particles in the process. At threshold, $s_4 = 0$. The plus distributions are then of the form

$$\mathcal{D}_l(s_4) \equiv \left[\frac{\ln^l(s_4/M^2)}{s_4}\right]_+ , \tag{3}$$

where M is a hard scale relevant to the process, for example the mass m of a heavy quark or the transverse momentum of a jet. The plus distributions are defined through their integral with the parton distribution functions by

$$\int_0^{s_4\,max} ds_4\, \phi(s_4) \left[\frac{\ln^l(s_4/M^2)}{s_4}\right]_+ \equiv \int_0^{s_4\,max} ds_4 \frac{\ln^l(s_4/M^2)}{s_4}[\phi(s_4) - \phi(0)] + \frac{1}{l+1}\ln^{l+1}\left(\frac{s_{4\,max}}{M^2}\right)\phi(0)\,. \tag{4}$$

In PIM kinematics, x_{th} is usually denoted as $1-x$ or $1-z$, with $z = Q^2/s \to 1$ at threshold. Then the plus distributions are of the form

$$\mathcal{D}_l(z) \equiv \left[\frac{\ln^l(1-z)}{1-z}\right]_+ \tag{5}$$

defined by

$$\int_{z_{min}}^1 dz\, \phi(z) \left[\frac{\ln^l(1-z)}{1-z}\right]_+ \equiv \int_{z_{min}}^1 dz \frac{\ln^l(1-z)}{1-z}[\phi(z) - \phi(1)] + \frac{1}{l+1}\ln^{l+1}(1-z_{min})\phi(1)\,. \tag{6}$$

The highest powers of these distributions in the nth-order corrections are the leading logarithms (LL) with $l = 2n - 1$, the second highest are the next-to-leading logarithms (NLL) with $l = 2n - 2$, etc. (note that the counting of logarithms is different in the exponent and in the fixed-order expansions). These logarithms can be resummed in principle to all orders in perturbation theory.

2.1. Exponentiation

The resummation of threshold logarithms is performed in moment space. By taking moments, divergent distributions in $1 - z$ (or s_4) produce powers of $\ln N$, with N the moment variable:

$$\int_0^1 dz\, z^{N-1} \left[\frac{\ln^m(1-z)}{1-z}\right]_+ = \frac{(-1)^{m+1}}{m+1} \ln^{m+1} N + \mathcal{O}\left(\ln^{m-1} N\right). \qquad (7)$$

If we define moments of the partonic cross section by $\hat{\sigma}(N) = \int dz\, z^{N-1} \hat{\sigma}(z)$ (PIM) or by $\hat{\sigma}(N) = \int (ds_4/s)\, e^{-Ns_4/s} \hat{\sigma}(s_4)$ (1PI), then the logarithms of N that appear in $\hat{\sigma}(N)$ exponentiate.

The resummation follows from the factorization properties of the cross section. We begin the derivation of the resummed cross section by first writing a factorized form for the moment-space infrared-regularized parton-parton scattering cross section, $\sigma_{f_1 f_2 \to F}(N, \epsilon)$, which factorizes as the hadronic cross section

$$\sigma_{f_1 f_2 \to F}(N, \epsilon) = \tilde{\phi}_{f_1/f_1}(N, \mu_F, \epsilon)\, \tilde{\phi}_{f_2/f_2}(N, \mu_F, \epsilon)\, \hat{\sigma}_{f_1 f_2 \to F}(N, \mu_F, \mu_R), \qquad (8)$$

with the moments of ϕ given by $\tilde{\phi}(N) = \int_0^1 dx\, x^{N-1} \phi(x)$. We factorize the initial-state collinear divergences, regularized by ϵ, into the parton distribution functions, ϕ, which are expanded to the same order in α_s as the partonic cross section, and we thus obtain the perturbative expansion for the infrared-safe partonic short-distance function $\hat{\sigma}$.

The partonic short-distance function $\hat{\sigma}$ still has sensitivity to soft-gluon dynamics through its N dependence. We then refactorize the moments of the cross section as [4, 8]

$$\sigma_{f_1 f_2 \to F}(N, \epsilon) = \tilde{\psi}_{f_1/f_1}(N, \mu_F, \epsilon)\, \tilde{\psi}_{f_2/f_2}(N, \mu_F, \epsilon)$$
$$\times\, H_{IL}^{f_1 f_2 \to F}(\alpha_s(\mu_R))\, \tilde{S}_{LI}^{f_1 f_2 \to F}\left(\frac{M}{N\mu_F}, \alpha_s(\mu_R)\right) \prod_j \tilde{J}_j(N, \mu_F, \epsilon) + \mathcal{O}(1/N), \qquad (9)$$

where ψ are center-of-mass distributions that absorb the universal collinear singularities from the incoming partons, H_{IL} are N-independent hard components which describe the hard-scattering, S_{LI} is a soft gluon function associated with non-collinear soft gluons, and J are functions that absorb the collinear singularities from massless partons, if any, in the final state.

H and S are matrices in color space and we sum over the color indices I and L that describe the color structure of the hard scattering. The hard-scattering function involves contributions from the amplitude of the process and the complex conjugate of the amplitude, $H_{IL} = h_L^*\, h_I$. The soft function S_{LI} represents the coupling of soft gluons to the partons in the scattering. The color tensors of the hard scattering connect together the eikonal lines to which soft gluons couple. One can construct an eikonal operator describing soft-gluon emission and write a dimensionless eikonal cross section, which describes the emission of soft gluons by the eikonal lines [4, 6–8].

Comparing Eqs. (8) and (9), we see that the moments of the short-distance partonic cross section are given by

$$\begin{aligned}\hat{\sigma}_{f_1 f_2 \to F}(N, \mu_F, \mu_R) &= \frac{\tilde{\psi}_{f_1/f_1}(N, \mu_F, \epsilon)\, \tilde{\psi}_{f_2/f_2}(N, \mu_F, \epsilon)}{\tilde{\phi}_{f_1/f_1}(N, \mu_F, \epsilon)\, \tilde{\phi}_{f_2/f_2}(N, \mu_F, \epsilon)} H_{IL}^{f_1 f_2 \to F}\left(\alpha_s(\mu_R)\right) \\ &\times \tilde{S}_{LI}^{f_1 f_2 \to F}\left(\frac{M}{N\mu_F}, \alpha_s(\mu_R)\right) \prod_j \tilde{J}_j\left(N, \mu_F, \epsilon\right) . \end{aligned} \tag{10}$$

All the factors in Eq. (10) are gauge and factorization scale dependent. The constraint that the product of these factors must be independent of the gauge and factorization scale results in the exponentiation of logarithms of N in ψ/ϕ and S_{LI} [4, 5].

The soft matrix S_{LI} depends on N through the ratio $M/(N\mu_F)$, and it requires renormalization as a composite operator. Its N-dependence can thus be resummed by renormalization group analysis [17–20]. However, the product $H_{IL}S_{LI}$ needs no overall renormalization, because the UV divergences of S_{LI} are balanced by those of H_{IL}. Thus, we have [4, 8]

$$\begin{aligned} H^0_{IL} &= \prod_{i=a,b} Z_i^{-1} \left(Z_S^{-1}\right)_{IC} H_{CD} \left[\left(Z_S^{\dagger}\right)^{-1}\right]_{DL} , \\ S^0_{LI} &= (Z_S^{\dagger})_{LB} S_{BA} Z_{S,AI} , \end{aligned} \tag{11}$$

where H^0 and S^0 denote the unrenormalized quantities, Z_i is the renormalization constant of the ith incoming partonic field, and Z_S is a matrix of renormalization constants, which describe the renormalization of the soft function. Z_S is defined to include the wave function renormalization necessary for the outgoing eikonal lines that represent any heavy quarks.

From Eq. (11), we see that the soft function S_{LI} satisfies the renormalization group equation [4, 6–8]

$$\left(\mu \frac{\partial}{\partial \mu} + \beta(g_s) \frac{\partial}{\partial g_s}\right) S_{LI} = -(\Gamma_S^{\dagger})_{LB} S_{BI} - S_{LA} (\Gamma_S)_{AI} , \tag{12}$$

where β is the QCD beta function and $g_s^2 = 4\pi\alpha_s$. Γ_S is an anomalous dimension matrix that is calculated in the eikonal approximation by explicit renormalization of the soft function. In a minimal subtraction renormalization scheme and with $\epsilon = 4 - n$, where n is the number of space-time dimensions, the soft anomalous dimension matrix is given at one loop by

$$\Gamma_S^{(1l)}(g_s) = -\frac{g_s}{2} \frac{\partial}{\partial g_s} \mathrm{Res}_{\epsilon \to 0} Z_S(g_s, \epsilon) . \tag{13}$$

The process-dependent matrices Γ_S have been calculated at one loop for all $2 \to 2$ partonic processes; a compilation of results is given in [8]. In processes with trivial or simple color structure Γ_S is simply a function (1×1 matrix) while in processes with complex color structure it is a non-trivial matrix in color exchange. For quark-(anti)quark scattering, Γ_S

is a 2×2 matrix [4, 18]; for quark-gluon scattering it is a 3×3 matrix [7]; for gluon-gluon scattering it is an 8×8 matrix [7]. Complete two-loop calculations of soft anomalous dimensions for processes with massless quarks have appeared in [21]. Selected two-loop results for heavy quark production appeared in [22]. We present a sample one-loop calculation in Section 4.1 and a sample two-loop calculation in Section 4.2, both with outgoing massive quarks (see [23]).

The exponentiation of logarithms of N in the ratios ψ/ϕ and in the functions J in Eq. (10), together with the solution of the renormalization group equation (12), provide us with the complete expression for the resummed partonic cross section in moment space [4, 6–8, 24, 25]

$$
\begin{aligned}
\hat{\sigma}^{res}(N) &= \exp\left[\sum_i E^{f_i}(N_i)\right] \exp\left[\sum_j E'^{f_j}(N_j)\right] \\
&\times \exp\left[\sum_i 2\int_{\mu_F}^{\sqrt{s}} \frac{d\mu}{\mu}\gamma_{f_i/f_i}\left(\alpha_s(\mu)\right)\right] \exp\left[2\, d_{\alpha_s}\int_{\mu_R}^{\sqrt{s}} \frac{d\mu}{\mu}\beta\left(\alpha_s(\mu)\right)\right] \\
&\times \mathrm{Tr}\left\{H^{f_1 f_2\to F}\left(\alpha_s(\mu_R)\right) \exp\left[\int_{\sqrt{s}}^{\sqrt{s}/\tilde{N}_j} \frac{d\mu}{\mu}\Gamma_S^{\dagger\, f_1 f_2\to F}\left(\alpha_s(\mu)\right)\right]\right. \\
&\times \left.\tilde{S}^{f_1 f_2\to F}\left(\alpha_s\left(\frac{\sqrt{s}}{\tilde{N}_j}\right)\right) \exp\left[\int_{\sqrt{s}}^{\sqrt{s}/\tilde{N}_j} \frac{d\mu}{\mu}\Gamma_S^{f_1 f_2\to F}\left(\alpha_s(\mu)\right)\right]\right\}. \qquad (14)
\end{aligned}
$$

The sums over $i = 1, 2$ run over incoming partons. The sum over j is over massless partons, if any, in the final state at lowest order. The resummed expression is valid for either 1PI or PIM kinematics. In 1PI kinematics $N_i = N(-t_i/M^2)$, where t_i denotes t or u, and $N_j = N(s/M^2)$, while in PIM kinematics $N_i = N_j = N$. Also $\tilde{N} = Ne^{\gamma_E}$, with γ_E the Euler constant.

The first exponent in Eq. (14) arises from the exponentiation of logarithms of N in the ratios ψ/ϕ of Eq. (10), and is given in the $\overline{\text{MS}}$ scheme by

$$
E^{f_i}(N_i) = -\int_0^1 dz \frac{z^{N_i-1}-1}{1-z}\left\{\int_{(1-z)^2}^1 \frac{d\lambda}{\lambda} A_i\left(\alpha_s(\lambda s)\right) + \nu_i\left[\alpha_s((1-z)^2 s)\right]\right\}, \qquad (15)
$$

with $A_i(\alpha_s) = \sum_{n=1}^{\infty}(\alpha_s/\pi)^n A_i^{(n)}$. At one loop, $A_i^{(1)} = C_i$ which is $C_F = (N_c^2 - 1)/(2N_c)$ for a quark or antiquark and $C_A = N_c$ for a gluon, with N_c the number of colors, while $A_i^{(2)} = C_i K/2$ with $K = C_A(67/18 - \zeta_2) - 5n_f/9$ [26], where n_f is the number of quark flavors and $\zeta_2 = \pi^2/6$. Also $\nu_i = \sum_{n=1}^{\infty}(\alpha_s/\pi)^n \nu_i^{(n)}$, with $\nu_i^{(1)} = C_i$.

The second exponent in Eq. (14) arises from the exponentiation of logarithms of N in the functions J_j of Eq. (10), and is given by

$$
\begin{aligned}
E'^{f_j}(N_j) &= \int_0^1 dz \frac{z^{N_j-1}-1}{1-z}\left\{\int_{(1-z)^2}^{1-z} \frac{d\lambda}{\lambda} A_j\left(\alpha_s\left(\lambda s\right)\right) - B_j\left[\alpha_s((1-z)s)\right]\right. \\
&\left. - \nu_j\left[\alpha_s((1-z)^2 s)\right]\right\}. \qquad (16)
\end{aligned}
$$

Here $B_j = \sum_{n=1}^{\infty}(\alpha_s/\pi)^n B_j^{(n)}$ with $B_j^{(1)}$ equal to $3C_F/4$ for quarks and $\beta_0/4$ for gluons, where $\beta_0 = (11C_A - 2n_f)/3$ is the lowest-order β function.

The third exponent in Eq. (14) controls the factorization scale dependence of the cross section, and γ_{f_i/f_i} is the moment-space anomalous dimension of the $\overline{\rm MS}$ density ϕ_{f_i/f_i}. The β function in the fourth exponent controls the renormalization scale dependence of the cross section. The constant d_{α_s} takes the value k if the Born cross section is of order α_s^k. Explicit expressions for the functions in these four exponents, and related references, are assembled for convenience in Appendix A of Ref. [27].

As noted before, both H and S are process-dependent matrices in color space and thus the trace is taken in Eq. (14). At lowest order, the trace of the product of H and S reproduces the Born cross section. The evolution of the soft function S follows from its renormalization group equation, (12), and is given in terms of the soft anomalous dimension matrix Γ_S.

2.2. NNNLO Expansions

The exponentials in the resummed partonic cross section can be expanded to any fixed order in α_s and then inverted to momentum space to provide explicit results for the higher-order corrections. A fixed-order expansion avoids using a prescription to regulate the infrared singularities in the exponents and thus no prescription is needed to deal with these in this approach (see discussion in Ref. [28]).

We now expand the resummed cross section, Eq. (14), in 1PI kinematics through NNNLO. We provide results here for the case where Γ_S are trivial (1×1) color matrices. Explicit expressions for the more general case are found through NNLO in [24] and through NNNLO in [25].

At NLO, we find the expression for the soft-gluon corrections

$$\hat{\sigma}^{(1)} = \sigma^B \frac{\alpha_s(\mu_R^2)}{\pi} \{c_3\, \mathcal{D}_1(s_4) + c_2\, \mathcal{D}_0(s_4) + c_1\, \delta(s_4)\} \tag{17}$$

where σ^B is the leading-order (LO) term, the LL coefficient is

$$c_3 = \sum_i 2\, C_i - \sum_j C_j\,, \tag{18}$$

with $C_q = C_F$ and $C_g = C_A$, and the NLL coefficient c_2 is defined by $c_2 = c_2^\mu + T_2$, with

$$c_2^\mu = -\sum_i C_i \ln\left(\frac{\mu_F^2}{M^2}\right) \tag{19}$$

denoting the terms involving logarithms of the factorization scale, and

$$\begin{aligned} T_2 \;&=\; 2{\rm Re}\Gamma_S^{(1)} - \sum_i \left[C_i + 2\, C_i\, \ln\left(\frac{-t_i}{M^2}\right) + C_i \ln\left(\frac{M^2}{s}\right)\right] \\ &\quad - \sum_j \left[B_j^{(1)} + C_j + C_j\, \ln\left(\frac{M^2}{s}\right)\right] \end{aligned} \tag{20}$$

denoting the scale-independent terms. Again, t_i denotes t or u, the sums over i run over incoming partons, and the sums over j run over any massless partons in the final state at LO.

We write the NLO $\delta(s_4)$ terms as $c_1 = c_1^\mu + T_1$, where

$$c_1^\mu = \sum_i \left[C_i \ln\left(\frac{-t_i}{M^2}\right) - \gamma_i^{(1)}\right] \ln\left(\frac{\mu_F^2}{M^2}\right) + d_{\alpha_s} \frac{\beta_0}{4} \ln\left(\frac{\mu_R^2}{M^2}\right) \tag{21}$$

denotes the terms involving logarithms of the factorization and renormalization scales. Here $\gamma_q^{(1)} = 3C_F/4$ and $\gamma_g^{(1)} = \beta_0/4$, and T_1 denotes virtual terms that cannot be derived from the resummation formalism but can be determined by matching to a full NLO calculation for any specified process.

At NNLO, the soft-gluon corrections are

$$\begin{aligned}
\hat{\sigma}^{(2)} &= \sigma^B \frac{\alpha_s^2(\mu_R^2)}{\pi^2} \left\{ \frac{1}{2} c_3^2 \, \mathcal{D}_3(s_4) + \left[\frac{3}{2} c_3 \, c_2 - \frac{\beta_0}{4} c_3 + \sum_j C_j \frac{\beta_0}{8}\right] \mathcal{D}_2(s_4) \right. \\
&+ \left. \left[c_3 \, c_1 + c_2^2 - \zeta_2 \, c_3^2 - \frac{\beta_0}{2} T_2 + \frac{\beta_0}{4} c_3 \ln\left(\frac{\mu_R^2}{M^2}\right) + c_3 \frac{K}{2} - \sum_j \frac{\beta_0}{4} B_j^{(1)}\right] \mathcal{D}_1(s_4) \right\} \\
&+ \cdots
\end{aligned} \tag{22}$$

where we show explicitly results through next-to-next-to-leading logarithms (NNLL). For a complete expression see [24,25].

At NNNLO, the soft-gluon corrections are

$$\begin{aligned}
\hat{\sigma}^{(3)} &= \sigma^B \frac{\alpha_s^3(\mu_R^2)}{\pi^3} \left\{ \frac{1}{8} c_3^3 \, \mathcal{D}_5(s_4) + \left[\frac{5}{8} c_3^2 \, c_2 - \frac{5}{24} \beta_0 \, c_3^2 + \frac{5}{48} c_3 \, \beta_0 \sum_j C_j\right] \mathcal{D}_4(s_4) \right. \\
&+ \left[c_3 \, c_2^2 + \frac{1}{2} c_3^2 \, c_1 - \zeta_2 \, c_3^3 + \frac{\beta_0^2}{12} c_3 - \frac{\beta_0}{3} c_3 c_2 - \frac{\beta_0}{2} c_3 T_2 + \frac{\beta_0}{4} c_3^2 \ln\left(\frac{\mu_R^2}{M^2}\right) \right. \\
&+ \left. \left. c_3^2 \frac{K}{2} + c_2 \frac{\beta_0}{6} \sum_j C_j - c_3 \sum_j \frac{\beta_0}{4} B_j^{(1)} - \sum_j C_j \frac{3\beta_0^2}{48}\right] \mathcal{D}_3(s_4) \right\} + \cdots
\end{aligned} \tag{23}$$

where again we show explicitly results through NNLL. The complete expression is given in [25].

In PIM kinematics we simply replace s_4 by $1 - z$, set $s = M^2$, and drop the terms with $\ln(-t_i/M^2)$ in the above formulas.

The NNNLO master equation, (23), gives the structure of the soft corrections and can provide the full soft corrections explicitly if all the necessary two-loop and three-loop quantities are known. For processes with non-trivial color structure we are currently limited to NLL or NNLL accuracy. For processes with trivial color structure, such as $b\bar{b} \to H$ [27,29], all soft-gluon corrections have been determined through NNNLO. Below, the term "N$^{(n)}$LO-N$^{(l)}$LL" means that the soft-gluon contributions through N$^{(l)}$LL accuracy to the n-th order QCD corrections have been included.

3. Applications of Large-x Resummations

In this section we provide some calculations that are applications of the large-x resummation formalism to processes of interest at the Tevatron and the LHC. We present results for

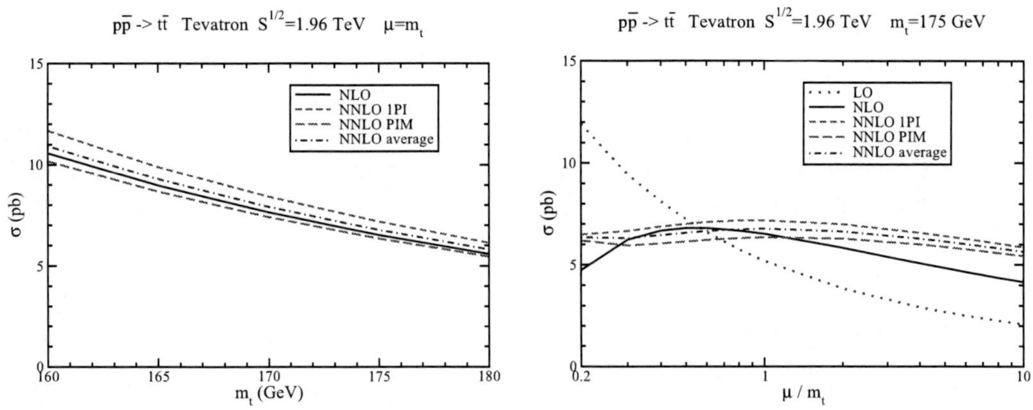

Figure 1. The $t\bar{t}$ total cross sections in $p\bar{p}$ collisions at the Tevatron with $\sqrt{S} = 1.96$ TeV as functions of m_t (left) and μ/m_t (right).

top-antitop pair production, single top quark production, W-boson production, and Higgs boson production via $b\bar{b} \to H$.

3.1. $t\bar{t}$ Production

The top quark, the heaviest known elementary particle, was discovered in $p\bar{p}$ collisions at Run I of the Tevatron in 1995 [30,31]. More recent measurements at Run II have increased the accuracy of the top mass and cross section measurements (for a review see [32, 33]) and thus require accurate theoretical calculations of top production cross sections and differential distributions. The main partonic channels in $t\bar{t}$ production are $q\bar{q} \to t\bar{t}$, which is dominant at the Tevatron, and $gg \to t\bar{t}$, which will be dominant at the LHC.

The latest calculation for top-antitop pair hadroproduction includes NNLO soft-gluon corrections to the double differential cross section [34]. Near threshold the soft-gluon corrections dominate the cross section at the Tevatron and contribute sizable enhancements. The form of the corrections and their numerical values depend crucially on the kinematics chosen to describe the process. The NNLO soft corrections were calculated fully to NNLL in both 1PI and PIM kinematics [28, 35]. In addition a good approximation for the next-to-next-to-next-to-leading logarithms (NNNLL) was provided in [34]. The best theoretical result for the cross section is the average of the NNLO-NNNLL cross sections in the two different kinematics [34].

In Fig. 1, we present the NLO and approximate NNLO-NNNLL $t\bar{t}$ cross sections at the Tevatron with $\sqrt{S} = 1.96$ TeV using the MRST2002 [36] parton densities. On the left we plot the cross sections as functions of m_t, the top quark mass, for $\mu = m_t$, where μ denotes the factorization and renormalization scales which we have set equal to each other. On the right we plot the cross sections as functions of μ/m_t with $m_t = 175$ GeV. The results are given in both 1PI and PIM kinematics together with their average. The NLO cross section depends less on μ than the LO cross section, as expected. The NNLO-NNNLL cross sections exhibit even less dependence on μ, approaching the scale independence of a true physical cross section. They change by less than 3% in the range $m_t/2 < \mu < 2m_t$. For a top mass of 175 GeV the average of the NNLO-NNNLL 1PI and PIM results is 6.77 ± 0.42

pb, where the uncertainty indicated is from the kinematics. Including all sources of uncertainty (kinematics, scale variation, and uncertainty from the parton distribution functions) we may write the cross section as 6.8 ± 0.6 pb. This theoretical result is in agreement with the latest experimental result for the cross section at the Tevatron [37, 38]. Finally, we note that NNNLO soft-gluon contributions in the $q\bar{q} \to t\bar{t}$ channel were presented in [25]. These NNNLO-NNLL corrections further stabilize the scale dependence of the cross section at the Tevatron.

3.2. Single Top Quark Production

Single top quark production provides a way to directly measure electroweak properties of the top quark, such as the V_{tb} CKM matrix element. It also allows a deeper study of electroweak theory since the top quark mass is of the same order of magnitude as the electroweak symmetry breaking scale, and may be useful in the discovery of new physics. Therefore it is crucial to have accurate theoretical predictions for the cross section.

The cross section for single top quark production is less than the $t\bar{t}$ cross section and the backgrounds to the production processes make the extraction of the single top signal challenging. Intensive searches for single top quark events at the Tevatron have recently produced evidence of such events [39,40]. The LHC has good potential for observation and further analysis of single top events.

Single top quarks can be produced through three distinct partonic processes. One is t-channel production, $qb \to q't$ and $\bar{q}b \to \bar{q}'t$, via the exchange of a space-like W boson, a second is s-channel production, $q\bar{q}' \to \bar{b}t$, via the exchange of a time-like W boson, and a third is associated tW production, $bg \to tW^-$.

The threshold corrections to single top production have been calculated for both the Tevatron and LHC colliders through NNNLO [41–43]. At the Tevatron the t-channel process is numerically dominant, but the higher-order corrections are relatively small. The s-channel is smaller, but receives large corrections and it was shown that the threshold soft-gluon corrections dominate the cross section. Associated tW production is quite minor, although it also has large K factors, defined as the ratios of the higher-order cross sections to the LO cross section. At the LHC the t channel is again dominant, but the second largest channel is tW production; the s channel is numerically the smallest. Below we provide some numerical results for all three channels at both the Tevatron and the LHC colliders using the MRST2004 parton densities [44]. We add the soft-gluon corrections through NNNLO to the complete NLO cross section [45,46].

We begin with single top production at the Tevatron [41] with $\sqrt{S} = 1.96$ TeV. For t-channel production, the NNNLO-NLL cross section is $\sigma^{t-\text{channel}}(m_t = 175\,\text{GeV}) = 1.08^{+0.02}_{-0.01} \pm 0.06$ pb, where the first uncertainty is from variation of the factorization and renormalization scales, μ_F and μ_R, between $m_t/2$ and $2m_t$, and the second is due to the parton distribution functions. For the s channel, the corresponding cross section is $\sigma^{s-\text{channel}}(m_t = 175\,\text{GeV}) = 0.49 \pm 0.02 \pm 0.01$ pb. Finally, in the tW channel $\sigma^{tW}(m_t = 175\,\text{GeV}) = 0.13 \pm 0.02 \pm 0.02$ pb. We note that the cross sections for antitop production at the Tevatron are identical to those for single top production in each channel.

In Fig. 2 we plot the cross section and the K factors for single top quark production at the Tevatron in the s channel setting both the factorization and renormalization scales to $\mu =$

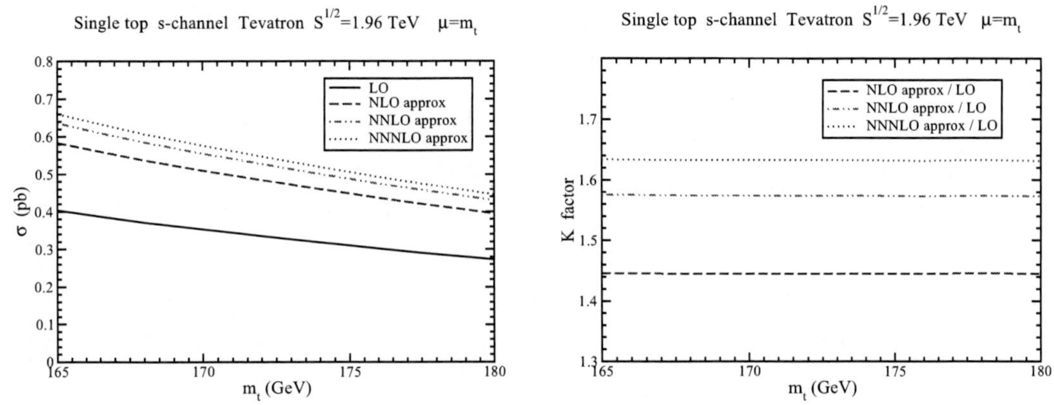

Figure 2. The cross section (left) and K factors (right) for single top quark production at the Tevatron in the s channel. Here $\mu = \mu_F = \mu_R = m_t$.

m_t. We plot the LO cross section and the approximate NLO, NNLO, and NNNLO cross sections at NLL accuracy. The K factors are quite large, thus showing that the corrections provide a big enhancement to the cross section.

We continue with single top production at the LHC [42] with $\sqrt{S} = 14$ TeV. For the t channel the threshold corrections are not a good approximation of the complete corrections. The NLO cross section for top production $\sigma_{\rm top}^{t-{\rm channel}}(m_t = 175\,{\rm GeV}) = 146 \pm 4 \pm 3$ pb. For antitop production the corresponding result is $\sigma_{\rm antitop}^{t-{\rm channel}}(m_t = 175\,{\rm GeV}) = 89 \pm 3 \pm 2$ pb. For the s channel, the soft-gluon corrections are relatively large and the soft-gluon approximation is good. The NNNLO-NLL cross section is $\sigma_{\rm top}^{s-{\rm channel}}(m_t = 175\,{\rm GeV}) = 7.23^{+0.53}_{-0.45} \pm 0.13$ pb for single top production and $\sigma_{\rm antitop}^{s-{\rm channel}}(m_t = 175\,{\rm GeV}) = 4.03^{+0.10}_{-0.12} \pm 0.10$ pb for single antitop production. Finally, for tW production the cross section is $\sigma^{tW}(m_t = 175\,{\rm GeV}) = 41.1 \pm 4.1 \pm 1.0$ pb, which is identical to that for associated antitop production. In Fig. 3 we plot the cross section and K factors for associated tW production at the LHC setting the scales to $\mu = m_t$. As seen from the plots, the soft-gluon corrections are large for this process.

3.3. W-boson Production at Large Transverse Momentum

W-boson production in hadron colliders can be used in testing the Standard Model and in estimating backgrounds to Higgs production and new physics. Precise calculations for W production at large transverse momentum, Q_T, are needed to identify signals of new physics which may be expected to enhance the Q_T distribution at high Q_T.

Analytical NLO calculations of the cross section for W production at large transverse momentum were presented in Refs. [47, 48], where numerical results were also presented for the Fermilab Tevatron. Numerical NLO results for W production at the LHC were more recently presented in [49]. The NLO corrections enhance the Q_T distribution of the W boson and they reduce the factorization and renormalization scale dependence of the cross section.

A recent theoretical study [50] included soft-gluon corrections through NNLO, which

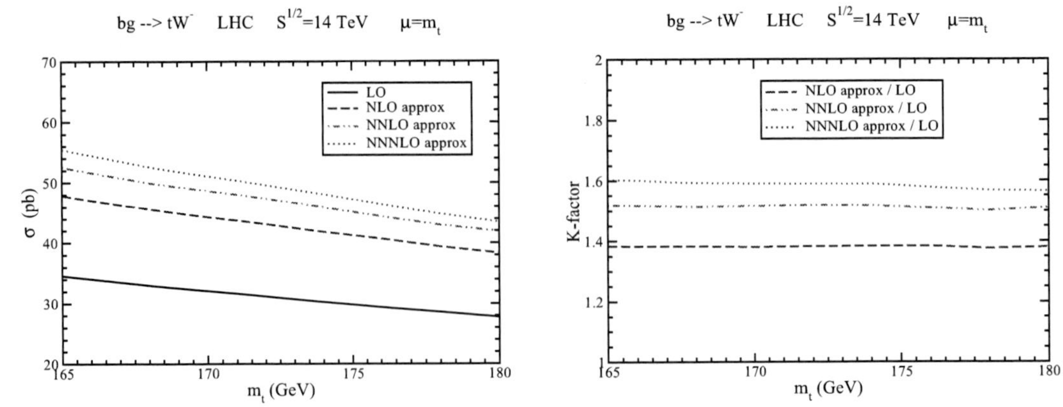

Figure 3. The cross section (left) and K factors (right) for associated tW production at the LHC. Here $\mu = \mu_F = \mu_R = m_t$.

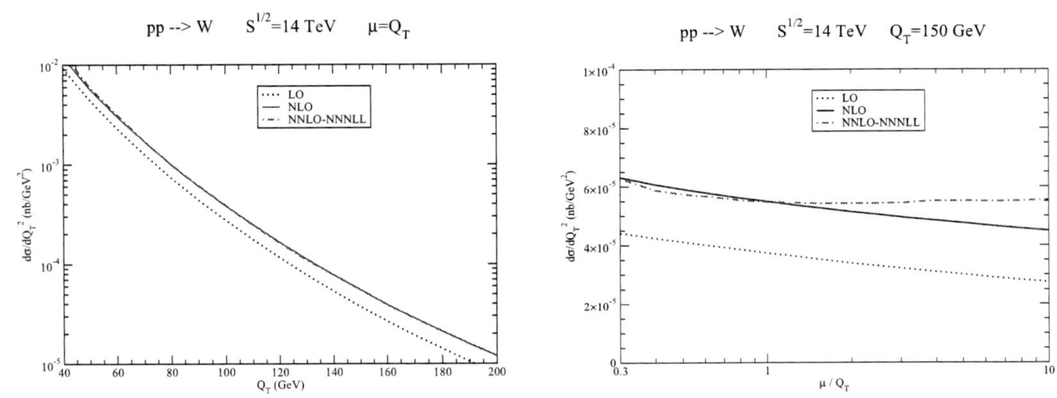

Figure 4. The differential cross section, $d\sigma/dQ_T^2$, for W production at the LHC with $\mu = Q_T$ (left) and $Q_T = 150$ GeV (right).

provide additional enhancements and a further reduction of the scale dependence. The complete NNLL terms were calculated and an approximation for the NNNLL terms was derived at NNLO. Numerical results with these soft corrections were calculated for W production at the Tevatron [50] and the LHC [49].

Here we discuss W production at large transverse momentum at the LHC with $\sqrt{S} = 14$ TeV using the MRST2002 parton densities [36]. The LO partonic processes for the production of a W boson and a parton are $qg \to Wq$ and $q\bar{q} \to Wg$. The electroweak coupling $\alpha(M_Z^2)$ is evaluated at the mass of the Z boson, and standard values [51] are used for the various electroweak parameters. In the numerical results we present the sum of cross sections for W^- and W^+ production. The W bosons at the LHC will be detected primarily through their leptonic decay products e.g., $W^- \to \ell\bar{\nu}_\ell$, therefore the cross sections presented here should be multiplied by the appropriate branching ratios.

In Fig. 4 (left plot) we plot the transverse momentum distribution, $d\sigma/dQ_T^2$, at high Q_T for W production at the LHC. We set $\mu_F = \mu_R = Q_T$ and denote this common scale by

μ. We plot LO, NLO, and NNLO-NNNLL results using the corresponding parton densities. As seen from the plot, the NLO corrections provide a significant enhancement of the LO Q_T distribution. The NNLO-NNNLL corrections provide a rather small further enhancement of the Q_T distribution. However, the NNLO-NNNLL corrections can be much bigger for other choices of factorization and renormalization scales. The NLO corrections increase the LO result by about 30% to 50% in the Q_T range shown. In contrast, the NNLO-NNNLL/NLO ratio for this scale is rather small. Part of the reason for this is that the NNLO parton distribution functions are significantly smaller than the NLO pdf.

On the plot on the right in Fig. 4 we show the scale dependence of $d\sigma/dQ_T^2$ for $Q_T =$ 150 GeV versus μ/Q_T over two orders of magnitude. It is interesting to note that the scale dependence of the cross section is not reduced when the NLO corrections are included. This is due to the fact that the cross section is dominated by the process $qg \to Wq$. The gluon density in the proton, at fixed x less than $\sim$0.01, increases rapidly with scale. Thus, the μ_R and μ_F dependencies cancel one another to a large extent. However, we have an improvement in the scale variation when the NNLO-NNNLL corrections are added. The NNLO-NNNLL result displays very little scale dependence.

3.4. Higgs Boson Production via $b\bar{b} \to H$

The search for the Higgs boson [52] is one of the most important goals at the Tevatron and the LHC colliders [53]. The main Standard Model production channel at these colliders is $gg \to H$. However, the channel $b\bar{b} \to H$ can be competitive in the Minimal Supersymmetric Standard Model at high $\tan\beta$, with $\tan\beta$ the ratio of the vacuum expectation values for the two Higgs doublets. The complete NNLO QCD corrections for this process were calculated in [54].

Complete expressions for the soft-gluon corrections at NNNLO were presented in [27, 29]. However, it is known at NNLO that the soft corrections alone are not a good approximation of the full corrections [27, 54]. Purely collinear terms [27, 55, 56] have to be included to provide an accurate calculation. An approximation for the collinear terms through NNLL accuracy at NNNLO was provided in [27].

We now present numerical results for $b\bar{b} \to H$ at the Tevatron and the LHC [27] using the MRST2006 parton densities [57]. Figure 5 shows the K factors for Higgs production via $b\bar{b} \to H$ at the Tevatron (left) and the LHC (right), with $\mu = m_H$. The complete NLO corrections increase the LO result by around 60% at both the Tevatron and the LHC. Inclusion of the complete NNLO corrections futher increases the cross section: the NNLO K factor is around 1.9 at the Tevatron and 1.8 at the LHC. By including at NNNLO the sum of the complete soft-gluon corrections and the collinear approximate NNLL corrections (S+NNLLapp), we find further enhancement. From the study of the contributions of the soft and collinear terms at NLO and NNLO at both the Tevatron and the LHC we expect that the NNNLO S+NNLCapp curve provides a good approximation of the complete NNNLO cross section. The NNNLO S+NNLCapp K factor is between 2.06 and 2.01 at the Tevatron and between 1.95 and 1.87 at the LHC for Higgs masses ranging between 110 and 180 GeV, which is a significant addition to the NNLO result.

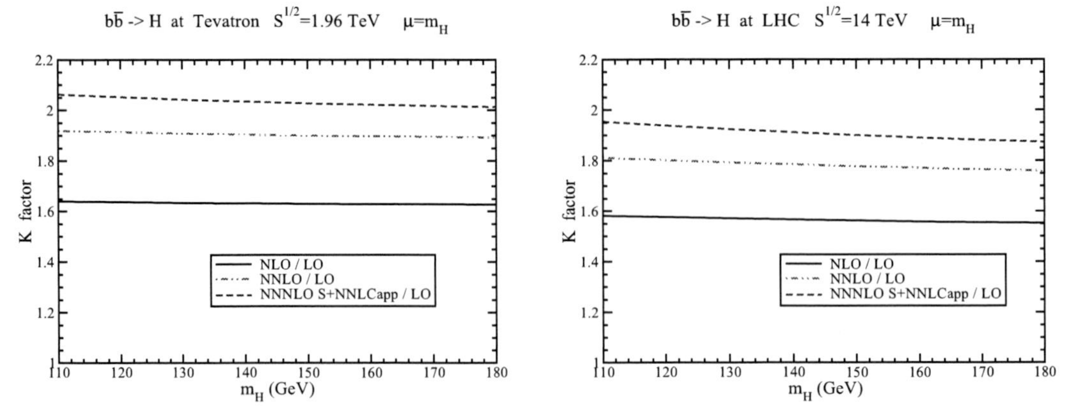

Figure 5. The K factors for $b\bar{b} \to H$ at the Tevatron (left) and the LHC (right). Here $\mu = \mu_F = \mu_R = m_H$.

4. Loop Calculations in the Eikonal Approximation

The soft-gluon resummation formalism, and in particular the calculation of the soft anomalous dimension matrices, employs the use of the eikonal approximation in loop diagrams. The eikonal approximation is valid for descibing the emission of soft gluons from partons in the hard scattering. The approximation leads to a simplified form of the Feynman rules by removing the Dirac matrices from the calculation. When the gluon momentum goes to zero, the Feynman rules for the quark propagator and quark-gluon vertex in Figure 6 simplify as follows:

$$\bar{u}(p)\,(-ig_sT_F^c)\,\gamma^\mu \frac{i(\not{p}+\not{k}+m)}{(p+k)^2-m^2+i\epsilon} \to \bar{u}(p)\,g_sT_F^c\,\gamma^\mu \frac{\not{p}+m}{2p\cdot k+i\epsilon} = \bar{u}(p)\,g_sT_F^c\,\frac{v^\mu}{v\cdot k+i\epsilon} \tag{24}$$

with v a dimensionless vector, $p \propto v$, and T_F^c the generators of SU(3) in the fundamental representation.

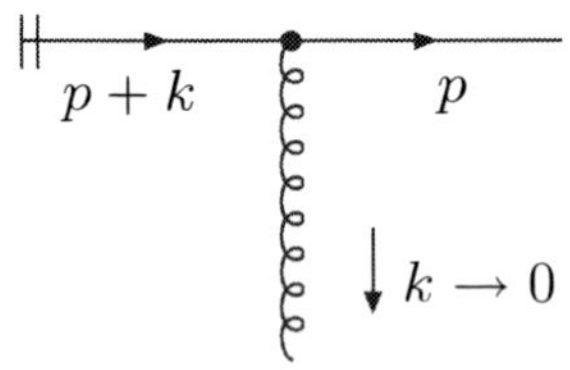

Figure 6. Eikonal approximation.

The ultraviolet poles in loop diagrams involving eikonal lines are particularly important as they play a direct role in the renormalization group evolution equations that are used in threshold resummations [4, 6, 7] (see Eq. (13)).

Below we give examples of a one-loop and a two-loop calculation for diagrams involving eikonal lines representing massive quarks. For the calculation we use the Feynman gauge, and we use dimensional regularization with $n = 4 - \epsilon$ dimensions.

4.1. One-Loop Calculation

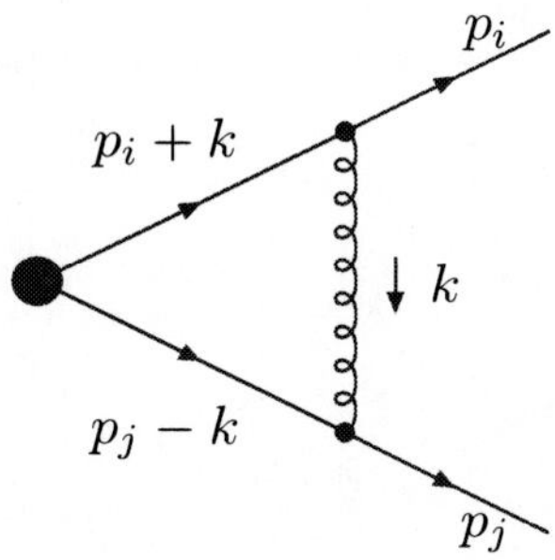

Figure 7. One-loop eikonal diagram with outgoing massive quarks.

In this subsection we calculate the integral I_{1l} for the one-loop diagram in Fig. (7) with eikonal lines representing outgoing massive quarks. This one-loop integral is given by

$$I_{1l} = g_s^2 \int \frac{d^n k}{(2\pi)^n} \frac{(-i) g^{\mu\nu}}{k^2} \frac{v_i^\mu}{v_i \cdot k} \frac{(-v_j^\nu)}{(-v_j \cdot k)} \, . \tag{25}$$

Using Feynman parameterization, this integral can be rewritten as

$$I_{1l} = -2ig_s^2 \frac{v_i \cdot v_j}{(2\pi)^n} \int_0^1 dx \int_0^{1-x} dy \int \frac{d^n k}{\left[xk^2 + y v_i \cdot k + (1-x-y) v_j \cdot k\right]^3} \, . \tag{26}$$

After several manipulations, Eq. (26) becomes

$$\begin{aligned} I_{1l} &= \frac{\alpha_s}{\pi} (-1)^{-1-\epsilon/2} \, 2^{5\epsilon/2} \, \pi^{\epsilon/2} \, \Gamma\left(1 + \frac{\epsilon}{2}\right) (1+\beta^2) \int_0^1 dx \, x^{-1+\epsilon} (1-x)^{-1-\epsilon} \\ &\times \left\{ \int_0^1 dz \left[4z\beta^2(1-z) + 1 - \beta^2\right]^{-1} - \frac{\epsilon}{2} \int_0^1 dz \frac{\ln\left[4z\beta^2(1-z) + 1 - \beta^2\right]}{4z\beta^2(1-z) + 1 - \beta^2} + \mathcal{O}\left(\epsilon^2\right) \right\} \end{aligned} \tag{27}$$

where here $\beta = \sqrt{1 - 4m^2/s}$, with m the quark mass, and we have used the relations $v_i \cdot v_j = (1+\beta^2)/2$ and $v_i^2 = v_j^2 = (1-\beta^2)/2$.

The integral over x in Eq. (27) contains both ultraviolet (UV) and infrared (IR) singularities. We isolate the UV singularities and find that

$$\int_0^1 dx \, x^{-1+\epsilon} \, (1-x)^{-1-\epsilon} = \frac{1}{\epsilon} + \text{IR}. \tag{28}$$

After calculating the integrals over z in Eq. (27), we find that the UV poles and constant terms of I_{1l} are

$$\begin{aligned} I_{1l}^{UV} &= \frac{\alpha_s}{\pi} \frac{(1+\beta^2)}{2\beta} \left\{ \frac{1}{\epsilon} \ln\left(\frac{1-\beta}{1+\beta}\right) + \frac{1}{2} \left(4 \ln 2 + \ln \pi - \gamma_E - i\pi\right) \ln\left(\frac{1-\beta}{1+\beta}\right) \right. \\ &\left. + \frac{1}{4} \ln^2(1+\beta) - \frac{1}{4} \ln^2(1-\beta) - \frac{1}{2} \text{Li}_2\left(\frac{1+\beta}{2}\right) + \frac{1}{2} \text{Li}_2\left(\frac{1-\beta}{2}\right) \right\} . \end{aligned} \tag{29}$$

Complete one-loop calculations for heavy quark production in axial gauge were presented in Ref. [4].

4.2. Two-Loop Calculation

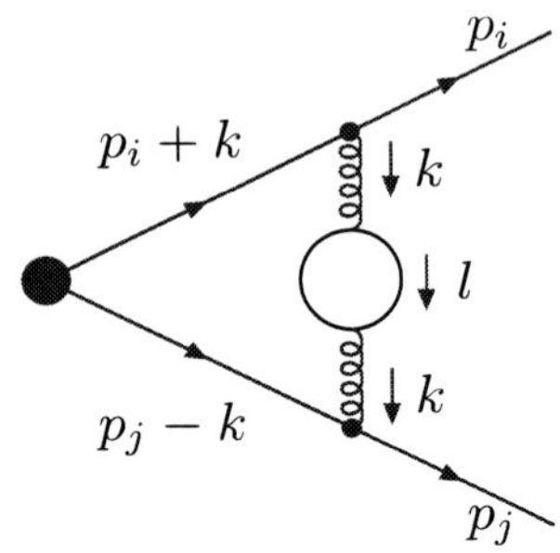

Figure 8. Two-loop eikonal diagram, involving a quark loop, with outgoing massive quarks.

In this subsection we calculate the two-loop integral I_{2l} for the quark-loop diagram in Fig. (8) given by

$$\begin{aligned} I_{2l} &= (-1)n_f g_s^4 \int \frac{d^n k}{(2\pi)^n} \frac{d^n l}{(2\pi)^n} \frac{v_i^\mu}{v_i \cdot k} \frac{(-v_j^\rho)}{(-v_j \cdot k)} \frac{(-i)g^{\mu\nu}}{k^2} \frac{(-i)g^{\rho\sigma}}{k^2} \\ &\times \mathrm{Tr}\left[-i\gamma^\nu \frac{i\not{l}}{l^2}(-i)\gamma^\sigma i \frac{(\not{l}-\not{k})}{(l-k)^2}\right] . \end{aligned} \tag{30}$$

After a few manipulations involving the trace we can write this integral as

$$I_{2l} = -n_f \frac{n g_s^4}{(2\pi)^{2n}} \left[I_{2l}^a + I_{2l}^b + I_{2l}^c + I_{2l}^d + I_{2l}^e\right] \tag{31}$$

where

$$I_{2l}^a = v_i \cdot v_j \int \frac{d^n k}{v_i \cdot k \, v_j \cdot k \, k^4} \int \frac{d^n l}{(l-k)^2} \tag{32}$$

$$I_{2l}^b = -v_i \cdot v_j \int \frac{d^n k}{v_i \cdot k \, v_j \cdot k \, k^4} \int d^n l \frac{l \cdot k}{l^2 (l-k)^2} \tag{33}$$

$$I_{2l}^c = -2 \int \frac{d^n k}{v_i \cdot k \, v_j \cdot k \, k^4} \int d^n l \frac{v_i \cdot l \, v_j \cdot l}{l^2 (l-k)^2} \tag{34}$$

$$I_{2l}^d = \int \frac{d^n k}{v_i \cdot k \, k^4} \int d^n l \frac{v_i \cdot l}{l^2 (l-k)^2} \tag{35}$$

$$I_{2l}^e = \int \frac{d^n k}{v_j \cdot k \, k^4} \int d^n l \frac{v_j \cdot l}{l^2 (l-k)^2} . \tag{36}$$

We begin with the evaluation of I_{2l}^a. Since

$$\int \frac{d^n l}{(l-k)^2} = 0 \tag{37}$$

we find $I_{2l}^a = 0$.

Next we evaluate I_{2l}^b. Using Feynman parameterization, we find

$$\int d^n l \frac{l^\mu}{l^2(l-k)^2} = i\pi^{(5-\epsilon)/2}\, 2^{-2+\epsilon}\, \Gamma\left(\frac{\epsilon}{2}\right) \frac{\Gamma\left(1-\frac{\epsilon}{2}\right)}{\Gamma\left(\frac{3}{2}-\frac{\epsilon}{2}\right)} \left(k^2\right)^{-\epsilon/2} k^\mu \tag{38}$$

therefore

$$I_{2l}^b = -i\pi^{(5-\epsilon)/2}\, 2^{-2+\epsilon}\, v_i \cdot v_j\, \Gamma\left(\frac{\epsilon}{2}\right) \frac{\Gamma\left(1-\frac{\epsilon}{2}\right)}{\Gamma\left(\frac{3}{2}-\frac{\epsilon}{2}\right)} \int \frac{d^n k}{v_i \cdot k\; v_j \cdot k\; (k^2)^{1+\epsilon/2}}\,. \tag{39}$$

The k integral in the above expression is

$$\begin{aligned}\int \frac{d^n k}{v_i \cdot k\; v_j \cdot k\; (k^2)^{1+\epsilon/2}} &= i\pi^{2-\epsilon/2}\, 2^{2+2\epsilon}\, (-1)^{-1-\epsilon} \frac{\Gamma(1+\epsilon)}{\Gamma\left(1+\frac{\epsilon}{2}\right)} \\ &\times \int_0^1 dx\, x^{-1+2\epsilon}(1-x)^{-1-2\epsilon} \int_0^1 dy \left[-2\beta^2 y^2 + 2\beta^2 y + \frac{1-\beta^2}{2}\right]^{-1-\epsilon}. \end{aligned} \tag{40}$$

The integral over x in Eq. (40) contains both UV and IR singularities. We isolate the UV singularities and find that

$$\int_0^1 dx\, x^{-1+2\epsilon}(1-x)^{-1-2\epsilon} = \frac{1}{2\epsilon} + \mathrm{IR}\,. \tag{41}$$

The integral over y is given in terms of hypergeometric functions ${}_2F_1\left(-\epsilon, 1+\epsilon, 1-\epsilon, \frac{1\pm\beta}{2}\right)$ which can be expanded in powers of ϵ. After some calculation, we find

$$\begin{aligned}\int \frac{d^n k}{v_i \cdot k\; v_j \cdot k\; (k^2)^{1+\epsilon/2}} &= \frac{2i\pi^2}{\beta\,\epsilon} \ln\left(\frac{1-\beta}{1+\beta}\right) + \frac{2i\pi^2}{\beta}\left[\mathrm{Li}_2\left(\frac{2}{1+\beta}\right) - \mathrm{Li}_2\left(\frac{2}{1-\beta}\right)\right. \\ &\left. + \ln^2(1+\beta) - \ln^2(1-\beta) + \frac{1}{2}(6\ln 2 - \ln\pi - \gamma_E)\ln\left(\frac{1-\beta}{1+\beta}\right)\right]. \end{aligned} \tag{42}$$

Assembling everything together we find the result for the UV poles of I_{2l}^b,

$$\begin{aligned} I_{2l}^{b\,UV} &= \pi^4 \frac{(1+\beta^2)}{\beta}\left\{\frac{1}{\epsilon^2}\ln\left(\frac{1-\beta}{1+\beta}\right) + \frac{1}{\epsilon}\left[\mathrm{Li}_2\left(\frac{2}{1+\beta}\right) - \mathrm{Li}_2\left(\frac{2}{1-\beta}\right)\right.\right. \\ &\left.\left. + \ln^2(1+\beta) - \ln^2(1-\beta) + (1 + 3\ln 2 - \ln\pi - \gamma_E)\ln\left(\frac{1-\beta}{1+\beta}\right)\right]\right\}. \end{aligned} \tag{43}$$

We continue with the evaluation of I_{2l}^c. Now

$$\begin{aligned}\int d^n l \frac{l^\mu l^\nu}{l^2(l-k)^2} &= i\pi^{2-\epsilon/2}\, \Gamma\left(\frac{\epsilon}{2}\right) \Gamma\left(1-\frac{\epsilon}{2}\right) \frac{\Gamma\left(3-\frac{\epsilon}{2}\right)}{\Gamma(4-\epsilon)} (k^2)^{-\epsilon/2} k^\mu k^\nu \\ &+ \frac{i\pi^{2-\epsilon/2}}{2} g^{\mu\nu} \Gamma\left(-1+\frac{\epsilon}{2}\right) \frac{\left(\Gamma\left(2-\frac{\epsilon}{2}\right)\right)^2}{\Gamma(4-\epsilon)} (k^2)^{1-\epsilon/2} \end{aligned} \tag{44}$$

and after a few manipulations we find

$$I_{2l}^{c} = -i\pi^{2-\epsilon/2}\Gamma\left(-1+\frac{\epsilon}{2}\right)\frac{\left(\Gamma\left(2-\frac{\epsilon}{2}\right)\right)^{2}}{\Gamma(4-\epsilon)}v_i \cdot v_j \int \frac{d^n k}{v_i \cdot k \, v_j \cdot k \, (k^2)^{1+\epsilon/2}} \, . \tag{45}$$

The integral over k was evaluated before for I_{10}^{b}, Eq. (40). We thus find that the UV poles of I_{10}^{c} are

$$\begin{aligned} I_{2l}^{c\,UV} &= \pi^4 \frac{1+\beta^2}{3\beta}\left\{-\frac{1}{\epsilon^2}\ln\left(\frac{1-\beta}{1+\beta}\right)+\frac{1}{\epsilon}\left[-\mathrm{Li}_2\left(\frac{2}{1+\beta}\right)+\mathrm{Li}_2\left(\frac{2}{1-\beta}\right)\right.\right. \\ &\left.\left. - \ln^2(1+\beta)+\ln^2(1-\beta)+\left(-\frac{4}{3}-3\ln 2+\ln\pi+\gamma_E\right)\ln\left(\frac{1-\beta}{1+\beta}\right)\right]\right\} . \end{aligned} \tag{46}$$

Finally, we calculate I_{2l}^{d} and I_{2l}^{e}. We use Eq. (38) for the l integral and then find that the remaining integral over k vanishes, so $I_{2l}^{d} = I_{2l}^{e} = 0$.

Adding all the terms in Eq. (31), the final result for the UV poles of I_{2l} is

$$\begin{aligned} I_{2l}^{UV} &= -n_f \frac{\alpha_s^2}{\pi^2}\frac{(1+\beta^2)}{6\beta}\left\{\frac{1}{\epsilon^2}\ln\left(\frac{1-\beta}{1+\beta}\right)+\frac{1}{\epsilon}\left[\mathrm{Li}_2\left(\frac{2}{1+\beta}\right)-\mathrm{Li}_2\left(\frac{2}{1-\beta}\right)\right.\right. \\ &\left.\left. + \ln^2(1+\beta)-\ln^2(1-\beta)+\left(\frac{7}{12}+5\ln 2+\ln\pi-\gamma_E\right)\ln\left(\frac{1-\beta}{1+\beta}\right)\right]\right\} . \end{aligned} \tag{47}$$

More results for two-loop integrals with massive quarks will appear in [23].

5. Small-x Resummations

In the last forty years there has been a large effort trying to understand what are the correct effective degrees of freedom underlying the strong interaction at high energies. In scattering processes where the center-of-mass energy is much larger than any other scales the Balitsky-Fadin-Kuraev-Lipatov (BFKL) approach [10–14] emerges as the correct approach to describe the scattering. This framework relies upon t-channel "Reggeized" gluons interacting with each other via standard gluons in the s-channel and a gauge invariant three particle vertex. This simple structure is a consequence of using multi-Regge kinematics where gluon cascades are ordered in longitudinal components but with a random walk in transverse momenta. Although this simple iterative and linear structure must be modified at higher energies in order to introduce unitarization and non-linear corrections, there is a window at present and future colliders where the BFKL predictions hold.

In the leading logarithmic approximation (LLA) we resum terms of the form $(\alpha_s \ln s)^n$. Diagrams contributing to the running of the strong coupling do not appear and the coupling is a constant parameter. The factor needed to scale the energy in the logarithms is also free and the predictability of the LL approximation is limited. In the next-to-leading logarithmic approximation (NLLA) diagrams with an extra power in the coupling without introducing an extra logarithm in energy are considered. The coupling is allowed to run and the energy scale is determined.

In this contribution we discuss three aspects of the BFKL resummation program. In subsection 5.1. we review the relevant equations to describe final states at small values of

Bjorken x in Deep Inelastic Scattering (DIS). We introduce the concept of color coherence and the CCFM equation. We show the differences and similarities between the BFKL approach and the introduction of angular ordering in the case of jet rates. In subsection 5.2. we analyse in detail how to extend the region of applicability of the multi-Regge kinematics, the basic ingredient in the BFKL approach, to regions with collinear emissions. We will find an interesting structure in the higher-order corrections that can be resummed into a Bessel function of the first kind, which accounts for the double logarithms in tranverse scales. In subsection 5.3. we briefly explain the $SL(2, C)$ invariance associated to the BFKL Hamiltonian and how it shows up in the physics of multijet events, in particular in the production of Mueller-Navelet jets at a hadron collider.

5.1. QCD Coherence and Small-x Final States

In Quantum Electrodynamics coherence effects are responsible for the suppression of soft bremsstrahlung from electron-positron pairs. In QCD processes such as $g \to q\bar{q}$ any soft gluon emitted with an angle from one of the fermionic lines larger than the angle of emission in the $q\bar{q}$ pair will probe the total color charge of the pair. This charge is the same as the one from the parent gluon and the radiation takes place as if the soft gluon was emitted from it. This color coherence leads to the angular ordering of sequential gluon emissions.

In DIS, let us say that the $(i-1)$th emitted gluon from the proton has energy E_{i-1}. A gluon radiated from it with a fraction $(1-z_i)$ of its energy and a transverse momentum q_i will have an opening angle

$$\theta_i \approx \frac{q_i}{(1-z_i)E_{i-1}}, \tag{48}$$

with

$$z_i = \frac{E_i}{E_{i-1}}. \tag{49}$$

Color coherence leads to angular ordering with increasing opening angles towards the hard scale (the photon). Therefore, we have $\theta_{i+1} > \theta_i$, or

$$\frac{q_{i+1}}{1-z_{i+1}} > \frac{z_i q_i}{1-z_i}, \tag{50}$$

which in the limit $z_i, z_{i+1} \ll 1$ reduces to

$$q_{i+1} > z_i q_i. \tag{51}$$

In Ref. [58–61] the BFKL equation for the unintegrated structure function was obtained in a form suitable for the study of exclusive observables:

$$f_\omega(\boldsymbol{k}) = f_\omega^0(\boldsymbol{k}) + \bar{\alpha}_S \int \frac{d^2\boldsymbol{q}}{\pi q^2} \int_0^1 \frac{dz}{z} z^\omega \Delta_R(z,k)\Theta(q-\mu) f_\omega(\boldsymbol{q}+\boldsymbol{k}), \tag{52}$$

where μ is a collinear cutoff, $\boldsymbol{q}$ is the transverse momentum of the emitted gluon, and the gluon Regge factor is

$$\Delta_R(z_i, k_i) = \exp\left[-\bar{\alpha}_S \ln\frac{1}{z_i}\ln\frac{k_i^2}{\mu^2}\right], \tag{53}$$

with $k_i \equiv |\boldsymbol{k}_i|$, and $\bar{\alpha}_S \equiv \alpha_S N_c/\pi$. Under iteration, this expression generates real gluon emissions with all the virtual corrections summed to all orders. Since f_ω is an inclusive structure function, it includes the sum over all final states and the μ-dependence cancels between the real and virtual contributions.

The structure function is defined by integrating over all $\mu^2 \leq q_i^2 \leq Q^2$, *i.e.*

$$F_{0\omega}(Q,\mu) \equiv \Theta(Q-\mu) + \sum_{r=1}^{\infty} \int_{\mu^2}^{Q^2} \prod_{i=1}^{r} \frac{d^2\boldsymbol{q}_i}{\pi q_i^2} dz_i \frac{\bar{\alpha}_S}{z_i} z_i^\omega \Delta_R(z_i,k_i), \tag{54}$$

with i real gluon emissions in each iteration of the kernel. The contributions from a fixed number r of emitted gluons is

$$F_{0\omega}(Q) = \int_0^1 dx\, x^\omega F_0(x,Q) = 1 + \sum_{r=1}^{\infty} F_{0\omega}^{(r)}(Q). \tag{55}$$

In Ref. [59] the perturbative expansion for the $F_{0\omega}^{(r)}(Q,\mu)$

$$F_{0\omega}^{(r)}(Q,\mu) = \sum_{n=r}^{\infty} C_0^{(r)}(n;T) \frac{\bar{\alpha}_S^n}{\omega^n}, \tag{56}$$

was obtained with $T \equiv \ln(Q/\mu)$. Then we have

$$F_{0\omega}(Q) \equiv \sum_{i=0}^{\infty} F_{0\omega}^{(i)}(Q) = \left(\frac{Q^2}{\mu^2}\right)^{\bar{\gamma}}, \tag{57}$$

where $\bar{\gamma}$ is the BFKL anomalous dimension. It was pointed out that coherence effects significantly modify the individual $F_{0\omega}^{(r)}(Q)$ whilst preserving the sum $F_{0\omega}(Q)$, and care must be taken to account properly for coherence in the calculation of associated distributions.

Modifying the BFKL formalism to account for coherence [58–61], $F_{0\omega}(Q,\mu)$ becomes

$$F_\omega(Q,\mu) = \Theta(Q-\mu) + \sum_{r=1}^{\infty} \int_0^{Q^2} \prod_{i=1}^{r} \frac{d^2\boldsymbol{q}_i}{\pi q_i^2} dz_i \frac{\bar{\alpha}_S}{z_i} z_i^\omega \Delta(z_i,q_i,k_i)\Theta(q_i - z_{i-1}q_{i-1}), \tag{58}$$

where $\Delta_R(z_i,k_i)$ is substituted by

$$\Delta(z_i,q_i,k_i) = \exp\left[-\bar{\alpha}_S \ln\frac{1}{z_i} \ln\frac{k_i^2}{z_i q_i^2}\right]; \quad k_i > q_i, \tag{59}$$

and for the first emission we take $q_0 z_0 = \mu$. The expansion of $F_\omega^{(r)}(Q)$ is now

$$F_\omega^{(r)}(Q) = \sum_{n=r}^{\infty} \sum_{m=1}^{n} C^{(r)}(n,m;T) \frac{\bar{\alpha}_S^n}{\omega^{2n-m}}. \tag{60}$$

A collinear cutoff is needed only on the emission of the first gluon because subsequent collinear emissions are regulated by the angular ordering constraint.

The rates for the emission of a fixed number of resolved gluons, with a transverse momentum larger than a resolution scale μ_R, together with any number of unresolved ones, were calculated in Ref. [62] in the LLA, to third order in $\bar{\alpha}_S$. μ_R is constrained by the collinear cutoff and the hard scale, $\mu \ll \mu_R \ll Q$. For the n-jet rate all the graphs with n resolved gluons and any number of unresolvable ones were considered. Expanding the Regge factors to $\mathcal{O}(\bar{\alpha}_S^3)$ we find that the jet rates both in the multi-Regge (BFKL) approach and in the coherent (CCFM) approach are the same:

$$\text{0 jet} = \frac{(2\bar{\alpha}_S)}{\omega}S + \frac{(2\bar{\alpha}_S)^2}{\omega^2}\left[\frac{S^2}{2}\right] + \frac{(2\bar{\alpha}_S)^3}{\omega^3}\left[\frac{S^3}{6}\right], \tag{61}$$

$$\text{1 jet} = \frac{(2\bar{\alpha}_S)}{\omega}T + \frac{(2\bar{\alpha}_S)^2}{\omega^2}\left[TS - \frac{1}{2}T^2\right] + \frac{(2\bar{\alpha}_S)^3}{\omega^3}\left[\frac{1}{3}T^3 - \frac{1}{2}T^2 S + \frac{1}{2}TS^2\right], \tag{62}$$

$$\text{2 jet} = \frac{(2\bar{\alpha}_S)^2}{\omega^2}\left[T^2\right] + \frac{(2\bar{\alpha}_S)^3}{\omega^3}\left[T^2 S - \frac{7}{6}T^3\right], \tag{63}$$

$$\text{3 jet} = \frac{(2\bar{\alpha}_S)^3}{\omega^3}\left[T^3\right], \tag{64}$$

with $T = \ln(Q/\mu_R)$ and $S = \ln(\mu_R/\mu)$. When coherence is introduced the singularities at $\omega \sim 0$ are stronger than in the BFKL approach but the extra logarithms cancel in the sum of all the graphs needed for the jet rates. The net effect is that the final results are the same as those obtained without coherence [62]. This is true to all orders in the coupling [63] since a generating function for the jet multiplicity distribution was obtained in [64]. Within the multi-Regge kinematics the r-jet rate reads

$$R_\omega^{(n\ \text{jet})}(Q,\mu_R) = \frac{F_\omega^{(n\ \text{jet})}(Q,\mu_R,\mu)}{F_\omega(Q,\mu)} = \frac{1}{n!}\frac{\partial^n}{\partial u^n}R_\omega(u,T)\bigg|_{u=0}, \tag{65}$$

where the jet-rate generating function R_ω is given by

$$R_\omega(u,T) = \exp\left(-\frac{2\bar{\alpha}_s}{\omega}T\right)\left[1 + (1-u)\frac{2\bar{\alpha}_s}{\omega}T\right]^{\frac{u}{1-u}}. \tag{66}$$

The same generating function is obtained when coherence is considered. The mean number of jets and the mean square fluctuation in this number are

$$\langle n\rangle = \frac{\partial}{\partial u}R_\omega(u,T)\bigg|_{u=1} = \frac{2\bar{\alpha}_s}{\omega}T + \frac{1}{2}\left(\frac{2\bar{\alpha}_s}{\omega}T\right)^2, \tag{67}$$

$$\langle n^2\rangle - \langle n\rangle^2 = \frac{2\bar{\alpha}_s}{\omega}T + \frac{3}{2}\left(\frac{2\bar{\alpha}_s}{\omega}T\right)^2 + \frac{2}{3}\left(\frac{2\bar{\alpha}_s}{\omega}T\right)^3. \tag{68}$$

In general, the pth central moment of the jet multiplicity distribution is a polynomial in $\bar{\alpha}_s T/\omega$ of degree $2p-1$, indicating that the distribution becomes relatively narrow in the limit of very small x and large Q/μ_R [64].

In Ref. [65, 66] the subject was developed even further and all subleading logarithms of Q^2/μ_R^2 were included to calculate the jet multiplicity in Higgs production at the LHC. In Ref. [65] they extended the results from a $[\bar{\alpha}_S \ln(1/x)\ln(Q^2/\mu_R^2)]^n$ resummation to a $[\bar{\alpha}_s \ln(1/x)]^n[\ln(Q^2/\mu_R^2)]^m$ one with $0 < m \leq n$, proving that the quadratic and cubic forms of the mean and the variance remain valid. It has also been shown that for any sufficiently inclusive observables the CCFM formalism leads to the same results as the BFKL equation [67]. The key idea to understand this result comes if we try to obtain the results of Ref. [59] from those in the previous section in the limit $\mu_R \to 0$. To get the right solution we should consider subleading terms $\bar{\alpha}_s \ln^2(Q/\mu_R)$ which must be resummed when taking the limit $\mu_R \to 0$ to obtain a continuous transition from the case where BFKL and CCFM results are equivalent, to that of them being different. If we also consider the effects of introducing the $z \to 1$ divergent part of the splitting function in the CCFM approach we will see that this leads to all BFKL and CCFM final-state properties being identical in the $[\bar{\alpha}_s \ln(1/x)\ln(Q^2/\mu_R^2)]^n$ approximation [67]. Recent reviews devoted to the implementation of CCFM in Monte Carlo event generators can be found in, *e.g.*, [68–71]. An approach which has the potential to apply BFKL in the NLLA to DIS phenomenology is that in Ref. [72–75]. In the NLLA approximation it is important to carefully take into account k_t factorization [76].

5.2. Improving the Collinear Region of Multi-Regge Kinematics

In this section we revisit the approach of Ref. [77] where the multi-Regge kinematics was extended to include collinear contributions present to all orders in the BFKL formalism. In Ref. [78] the structure in transverse momentum space of the double logarithms resummed was explicitly extracted. A new renormalization group (RG)-improved kernel was obtained which does not mix transverse with longitudinal momentum components.

In the $\overline{\text{MS}}$ renormalisation scheme, the BFKL kernel in the NLLA acting on a smooth function [15, 16] is

$$\int d^2\vec{q}_2\, \mathcal{K}(\vec{q}_1,\vec{q}_2)\, f\left(q_2^2\right) =$$

$$\int \frac{d^2\vec{q}_2}{\left|q_1^2-q_2^2\right|}\left\{\left[\bar{\alpha}_s + \bar{\alpha}_s^2\left(\mathcal{S} - \frac{\beta_0}{4N_c}\ln\left(\frac{\left|q_1^2-q_2^2\right|^2}{\max\left(q_1^2,q_2^2\right)\mu^2}\right)\right)\right]\right.$$
$$\left.\times\left(f\left(q_2^2\right) - 2\frac{\min\left(q_1^2,q_2^2\right)}{\left(q_1^2+q_2^2\right)} f\left(q_1^2\right)\right) - \frac{\bar{\alpha}_s^2}{4}\left(\mathcal{T}\left(q_1^2,q_2^2\right) + \ln^2\left(\frac{q_1^2}{q_2^2}\right)\right) f\left(q_2^2\right)\right\}, \quad (69)$$

where $\beta_0 = (11N_c - 2n_f)/3$, $\mathcal{S} = (4 - \pi^2 + 5\beta_0/N_c)/12$, and $\mathcal{T}(q_1^2, q_2^2)$ can be found in Ref. [15]. The collinear structure can be obtained acting on the eigenfunctions in the LLA, *i.e.*

$$\int d^2\vec{q}_2\, \mathcal{K}(\vec{q}_1,\vec{q}_2)\left(\frac{\bar{\alpha}_s\left(q_2^2\right)}{\bar{\alpha}_s\left(q_1^2\right)}\right)^{-\frac{1}{2}}\left(\frac{q_2^2}{q_1^2}\right)^{\gamma-1} = \bar{\alpha}_s\left(q_1^2\right)\chi_0(\gamma) + \bar{\alpha}_s^2\chi_1(\gamma). \quad (70)$$

Here we have

$$\chi_0(\gamma) = 2\psi(1) - \psi(\gamma) - \psi(1-\gamma), \quad (71)$$

$$\chi_1(\gamma) = \mathcal{S}\chi_0(\gamma) + \frac{1}{4}\left(\psi''(\gamma) + \psi''(1-\gamma)\right) - \frac{1}{4}\left(\phi(\gamma) + \phi(1-\gamma)\right) \quad (72)$$
$$- \frac{\pi^2\cos(\pi\gamma)}{4\sin^2(\pi\gamma)(1-2\gamma)}\left(3 + \left(1 + \frac{n_f}{N_c^3}\right)\frac{(2+3\gamma(1-\gamma))}{(3-2\gamma)(1+2\gamma)}\right) + \frac{3}{2}\zeta_3 - \frac{\beta_0}{8N_c}\chi_0^2(\gamma),$$

with $\psi(\gamma) = \Gamma'(\gamma)/\Gamma(\gamma)$ and

$$\phi(\gamma) + \phi(1-\gamma) = \sum_{m=0}^{\infty}\left(\frac{1}{\gamma+m} + \frac{1}{1-\gamma+m}\right)\left(\psi'\left(\frac{2+m}{2}\right) - \psi'\left(\frac{1+m}{2}\right)\right). \quad (73)$$

The pole structure around $\gamma = 0, 1$ is

$$\chi_0(\gamma) \simeq \frac{1}{\gamma} + \{\gamma \to 1-\gamma\}, \quad (74)$$

$$\chi_1(\gamma) \simeq \frac{\mathrm{a}}{\gamma} + \frac{\mathrm{b}}{\gamma^2} - \frac{1}{2\gamma^3} + \{\gamma \to 1-\gamma\} \quad (75)$$

with

$$\mathrm{a} = \frac{5}{12}\frac{\beta_0}{N_c} - \frac{13}{36}\frac{n_f}{N_c^3} - \frac{55}{36}, \quad \mathrm{b} = -\frac{1}{8}\frac{\beta_0}{N_c} - \frac{n_f}{6N_c^3} - \frac{11}{12}. \quad (76)$$

The cubic poles stem from ψ'' and compensate for the equivalent terms appearing when the Regge-like energy scale $s_0 = q_1 q_2$ is shifted to the DIS choice $s_0 = q_{1,2}^2$. Higher order terms beyond the NLLA, not compatible with RG evolution, are also generated by this change of scale. The NLLA truncation of the perturbative expansion is then the reason why the gluon Green's function develops oscillations, where the Green's function can have negative values, in the q_1^2/q_2^2 ratio.

It is possible to remove the most dominant poles in γ-space incompatible with RG evolution by simply shifting the ω-pole present in the BFKL scale invariant eigenfunction. Here we focus on the scheme proposed in Ref. [77]:

$$\omega = \bar{\alpha}_s\left(1 + \left(\mathrm{a} + \frac{\pi^2}{6}\right)\bar{\alpha}_s\right)\left(2\psi(1) - \psi\left(\gamma + \frac{\omega}{2} - \mathrm{b}\,\bar{\alpha}_s\right) - \psi\left(1-\gamma+\frac{\omega}{2} - \mathrm{b}\,\bar{\alpha}_s\right)\right)$$
$$+ \bar{\alpha}_s^2\left(\chi_1(\gamma) + \left(\frac{1}{2}\chi_0(\gamma) - \mathrm{b}\right)\left(\psi'(\gamma) + \psi'(1-\gamma)\right) - \left(\mathrm{a} + \frac{\pi^2}{6}\right)\chi_0(\gamma)\right). \quad (77)$$

We can approximately solve this equation considering the ω-shift in the form

$$\omega = \bar{\alpha}_s\left(1 + \mathrm{A}\,\bar{\alpha}_s\right)\left(2\psi(1) - \psi\left(\gamma + \frac{\omega}{2} + \mathrm{B}\,\bar{\alpha}_s\right) - \psi\left(1-\gamma+\frac{\omega}{2} + \mathrm{B}\,\bar{\alpha}_s\right)\right), \quad (78)$$

which can be written as

$$\omega = \bar{\alpha}_s\left(1 + \mathrm{A}\bar{\alpha}_s\right)\sum_{m=0}^{\infty}\left(\frac{1}{\gamma+m+\frac{\omega}{2}+\mathrm{B}\,\bar{\alpha}_s} + \frac{1}{1-\gamma+m+\frac{\omega}{2}+\mathrm{B}\,\bar{\alpha}_s} - \frac{2}{m+1}\right). \quad (79)$$

The solution to this shift can be obtained by adding all the approximated solutions at the different poles plus a term related to the virtual contributions, *i.e.*

$$\omega = \sum_{m=0}^{\infty}\left\{-(1+2m+2\,\mathrm{B}\,\bar{\alpha}_s) + |\gamma+m+\mathrm{B}\,\bar{\alpha}_s|\left(1 + \frac{2\bar{\alpha}_s\left(1+\mathrm{A}\bar{\alpha}_s\right)}{\left(\gamma+m+\mathrm{B}\,\bar{\alpha}_s\right)^2}\right)^{\frac{1}{2}}\right.$$
$$\left. + |1-\gamma+m+\mathrm{B}\,\bar{\alpha}_s|\left(1 + \frac{2\bar{\alpha}_s\left(1+\mathrm{A}\bar{\alpha}_s\right)}{\left(1-\gamma+m+\mathrm{B}\,\bar{\alpha}_s\right)^2}\right)^{\frac{1}{2}} - \frac{2\bar{\alpha}_s\left(1+\mathrm{A}\bar{\alpha}_s\right)}{m+1}\right\}. \quad (80)$$

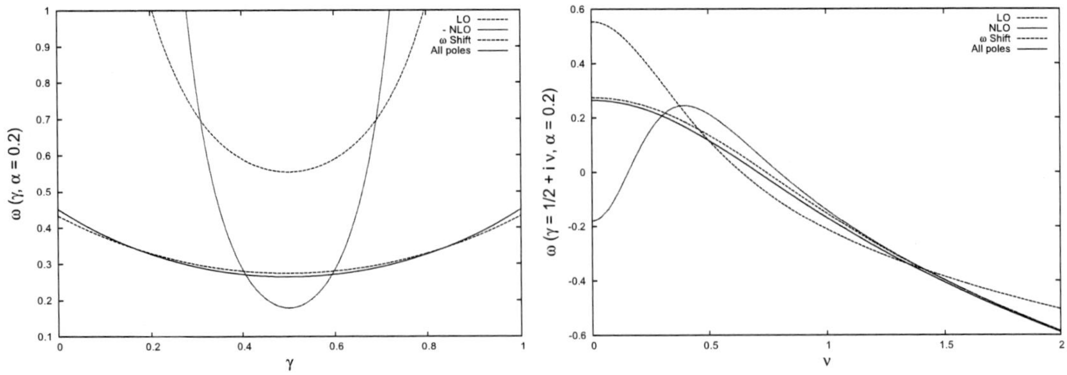

Figure 9. γ-representation of the LLA and NLLA kernels. The RG–improved kernel by a ω-shift is included together with the new "all-poles" approximation.

At the $\gamma = 0, 1$ poles this expansion generates the NLLA terms:

$$\omega \simeq \frac{\bar{\alpha}_s}{\gamma} + \bar{\alpha}_s^2 \left(\frac{\mathrm{A}}{\gamma} - \frac{\mathrm{B}}{\gamma^2} - \frac{1}{2\gamma^3} \right) + \{\gamma \to 1 - \gamma\} . \tag{81}$$

To match the original kernel at NLLA we set $\mathrm{A} = \mathrm{a}$ and $\mathrm{B} = -\mathrm{b}$ from Eq. (76).

We now include the full NLLA scale invariant kernel without double counting terms:

$$\begin{aligned} \omega &= \bar{\alpha}_s \chi_0(\gamma) + \bar{\alpha}_s^2 \chi_1(\gamma) \\ &+ \left\{ \sum_{m=0}^{\infty} \left[\left(\sum_{n=0}^{\infty} \frac{(-1)^n (2n)!}{2^n n! (n+1)!} \frac{\left(\bar{\alpha}_s + \mathrm{a}\, \bar{\alpha}_s^2\right)^{n+1}}{(\gamma + m - \mathrm{b}\, \bar{\alpha}_s)^{2n+1}} \right) \right. \right. \\ & \left. \left. - \frac{\bar{\alpha}_s}{\gamma + m} - \bar{\alpha}_s^2 \left(\frac{\mathrm{a}}{\gamma + m} + \frac{\mathrm{b}}{(\gamma+m)^2} - \frac{1}{2(\gamma+m)^3} \right) \right] + \{\gamma \to 1 - \gamma\} \right\} . \end{aligned} \tag{82}$$

This result reproduces the ω-shift very closely, see Fig. 9. The imaginary part of γ at the maximum of the NLLA scale invariant eigenvalue (middle plot of Fig. 9) is not zero and results in oscillations in the q_1^2/q_2^2 variable. These are eliminated when the RG–improved kernel is used, as it also happens for the "all–poles" kernel.

It is very important to note that in Eq. (82) the ω-space is decoupled from the γ-representation. In Ref. [78] an expression for the collinearly improved BFKL kernel which does not mix longitudinal with transverse degrees of freedom was found. The only modification needed in the full NLLA kernel to introduce the "all-poles" resummation is to remove the term

$$-\frac{\bar{\alpha}_s^2}{4} \frac{1}{(\vec{q} - \vec{k})^2} \ln^2 \left(\frac{q^2}{k^2} \right) \tag{83}$$

in the real emission kernel, $\mathcal{K}_r \left(\vec{q}, \vec{k} \right)$, and replace it with

$$\begin{aligned} \frac{1}{(\vec{q} - \vec{k})^2} & \left\{ \left(\frac{q^2}{k^2} \right)^{-\mathrm{b}\bar{\alpha}_s \frac{|k-q|}{k-q}} \sqrt{\frac{2\left(\bar{\alpha}_s + \mathrm{a}\, \bar{\alpha}_s^2\right)}{\ln^2\left(\frac{q^2}{k^2}\right)}} J_1 \left(\sqrt{2\left(\bar{\alpha}_s + \mathrm{a}\, \bar{\alpha}_s^2\right) \ln^2\left(\frac{q^2}{k^2}\right)} \right) \right. \\ & \left. - \bar{\alpha}_s - \mathrm{a}\, \bar{\alpha}_s^2 + \mathrm{b}\, \bar{\alpha}_s^2 \frac{|k-q|}{k-q} \ln \left(\frac{q^2}{k^2} \right) \right\} , \end{aligned} \tag{84}$$

with J_1 the Bessel function of the first kind. When the difference between the q^2 and k^2 scales is not very large then

$$J_1\left(\sqrt{2\bar{\alpha}_s \ln^2\left(\frac{q^2}{k^2}\right)}\right) \simeq \sqrt{\frac{\bar{\alpha}_s}{2}\ln^2\left(\frac{q^2}{k^2}\right)}, \quad (85)$$

and its influence is minimal, not affecting the "Regge–like" region. When the ratio of transverse momenta becomes larger then

$$J_1\left(\sqrt{2\bar{\alpha}_s \ln^2\left(\frac{q^2}{k^2}\right)}\right) \simeq \left(\frac{2}{\pi^2\bar{\alpha}_s \ln^2\left(\frac{q^2}{k^2}\right)}\right)^{\frac{1}{4}} \cos\left(\sqrt{2\bar{\alpha}_s \ln^2\left(\frac{q^2}{k^2}\right)} - \frac{3\pi}{4}\right) \quad (86)$$

compensating for the unphysical oscillations. This resummation of all-poles has been applied to extend the region of applicability of BFKL calculations in the NLLA in the case of electroproduction of light vector mesons in Ref. [79].

5.3. Conformal Signatures at the Large Hadron Collider: Azimuthal Angle

We now proceed to review the work of Ref. [80] where azimuthal angle decorrelations in inclusive dijet cross sections were studied analytically to include the NLLA to the BFKL kernel, while keeping the jet vertices at leading order. It was shown how the angular decorrelation for jets with a wide relative separation in rapidity largely decreases when higher order effects are considered.

Observables where BFKL effects should be dominant require a large enough center-of-mass energy, and two large and similar transverse scales. An example is the inclusive hadroproduction of two jets with large and similar transverse momenta and a large relative separation in rapidity, Y, the so-called Mueller-Navelet jets, first proposed in Ref. [81]. A rise with Y in the partonic cross section was predicted in agreement with the LLA hard Pomeron intercept. At the hadronic level, Mueller-Navelet jets are produced in a region where the parton distribution falls very quickly, reducing this rise. Small x resummation effects are very relevant if we investigate the azimuthal angle decorrelation of the pair of jets. BFKL enhances soft real emission as Y increases, reducing the angular correlation. This was first investigated in the LLA in Ref. [82–84]. The rate of decorrelation in the LLA lies quite below the experimental data [85–88] at the Tevatron. This motivates the NLLA discussion of this subsection.

We are interested in the cross section parton + parton $\longrightarrow$ jet + jet + soft emission, with the two jets having transverse momenta $\vec{q}_1$ and $\vec{q}_2$ and with a relative rapidity separation Y. The differential partonic cross section is

$$\frac{d\hat{\sigma}}{d^2\vec{q}_1 d^2\vec{q}_2} = \frac{\pi^2\bar{\alpha}_s^2}{2}\frac{f\left(\vec{q}_1,\vec{q}_2,\mathrm{Y}\right)}{q_1^2 q_2^2}. \quad (87)$$

It is useful to introduce a Mellin transform:

$$f\left(\vec{q}_1,\vec{q}_2,\mathrm{Y}\right) = \int \frac{d\omega}{2\pi i} e^{\omega \mathrm{Y}} f_\omega\left(\vec{q}_1,\vec{q}_2\right). \quad (88)$$

The solution to the BFKL equation in the LLA is

$$f_\omega\left(\vec{q}_1,\vec{q}_2\right) = \frac{1}{2\pi^2}\sum_{n=-\infty}^{\infty}\int_{-\infty}^{\infty} d\nu \left(q_1^2\right)^{-i\nu-\frac{1}{2}}\left(q_2^2\right)^{i\nu-\frac{1}{2}}\frac{e^{in(\theta_1-\theta_2)}}{\omega-\bar{\alpha}_s\chi_0\left(\left|n\right|,\nu\right)} \tag{89}$$

with

$$\chi_0\left(n,\nu\right) = 2\psi\left(1\right)-\psi\left(\frac{1}{2}+i\nu+\frac{n}{2}\right)-\psi\left(\frac{1}{2}-i\nu+\frac{n}{2}\right). \tag{90}$$

The nonforward BFKL equation corresponds to a Schrödinger-like equation with a holomorphically separable Hamiltonian where $-i\,\mathrm{Y}$ is the time variable. Both the holomorphic and antiholomorphic sectors are invariant under spin zero Möbius transformations with eigenfunctions carrying a conformal weight of the form $\gamma=\frac{1}{2}+i\nu+\frac{n}{2}$. In the principal series of the unitary representation, ν is real and $|n|$ the integer conformal spin [89]. In this way extracting information about n is equivalent to proving the conformal structure of high energy QCD.

We now integrate over the phase space of the two emitted gluons together with some general jet vertices, *i.e.*

$$\hat{\sigma}\left(\alpha_s,\mathrm{Y},p_{1,2}^2\right) = \int d^2\vec{q}_1\int d^2\vec{q}_2\,\Phi_{\mathrm{jet}_1}\left(\vec{q}_1,p_1^2\right)\Phi_{\mathrm{jet}_2}\left(\vec{q}_2,p_2^2\right)\frac{d\hat{\sigma}}{d^2\vec{q}_1 d^2\vec{q}_2}. \tag{91}$$

In the jet vertices only leading-order terms are kept:

$$\Phi^{(0)}_{\mathrm{jet}_i}\left(\vec{q},p_i^2\right) = \theta\left(q^2-p_i^2\right), \tag{92}$$

where p_i^2 corresponds to a resolution scale for the transverse momentum of the gluon jet. To extend this analysis it is needed to use the NLO jet vertices in Ref. [90, 91] where the definition of a jet is much more complex than Eq. (92). We then have

$$\hat{\sigma}\left(\alpha_s,\mathrm{Y},p_{1,2}^2\right) = \frac{\pi^2\bar{\alpha}_s^2}{2}\int d^2\vec{q}_1\int d^2\vec{q}_2\,\frac{\Phi^{(0)}_{\mathrm{jet}_1}\left(\vec{q}_1,p_1^2\right)}{q_1^2}\frac{\Phi^{(0)}_{\mathrm{jet}_2}\left(\vec{q}_2,p_2^2\right)}{q_2^2}f\left(\vec{q}_1,\vec{q}_2,\mathrm{Y}\right). \tag{93}$$

In a transverse momenta operator representation:

$$\langle\vec{q}|\,\nu,n\rangle = \frac{1}{\pi\sqrt{2}}\left(q^2\right)^{i\nu-\frac{1}{2}}e^{in\theta}, \tag{94}$$

the action of the NLO kernel, calculated in Ref. [92], is

$$\begin{aligned}\hat{K}\left|\nu,n\right\rangle &= \Bigg\{\bar{\alpha}_s\,\chi_0\left(\left|n\right|,\nu\right)+\bar{\alpha}_s^2\,\chi_1\left(\left|n\right|,\nu\right)\\ &+\bar{\alpha}_s^2\,\frac{\beta_0}{8N_c}\left[2\,\chi_0\left(\left|n\right|,\nu\right)\left(i\frac{\partial}{\partial\nu}+\log\mu^2\right)+\left(i\frac{\partial}{\partial\nu}\chi_0\left(\left|n\right|,\nu\right)\right)\right]\Bigg\}\left|\nu,n\right\rangle,\end{aligned} \tag{95}$$

where χ_1, for a general conformal spin, reads

$$\chi_1\left(n,\gamma\right) = \mathcal{S}\chi_0\left(n,\gamma\right)+\frac{3}{2}\zeta\left(3\right)-\frac{\beta_0}{8N_c}\chi_0^2\left(n,\gamma\right)$$

$$+ \frac{1}{4}\left[\psi''\left(\gamma+\frac{n}{2}\right)+\psi''\left(1-\gamma+\frac{n}{2}\right)-2\,\phi\left(n,\gamma\right)-2\,\phi\left(n,1-\gamma\right)\right]$$
$$- \frac{\pi^2\cos\left(\pi\gamma\right)}{4\sin^2\left(\pi\gamma\right)\left(1-2\gamma\right)}\left\{\left[3+\left(1+\frac{n_f}{N_c^3}\right)\frac{2+3\gamma\left(1-\gamma\right)}{\left(3-2\gamma\right)\left(1+2\gamma\right)}\right]\delta_{n0}\right.$$
$$\left. -\left(1+\frac{n_f}{N_c^3}\right)\frac{\gamma\left(1-\gamma\right)}{2\left(3-2\gamma\right)\left(1+2\gamma\right)}\delta_{n2}\right\}. \tag{96}$$

The function ϕ can be found in Ref. [92].

The jet vertices on the basis in Eq. (94) are:

$$\int d^2\vec{q}\,\frac{\Phi^{(0)}_{\text{jet}_1}\left(\vec{q},p_1^2\right)}{q^2}\left\langle\vec{q}\,|\nu,n\right\rangle = \frac{1}{\sqrt{2}}\frac{1}{\left(\frac{1}{2}-i\nu\right)}\left(p_1^2\right)^{i\nu-\frac{1}{2}}\delta_{n,0} \equiv c_1\left(\nu\right)\delta_{n,0}, \tag{97}$$

with the $c_2\left(\nu\right)$ projection of $\Phi^{(0)}_{\text{jet}_2}$ on $\left\langle n,\nu\right|\vec{q}\rangle$ being the complex conjugate of (97) with p_1^2 being replaced by p_2^2. The cross section can then be rewritten as

$$\hat{\sigma}\left(\alpha_s,\mathrm{Y},p_{1,2}^2\right) = \frac{\pi^2\bar{\alpha}_s^2}{2}\sum_{n=-\infty}^{\infty}\int_{-\infty}^{\infty}d\nu\, e^{\bar{\alpha}_s\chi_0(|n|,\nu)\mathrm{Y}}c_1\left(\nu\right)c_2\left(\nu\right)\delta_{n,0} \tag{98}$$
$$\times\left\{1+\bar{\alpha}_s^2\,\mathrm{Y}\left[\chi_1\left(|n|,\nu\right)+\frac{\beta_0}{4N_c}\left(\log\left(\mu^2\right)+\frac{i}{2}\frac{\partial}{\partial\nu}\log\left(\frac{c_1\left(\nu\right)}{c_2\left(\nu\right)}\right)+\frac{i}{2}\frac{\partial}{\partial\nu}\right)\chi_0\left(|n|,\nu\right)\right]\right\}.$$

For the LO jet vertices the logarithmic derivative in Eq. (98) is

$$-i\frac{\partial}{\partial\nu}\log\left(\frac{c_1\left(\nu\right)}{c_2\left(\nu\right)}\right) = \log\left(p_1^2p_2^2\right)+\frac{1}{\frac{1}{4}+\nu^2}. \tag{99}$$

If $\phi=\theta_1-\theta_2-\pi$, in the case of two equal resolution momenta, $p_1^2=p_2^2\equiv p^2$, the angular differential cross section can be expressed as

$$\frac{d\hat{\sigma}\left(\alpha_s,\mathrm{Y},p^2\right)}{d\phi} = \frac{\pi^3\bar{\alpha}_s^2}{2p^2}\frac{1}{2\pi}\sum_{n=-\infty}^{\infty}e^{in\phi}\mathcal{C}_n\left(\mathrm{Y}\right), \tag{100}$$

with

$$\mathcal{C}_n\left(\mathrm{Y}\right) = \int_{-\infty}^{\infty}\frac{d\nu}{2\pi}\frac{e^{\bar{\alpha}_s\left(p^2\right)\mathrm{Y}\left(\chi_0(|n|,\nu)+\bar{\alpha}_s\left(p^2\right)\left(\chi_1(|n|,\nu)-\frac{\beta_0}{8N_c}\frac{\chi_0(|n|,\nu)}{\left(\frac{1}{4}+\nu^2\right)}\right)\right)}}{\left(\frac{1}{4}+\nu^2\right)}. \tag{101}$$

The coefficient governing the energy dependence of the cross section corresponds to $n=0$:

$$\hat{\sigma}\left(\alpha_s,\mathrm{Y},p^2\right) = \frac{\pi^3\bar{\alpha}_s^2}{2p^2}\mathcal{C}_0\left(\mathrm{Y}\right). \tag{102}$$

We have chosen the resolution scale $p=30\,\mathrm{GeV}$, $n_f=4$ and $\Lambda_{\mathrm{QCD}}=0.1416$ GeV. The $n=0$ coefficient is directly related to the normalized cross section

$$\frac{\hat{\sigma}\left(\mathrm{Y}\right)}{\hat{\sigma}\left(0\right)} = \frac{\mathcal{C}_0\left(\mathrm{Y}\right)}{\mathcal{C}_0\left(0\right)}. \tag{103}$$

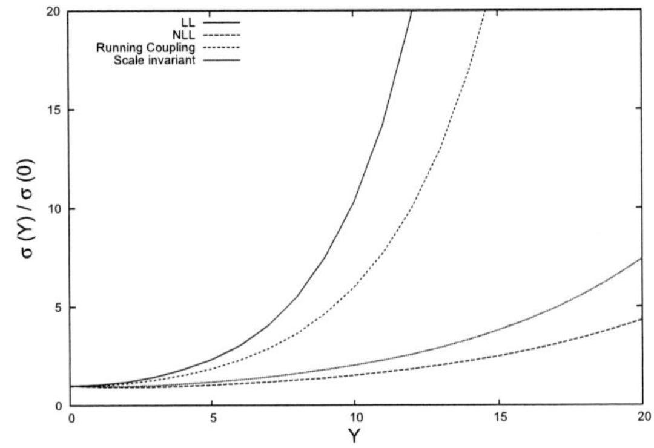

Figure 10. Evolution of the partonic cross section with the rapidity separation of the dijets.

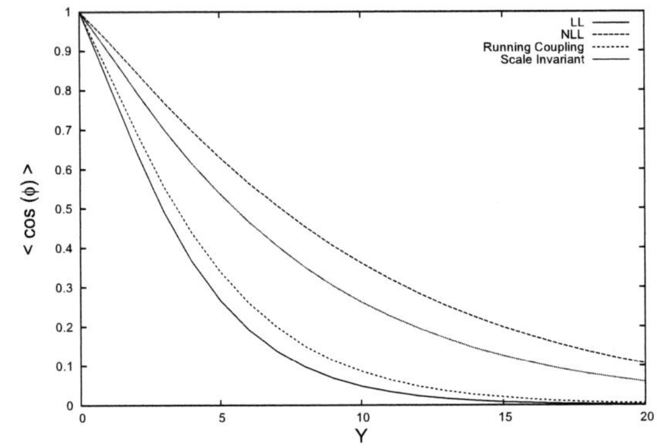

Figure 11. Dijet azimuthal angle decorrelation as a function of their separation in rapidity.

The rise with Y of this observable is shown in Fig. 10. Clearly the NLL intercept is very much reduced with respect to the LL case. The remaining coefficients with $n \geq 1$ all decrease with Y. Because of this, the angular correlations also diminish as the rapidity interval between the jets gets larger. This point can be studied in detail using the mean values

$$\langle \cos(m\phi) \rangle = \frac{\mathcal{C}_m(\mathrm{Y})}{\mathcal{C}_0(\mathrm{Y})}. \tag{104}$$

$\langle \cos(\phi) \rangle$ is calculated in Fig. 11. The NLL effects decrease the azimuthal angle decorrelation. This is the case for the running of the coupling and also for the scale invariant terms. This is encouraging from the phenomenological point of view given that the data at the Tevatron typically have lower decorrelation than predicted by LLA BFKL or LLA with running coupling. The difference in the decorrelation between LLA and NLLA is driven by the $n = 0$ conformal spin since the ratio

$$\frac{\langle \cos(\phi) \rangle^{\mathrm{NLLA}}}{\langle \cos(\phi) \rangle^{\mathrm{LLA}}} = \frac{\mathcal{C}_1^{\mathrm{NLLA}}(\mathrm{Y})\, \mathcal{C}_0^{\mathrm{LLA}}(\mathrm{Y})}{\mathcal{C}_0^{\mathrm{NLLA}}(\mathrm{Y})\, \mathcal{C}_1^{\mathrm{LLA}}(\mathrm{Y})}, \tag{105}$$

remains in the region

$$1.2 > \frac{\mathcal{C}_1^{\mathrm{NLLA}}(\mathrm{Y})}{\mathcal{C}_1^{\mathrm{LLA}}(\mathrm{Y})} > 1. \tag{106}$$

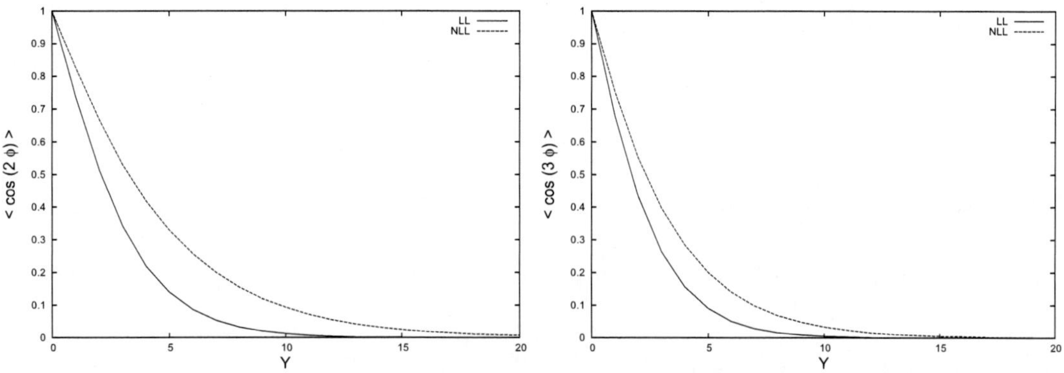

Figure 12. Dijet azimuthal angle decorrelation as a function of their separation in rapidity.

This is a consequence of the good convergence, in terms of asymptotic intercepts of the NLLA BFKL calculation, for conformal spins larger than zero. For completeness the $m = 2, 3$ cases for $\langle \cos(m\phi) \rangle$ are shown in Fig. 12. These distributions test the structure of the higher conformal spins. The methods of this subsection have been applied to phenomenology of dijets at the Tevatron and the LHC in [93, 94], and to the production of forward jets in DIS at HERA in [95].

6. Conclusion

The precision of perturbative QCD calculations will play a major role in the confidence of new physics discoveries, both at this generation of experiments, Tevatron and LHC, and in future experiments. The most available avenue of improving the precision of QCD is through resummation of large contributions. We have presented results for the resummation of large-x contributions and separately small-x contributions. In both cases, the large contributions arise from incomplete cancellations of virtual and real terms, and can be computed in the eikonal approximation.

We have shown that the inclusion of soft-gluon corrections to top quark production cross sections is essential to stabilize the unphysical scale variations in the order-by-order calculations. This is necessary for any sort of precision calculation of the top mass and production channels. Additionally, we have shown the importance of resummation on W production at large transverse momentum, and on Higgs production. Discovery of the Higgs boson is the last remaining test of the Standard Model and precision measurements of its properties is essential to proceed forward with beyond the Standard Model theories.

We have also presented a framework to include collinear effects into the BFKL formalism. This stabilizes the oscillatory behavior that arises when one moves away from the strict kinematic regime of validity. It was shown how this inclusion improves the prediction of Mueller-Navalet jets, jets with a large rapidity separation but similar transverse scales. This is a process which will be observed at the LHC where the BFKL formalism should flourish; an important test of the complex behavior of QCD. A comparison between the predictions steming from a pure BFKL analysis and one including QCD coherence in multijet final states in DIS has been also discussed in detail.

Acknowledgements

The work of N.K. was supported by the National Science Foundation under Grant No. PHY 0555372.

References

[1] J.C. Collins, D.E. Soper, and G. Sterman, in *Perturbative Quantum Chromodynamics*, ed. A.H. Mueller (World Scientific, Singapore, 1989), p. 1.

[2] G. Sterman, *Nucl. Phys. B* **281**, 310 (1987).

[3] S. Catani and L. Trentadue, *Nucl. Phys. B* **327**, 323 (1989).

[4] N. Kidonakis and G. Sterman, *Phys. Lett. B* **387**, 867 (1996); *Nucl. Phys. B* **505**, 321 (1997) [hep-ph/9705234].

[5] H. Contopanagos, E. Laenen, and G. Sterman, *Nucl. Phys. B* **484**, 303 (1997) [hep-ph/9604313].

[6] N. Kidonakis, G. Oderda, and G. Sterman, *Nucl. Phys. B* **525**, 299 (1998) [hep-ph/9801268].

[7] N. Kidonakis, G. Oderda, and G. Sterman, *Nucl. Phys. B* **531**, 365 (1998) [hep-ph/9803241].

[8] N. Kidonakis, *Int. J. Mod. Phys. A* **15**, 1245 (2000) [hep-ph/9902484].

[9] E. Laenen, G. Oderda, and G. Sterman, *Phys. Lett. B* **438**, 173 (1998) [hep-ph/9806467].

[10] L.N. Lipatov, *Sov. J. Nucl. Phys.* **23**, 338 (1976).

[11] V.S. Fadin, E.A. Kuraev, and L.N. Lipatov, *Phys. Lett. B* **60**, 50 (1975).

[12] E.A. Kuraev, L.N. Lipatov, and V.S. Fadin, *Sov. Phys. JETP* **44**, 443 (1976).

[13] E.A. Kuraev, L.N. Lipatov, V.S. Fadin, *Sov. Phys. JETP* **45**, 199 (1977).

[14] I.I. Balitsky and L.N. Lipatov, *Sov. J. Nucl. Phys.* **28**, 822 (1978).

[15] V.S. Fadin and L.N. Lipatov, *Phys. Lett. B* **429**, 127 (1998) [hep-ph/9802290].

[16] M. Ciafaloni and G. Camici, *Phys. Lett. B* **430**, 349 (1998) [hep-ph/9803389].

[17] G.P. Korchemsky and A.V. Radyushkin, *Phys. Lett. B* **171**, 459 (1986); *Nucl. Phys. B* **283**, 342 (1987).

[18] J. Botts and G. Sterman, *Nucl. Phys. B* **325**, 62 (1989).

[19] G.P. Korchemsky, *Phys. Lett. B* **325**, 459 (1994) [hep-ph/9311294].

[20] I.A. Korchemskaya and G.P. Korchemsky, *Nucl. Phys. B* **437**, 127 (1995) [hep-ph/9409446].

[21] S.M. Aybat, L.J. Dixon, and G. Sterman, *Phys. Rev. Lett.* **97**, 072001 (2006) [hep-ph/0606254]; *Phys. Rev. D***74**, 074004 (2006) [hep-ph/0607309].

[22] N. Kidonakis, hep-ph/0208056; in *Deep Inelastic Scattering, DIS 2003* (PNPI, Gatchina, 2003), p. 541 [hep-ph/0307145].

[23] N. Kidonakis and P. Stephens, in preparation.

[24] N. Kidonakis, *Int. J. Mod. Phys. A* **19**, 1793 (2004) [hep-ph/0303186]; *Mod. Phys. Lett. A* **19**, 405 (2004) [hep-ph/0401147].

[25] N. Kidonakis, *Phys. Rev. D* **73**, 034001 (2006) [hep-ph/0509079].

[26] J. Kodaira and L. Trentadue, *Phys. Lett.* **112B**, 66 (1982).

[27] N. Kidonakis, *Phys. Rev. D* (in press), arXiv:0711.0142 [hep-ph].

[28] N. Kidonakis, *Phys. Rev. D* **64**, 014009 (2001) [hep-ph/0010002].

[29] V. Ravindran, *Nucl. Phys. B* **752**, 173 (2006) [hep-ph/0603041].

[30] CDF Collaboration, *Phys. Rev. Lett.* **74**, 2626 (1995) [hep-ex/9503002].

[31] D0 Collaboration, *Phys. Rev. Lett.* **74**, 2632 (1995) [hep-ex/9503003].

[32] W. Wagner, *Rept. Prog. Phys.* **68**, 2409 (2005) [hep-ph/0507207].

[33] R. Kehoe, M. Narain, and A. Kumar, *Int. J. Mod. Phys. A* **23**, 353 (2008), arXiv:0712.2733 [hep-ex].

[34] N. Kidonakis and R. Vogt, *Phys. Rev. D* **68**, 114014 (2003) [hep-ph/0308222].

[35] N. Kidonakis, E. Laenen, S. Moch, and R. Vogt, *Phys. Rev. D* **64**, 114001 (2001) [hep-ph/0105041].

[36] A.D. Martin, R.G. Roberts, W.J. Stirling, and R.S. Thorne, *Eur. Phys. J. C* **28**, 455 (2003) [hep-ph/0211080].

[37] CDF Collaboration, *Phys. Rev. Lett.* **96**, 202002 (2006) [hep-ex/0603043]; *Phys. Rev. D* **74**, 072006 (2006) [hep-ex/0607035]; *Phys. Rev. D* **74**, 072005 (2006) [hep-ex/0607095]; *Phys. Rev. D* **76**, 072009 (2007), arXiv:0706.3790 [hep-ex].

[38] D0 Collaboration, *Phys. Rev. D* **74**, 112004 (2006) [hep-ex/0611002]; *Phys. Rev. D* **76**, 072007 (2007) [hep-ex/0612040]; *Phys. Rev. D* **76**, 092007 (2007), arXiv:0705.2788 [hep-ex]; *Phys. Rev. D* **76**, 052006 (2007), arXiv:0706.0458 [hep-ex].

[39] D0 Collaboration, *Phys. Rev. Lett.* **98**, 181802 (2007) [hep-ex/0612052].

[40] CDF Collaboration, *Conf. Note* 8964; *Conf. Note* 8968.

[41] N. Kidonakis, *Phys. Rev. D* **74**, 114012 (2006) [hep-ph/0609287].

[42] N. Kidonakis, *Phys. Rev. D* **75**, 071501 (2007) [hep-ph/0701080].

[43] N. Kidonakis, arXiv:0802.3381 [hep-ph].

[44] A.D. Martin, R.G. Roberts, W.J. Stirling, and R.S. Thorne, *Phys. Lett. B* **604**, 61 (2004) [hep-ph/0410230].

[45] B.W. Harris, E. Laenen, L. Phaf, Z. Sullivan, and S. Weinzierl, *Phys. Rev. D* **66**, 054024 (2002) [hep-ph/0207055].

[46] S.H. Zhu, *Phys. Lett. B* **524**, 283 (2002) [hep-ph/0109269]; (E) *B* **537**, 351 (2002).

[47] P.B. Arnold and M.H. Reno, *Nucl. Phys. B* **319**, 37 (1989); (E) *B* **330**, 284 (1990).

[48] R.J. Gonsalves, J. Pawlowski, and C.-F. Wai, *Phys. Rev. D* **40**, 2245 (1989); *Phys. Lett. B* **252**, 663 (1990).

[49] R.J. Gonsalves, N. Kidonakis, and A. Sabio Vera, *Phys. Rev. Lett.* **95**, 222001 (2005) [hep-ph/0507317].

[50] N. Kidonakis and A. Sabio Vera, *JHEP* **02**, 027 (2004) [hep-ph/0311266].

[51] Particle Data Group, S. Eidelman *et al.*, *Phys. Lett. B* **592**, 1 (2004).

[52] P.W. Higgs, *Phys. Rev. Lett* **12**, 132 (1964); *Phys. Rev. Lett.* **13**, 508 (1964); *Phys. Rev.* **145**, 1156 (1966); F. Englert and R. Brout, *Phys. Rev. Lett.* **13**, 321 (1964); G.S. Guralnik, C.R. Hagen, and T.W.B. Kibble, *Phys. Rev. Lett.* **13**, 585 (1964).

[53] The Higgs Working Group: Summary Report, in *Les Houches 2003, Physics at TeV colliders*, p. 1 [hep-ph/0406152], and references therein.

[54] R.V. Harlander and W.B. Kilgore, *Phys. Rev. D* **68**, 013001 (2003) [hep-ph/0304035].

[55] M. Kramer, E. Laenen, and M. Spira, *Nucl. Phys. B* **511**, 523 (1998) [hep-ph/9611272].

[56] N. Kidonakis, in *DPF 2004*, *Int. J. Mod. Phys. A* **20**, 3726 (2005) [hep-ph/0410116].

[57] A.D. Martin, W.J. Stirling, R.S. Thorne, and G. Watt, *Phys. Lett. B* **652**, 292 (2007) [arXiv:0706.0459].

[58] S. Catani, F. Fiorani, and G. Marchesini, *Phys. Lett. B* **234**, 339 (1990).

[59] G. Marchesini, *Nucl. Phys. B* **445**, 49 (1995) [hep-ph/9412327].

[60] M. Ciafaloni, *Nucl. Phys. B* **296**, 49 (1988).

[61] S. Catani, F. Fiorani, G. Marchesini, and G. Oriani, *Nucl. Phys. B* **361**, 645 (1991).

[62] J.R. Forshaw and A. Sabio Vera, *Phys. Lett. B* **440**, 141 (1998) [hep-ph/9806394].

[63] J.R. Forshaw, A. Sabio Vera, and B.R. Webber, *J. Phys. G* **25**, 1511 (1999) [hep-ph/9812318].

[64] B.R. Webber, *Phys. Lett. B* **444**, 81 (1998) [hep-ph/9810286].

[65] C. Ewerz and B.R. Webber, *JHEP* **04**, 022 (1999) [hep-ph/9904244].

[66] C. Ewerz and B.R. Webber, *JHEP* **08**, 019 (1999) [hep-ph/9907430].

[67] G.P. Salam, *JHEP* **03**, 009 (1999) [hep-ph/9902324].

[68] M. Dittmar *et al.*, arXiv:hep-ph/0511119.

[69] S. Alekhin *et al.*, arXiv:hep-ph/0601012.

[70] S. Alekhin *et al.*, arXiv:hep-ph/0601013.

[71] J.R. Andersen *et al.* [Small x Collaboration], *Eur. Phys. J. C* **48**, 53 (2006) [hep-ph/0604189].

[72] J.R. Andersen and A. Sabio Vera, *Phys. Lett. B* **567**, 116 (2003) [hep-ph/0305236].

[73] J.R. Andersen and A. Sabio Vera, *Nucl. Phys. B* **679**, 345 (2004) [hep-ph/0309331].

[74] J.R. Andersen and A. Sabio Vera, *Nucl. Phys. B* **699**, 90 (2004) [hep-th/0406009].

[75] J.R. Andersen and A. Sabio Vera, *JHEP* **01**, 045 (2005) [hep-ph/0411231].

[76] J. Bartels, A. Sabio Vera, and F. Schwennsen, *JHEP* **11**, 051 (2006) [hep-ph/0608154].

[77] G.P. Salam, *JHEP* **07**, 019 (1998) [hep-ph/9806482].

[78] A. Sabio Vera, *Nucl. Phys. B* **722**, 65 (2005) [hep-ph/0505128].

[79] F. Caporale, A. Papa, and A. Sabio Vera, *Eur. Phys. J. C* **53**, 525 (2008), arXiv:0707.4100 [hep-ph].

[80] A. Sabio Vera, *Nucl. Phys. B* **746**, 1 (2006) [hep-ph/0602250].

[81] A.H. Mueller and H. Navelet, *Nucl. Phys. B* **282**, 727 (1987).

[82] V. Del Duca and C.R. Schmidt, *Phys. Rev. D* **49**, 4510 (1994) [hep-ph/9311290].

[83] V. Del Duca and C.R. Schmidt, *Phys. Rev. D* **51**, 2150 (1995) [hep-ph/9407359].

[84] W.J. Stirling, *Nucl. Phys. B* **423**, 56 (1994) [hep-ph/9401266].

[85] S. Abachi *et al.* [D0 Collaboration], *Phys. Rev. Lett.* **77**, 595 (1996) [hep-ex/9603010].

[86] B. Abbott *et al.* [D0 Collaboration], *Phys. Rev. Lett.* **84**, 5722 (2000) [hep-ex/9912032].

[87] B. Abbott *et al.* [D0 Collaboration], *Phys. Rev. Lett.* **80**, 666 (1998) [hep-ex/9707016].

[88] F. Abe *et al.* [CDF Collaboration], *Phys. Rev. Lett.* **77**, 5336 (1996) [hep-ex/9609011]; (E) **78**, 4307 (1997).

[89] L.N. Lipatov, *Sov. Phys. JETP* **63**, 904 (1986) [*Zh. Eksp. Teor. Fiz.* **90**, 1536 (1986)].

[90] J. Bartels, D. Colferai and G.P. Vacca, *Eur. Phys. J. C* **24**, 83 (2002) [hep-ph/0112283].

[91] J. Bartels, D. Colferai and G.P. Vacca, *Eur. Phys. J. C* **29**, 235 (2003) [hep-ph/0206290].

[92] A.V. Kotikov and L.N. Lipatov, *Nucl. Phys. B* **582**, 19 (2000) [hep-ph/0004008].

[93] A. Sabio Vera and F. Schwennsen, *Nucl. Phys. B* **776**, 170 (2007) [hep-ph/0702158].

[94] C. Marquet and C. Royon, arXiv:0704.3409 [hep-ph].

[95] A. Sabio Vera and F. Schwennsen, *Phys. Rev. D* **77**, 014001 (2008), arXiv:0708.0549 [hep-ph].

In: The Physics of Quarks: New Research ISBN 978-1-60456-802-8
Editors: N.L. Watson and T.M. Grant , pp. 197-225 © 2009 Nova Science Publishers, Inc.

Chapter 9

SOLITONS AS BARYONS AND QUALITONS AS CONSTITUENT QUARKS IN TWO-DIMENSIONAL QCD

H. Blas [a] ***and H.L. Carrion*** [b]
[a] Instituto de Física
Universidade Federal de Mato Grosso
Av. Fernando Correa, s/n, Coxipó
78060-900, Cuiabá - MT - Brazil
[b] Instituto de Física, Universidade de São Paulo,
Caixa Postal 66318, 05315-970, São Paulo, SP, Brazil.

Abstract

We study the soliton type solutions arising in two-dimensional quantum chromodynamics (QCD_2). In bosonized QCD_2 these type of solutions emerge as describing baryons and quark solitons (excitations with "colored" states), respectively. The so-called generalized sine-Gordon model (GSG) arises as the low-energy effective action of bosonized QCD_2 for unequal quark mass parameters, and it has been shown that the relevant solitons describe the normal and exotic baryonic spectrum of QCD_2 [JHEP(03)(2007)(055)]. In the first part of this chapter we classify the soliton and kink type solutions of the sl(3) GSG model with three real fields, which corresponds to QCD_2 with three flavors. Related to the GSG model we consider the sl(3) affine Toda model coupled to matter fields (Dirac spinors) (ATM). The strong coupling sector is described by the $sl(3)$ GSG model which completely decouples from the Dirac spinors. In the spinor sector we are left with Dirac fields coupled to GSG fields. Based on the equivalence between the U(1) vector and topological currents, which holds in the ATM model, it has been shown the confinement of the spinors inside the solitons and kinks of the GSG model providing an extended hadron model for "quark" confinement [JHEP(01)(2007)(027)]. Moreover, it has been proposed that the constituent quark in QCD is a topological soliton. These qualitons (quark solitons), topological excitations with the quantum numbers of quarks, may provide an accurate description of what is meant by constituent quarks in QCD. In the second part of this chapter we discuss the appearance of these type of quark solitons in the context of bosonized QCD_2 (with $N_f = 1$ and N_c colors) and the relevance of the $sl(2)$ ATM model in order to describe the confinement of the color degrees of freedom. We have shown that QCD_2 has quark soliton solutions if the quark mass is sufficiently large.

1. Introduction

A useful theoretical laboratory for studying several problems in Quantum Chromodynamics is QCD in two dimensions [1, 2]. This theory can be written in bosonized form [3] for arbitrary numbers of colors N_c and flavors N_f [4]. It reflects accurately the phenomena of quark confinement and condensation in the vacuum that we expect to occur in QCD in four dimensions. In the low-energy and strong coupling limit ($e_c >> m_q$, e_c=coupling constant, m_q=quark mass) QCD_2 has finite-energy soliton solutions for arbitrary values of N_c and N_f that can be interpreted as baryons [2], in close analogy with the skyrmion interpretation of baryons as solitons in QCD_4 [5]. In this limit the static classical soliton which describes a baryon in QCD_2 turns out to be the ordinary sine-Gordon (SG) soliton. It has been shown that various aspects of the low-energy effective QCD_2 action with unequal quark masses can be described by the so-called (generalized) sine-Gordon model (GSG) [6].

Moreover, it has been proposed that the constituent quark in QCD_4 is a topological soliton [7]. These qualitons (quark solitons), topological excitations with the quantum numbers of quarks, may provide an accurate description of what is meant by constituent quarks in QCD. Related to this phenomenon, it has been found certain static soliton solutions to QCD_2 that have the quantum numbers of quarks [8]. They exist only for quarks heavier than the dimensional gauge coupling ($e_c << m_q$), and have infinite energy, corresponding to the presence of a string carrying the non-singlet color flux off to spatial infinity.

On the other hand, the sine-Gordon model (SG) has been studied over the decades due to its many properties and mathematical structures such as integrability and soliton solutions. It can be used as a toy model for non-perturbative quantum field theory phenomena. In this context, some extensions and modifications of the SG model deserve attention. An extension taking multi-frequency terms as the potential has been investigated in connection to various physical applications [9, 10, 11, 12]. Another extension defined for multi-fields is the so-called generalized sine-Gordon model (GSG) which has been found in the study of the strong/weak coupling sectors of the so-called $sl(N,\mathbb{C})$ affine Toda model coupled to matter fields (ATM) [14, 15]. In connection to these developments, the bosonization process of the multi-flavor massive Thirring model (GMT) provides the quantum version of the (GSG) model [16]. The GSG model provides a framework to obtain (multi-)soliton solutions for unequal mass parameters of the fermions in the GMT sector and study the spectrum and their interactions. The extension of this picture to the NC space-time has been addressed (see [17] and references therein).

It has been conjectured that the low-energy action of QCD_2 ($e >> m_q$, m_q quark mass and e gauge coupling) might be related to massive two dimensional integrable models, thus leading to the exact solution of the strong coupled QCD_2 [2]. In particular, it has been shown that the $sl(2)$ ATM model describes the low-energy spectrum of QCD_2 (1 flavor and N_c colors) and the exact computation of the string tension was performed [18]. A key role has been played by the equivalence between the Noether and topological currents at the quantum level. Moreover, one notice that the $SU(n)$ ATM theory [14, 15] is a $2D$ analogue of the chiral quark soliton model proposed to describe solitons in QCD_4 [19], provided that the pseudo-scalars lie in the Abelian subalgebra and certain kinetic terms are supplied for them.

Besides, coupled systems of scalar fields have been investigated by many authors [20,

21, 22, 23, 24, 25]. One of the motivations was the study of topological defects in relativistic field theories; since realistic theories involve more than one scalar field, the multi-field sine-Gordon theories with kink-type exact solutions deserve some attention. The interest in the study of the classical limit of string theory on determined backgrounds has recently been greatly stimulated in connection to integrability. It has been established that the classical string on $R \times S^2$ is essentially equivalent to the sine-Gordon integrable system [26]. More recently, on $R \times S^3$ background utilizing the Pohlmeyers reduction it has been obtained a family of classical string solutions called dyonic giant magnons which were associated with solitons of complex sine-Gordon equations [27]. String theory on $R \times S^{N-1}$ is classically equivalent to the so-called $SO(N)$ symmetric space sine-Gordon model (SSG) [28].

In the first part of this chapter we study the spectrum of solitons and kinks of the GSG model proposed in [14, 15, 16] and consider the closely related ATM model from which one gets the classical GSG model (cGSG) through a gauge fixing procedure. Some reductions of the GSG model to one-field theory lead to the usual SG model and to the so-called multi-frequency sine-Gordon models. In particular, the double (two-frequency) sine-Gordon model (DSG) appears in a reduction of the $sl(3, \mathbb{C})$ GSG model. The DSG theory is a nonintegrable quantum field theory with many physical applications [11, 12].

In the ATM model, once a convenient gauge fixing is performed by setting to constants some spinor bilinears, we are left with two sectors: the cGSG model which completely decouples from the spinors and a system of Dirac spinors coupled to the cGSG fields. In the references [29, 30] a $1+1$-dimensional bag model for quark confinement is considered, we follow their ideas and generalize for multi-flavor Dirac spinors coupled to cGSG solitons and kinks. The first reference considers a model similar to the $sl(2)$ ATM theory, and in the second one the DSG kink is proposed as an extended hadron model.

In the second part of this chapter we examine the quark soliton type solutions in QCD_2. Regarding this phenomenon several properties of the ATM model deserve careful consideration in view of the relationships with two-dimensional QCD. For simplicity we concentrate on the $sl(2)$ ATM model. So, in order to disentangle the quark solitons one needs to restore the heavy fields, i.e. the fields associated to the color degrees of freedom. This is done in two steps. First, by including N_c dynamical Dirac spinors coupled to the Toda field, second by breaking the chiral symmetry through certain bilinear terms in the scalar fields of the bosonized effective Lagrangian. In this way we arrive at a model similar to the one proposed in [8] in the regime when $m_q >> e_c$. We have shown that QCD_2 has quark soliton solutions if the quark mass is sufficiently large.

In the next section we define the $sl(3)$ GSG model and study its properties such as the vacuum structure and the soliton, kink and bounce type solutions. In section 3. we consider the $sl(3)$ affine Toda model coupled to matter and obtain the cGSG model through a gauge fixing procedure. It is discussed the physical soliton spectrum of the gauge fixed model. In section 4. the topological charges are introduced, as well as the idea of baryons as solitons (or kinks), and the quark confinement mechanism is discussed. In section 5. we examine the quark soliton solutions of QCD_2 and discuss the role played by the effective $sl(2)$ ATM model. The discussion section outlines the main results of this contribution and some lines of future research. In appendix A. we provide the zero curvature formulation of the $sl(3)$ ATM model.

2. The GSG Model

The generalized sine-Gordon model (GSG) related to $sl(N)$ is defined by [14, 15, 16]

$$S = \int d^2x \sum_{i=1}^{N_f} \Big[\frac{1}{2}(\partial_\mu \Phi_i)^2 + \mu_i\Big(\cos\beta_i\Phi_i - 1\Big)\Big]. \tag{1}$$

The Φ_i fields in (1) satisfy the constraints

$$\Phi_p = \sum_{i=1}^{N-1} \sigma_{pi}\Phi_i, \quad p = N, N+1, ..., N_f, \quad N_f = \frac{N(N-1)}{2}, \tag{2}$$

where σ_{pi} are some constant parameters and N_f is the number of positive roots of the Lie algebra $sl(N)$. In the context of the Lie algebraic construction of the GSG system these constraints arise from the relationship between the positive and simple roots of $sl(N)$. Thus, in (1) we have $(N-1)$ independent fields.

We will consider the $sl(3)$ case with two independent real fields $\varphi_{1,2}$, such that

$$\Phi_1 = 2\varphi_1 - \varphi_2; \quad \Phi_2 = 2\varphi_2 - \varphi_1; \quad \Phi_3 = r\,\varphi_1 + s\,\varphi_2, \quad s, r \in \mathbb{R} \tag{3}$$

which must satisfy the constraint

$$\beta_3\Phi_3 = \delta_1\beta_1\Phi_1 + \delta_2\beta_2\Phi_2, \quad \beta_i \equiv \beta_0\nu_i, \tag{4}$$

where β_0, ν_i, δ_1, δ_2 are some real numbers. Therefore, the $sl(3)$ GSG model can be regarded as three usual sine-Gordon models coupled through the linear constraint (4).

Taking into account (3)-(4) and the fact that the fields φ_1 and φ_2 are independent we may get the relationships

$$\nu_2\delta_2 = \rho_0\nu_1\delta_1 \quad \nu_3 = \frac{1}{r+s}(\nu_1\delta_1 + \nu_2\delta_2); \quad \rho_0 \equiv \frac{2s+r}{2r+s} \tag{5}$$

The $sl(3)$ model has a potential density

$$V[\varphi_i] = \sum_{i=1}^{3} \mu_i\Big(1 - \cos\beta_i\Phi_i\Big) \tag{6}$$

The GSG model has been found in the process of bosonization of the generalized massive Thirring model (GMT) [16]. The GMT model is a multiflavor extension of the usual massive Thirring model incorporating massive fermions with current-current interactions between them. In the $sl(3)$ construction of [16] the parameters δ_i depend on the couplings β_i and they satisfy certain relationship. This is obtained by assuming $\mu_i > 0$ and the zero of the potential given for $\Phi_i = \frac{2\pi}{\beta_i}n_i$, which substituted into (4) provides

$$n_1\delta_1 + n_2\delta_2 = n_3, \quad n_i \in \mathbb{Z} \tag{7}$$

The last relation combined with (5) gives

$$(2r+s)\frac{n_1}{\nu_1} + (2s+r)\frac{n_2}{\nu_2} = 3\,\frac{n_3}{\nu_3}. \tag{8}$$

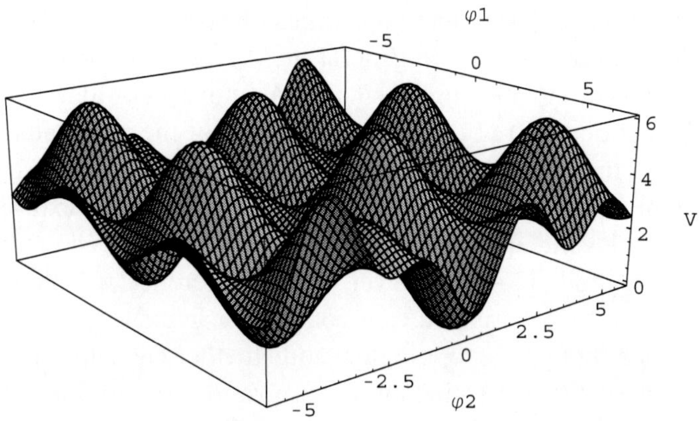

Figure 1. GSG potential V for the parameter values $\nu_1 = 1/2,\ \delta_1 = 2,\ \delta_2 = 1,\ \nu_2 = 1,\ r = s = 1,\ \beta_0 = 1,\ \mu_1 = \mu_2 = 1$.

The periodicity of the potential implies an infinitely degenerate ground state and then the theory supports topologically charged excitations. A typical potential is plotted in Fig. 1. The vacuum configuration is related to the fundamental weights (see sections 3., 4. and the Appendix). For the moment, consider the fields Φ_1 and Φ_2 and the vacuum lattice defined by

$$(\Phi_1\, ,\, \Phi_2) = \frac{2\pi}{\beta_0}(\frac{n_1}{\nu_1}\, ,\, \frac{n_2}{\nu_2}), \quad n_a \in \mathbb{Z}. \tag{9}$$

It is convenient to write the equations of motion in terms of the independent fields φ_1 and φ_2

$$\begin{aligned}
\partial^2\varphi_1 &= -\mu_1\beta_1\Delta_{11}\sin[\beta_1(2\varphi_1 - \varphi_2)] - \mu_2\beta_2\Delta_{12}\sin[\beta_2(2\varphi_2 - \varphi_1)] + \\
&\quad \mu_3\beta_3\Delta_{13}\sin[\beta_3(r\varphi_1 + s\varphi_2)] \qquad (10)\\
\partial^2\varphi_2 &= -\mu_1\beta_1\Delta_{21}\sin[\beta_1(2\varphi_1 - \varphi_2)] - \mu_2\beta_2\Delta_{22}\sin[\beta_2(2\varphi_2 - \varphi_1)] + \\
&\quad \mu_3\beta_3\Delta_{23}\sin[\beta_3(r\varphi_1 + s\varphi_2)], \qquad (11)
\end{aligned}$$

where

$$\begin{aligned}
A &= \beta_0^2\nu_1^2(4 + \delta^2 + \delta_1^2\rho_1^2 r^2), \quad B = \beta_0^2\nu_1^2(1 + 4\delta^2 + \delta_1^2\rho_1^2 s^2),\\
C &= \beta_0^2\nu_1^2(2 + 2\delta^2 + \delta_1^2\rho_1^2 r\, s),\\
\Delta_{11} &= (C - 2B)/\Delta, \quad \Delta_{12} = (B - 2C)/\Delta, \quad \Delta_{13} = (r\,B + s\,C)/\Delta,\\
\Delta_{21} &= (A - 2C)/\Delta, \quad \Delta_{22} = (C - 2A)/\Delta, \quad \Delta_{23} = (r\,C + s\,A)/\Delta\\
\Delta &= C^2 - AB, \quad \delta = \frac{\delta_1}{\delta_2}\rho_0, \quad \rho_1 = \frac{3}{2r + s}
\end{aligned}$$

Notice that the eqs. of motion (10)-(11) exhibit the symmetry

$$\varphi_1 \leftrightarrow \varphi_2, \quad \mu_1 \leftrightarrow \mu_2, \quad \nu_1 \leftrightarrow \nu_2, \quad \delta_1 \leftrightarrow \delta_2, \quad r \leftrightarrow s \tag{12}$$

Some type of coupled sine-Gordon models have been considered in connection to various interesting physical problems [31]. For example a system of two coupled SG models has been proposed in order to describe the dynamics of soliton excitations in deoxyribonucleic acid (DNA) double helices [32]. In general these type of equations have been solved by perturbation methods around decoupled sine-Gordon exact solitons.

The system of equations (10)-(11) for certain choice of the parameters r and s will be derived in section 3. in the context of the $sl(3)$ ATM type models, in which the fields φ_1 and φ_2 couple to some Dirac spinors in such a way that the model exhibits a local gauge invariance. The ATM relevant equations of motion have been solved using a hybrid of the Hirota and Dressing methods [33]. However, in this reference the physical spectrum of solitons and kinks of the theory, related to a convenient gauge fixing of the model, have not been discussed, even though the topological and Noether currents equivalence has been verified. The appearance of the so-called tau functions, in order to find soliton solutions in integrable models, is quite a general result in the both Dressing and Hirota approaches. In this section, we will find soliton and kink type solutions of the GSG model (10)-(11) and closely follow the spirit of the above hybrid method approach to find soliton solutions.

The general tau function for an $n-$soliton solution of the *gauge unfixed* ATM model has the form [33, 34]

$$
\begin{aligned}
\tau &= \sum_{p_1.....p_n=0}^{2} c_{p_1.....p_n} \exp[p_1\Gamma_{i_1}(z_1) + ... + p_n\Gamma_{i_n}(z_n)], \qquad (13)\\
& z_i = \gamma_i(x - v_i\, t), \quad c_{p_1.....p_n} \in \mathbb{C}
\end{aligned}
$$

In the following we discuss the developments of ref. [35]. Since the GSG model describes the strong coupling sector (soliton spectrum) of the ATM model [14, 15] then one can guess the following Ansatz for the tau functions of the GSG model

$$
e^{-i\beta_0\frac{\varphi_1}{2}} = \frac{\tau_1}{\tau_0}, \quad e^{-i\beta_0\frac{\varphi_2}{2}} = \frac{\tau_2}{\tau_0}, \qquad (14)
$$

where the tau functions τ_i $(i = 0, 1, 2)$ are assumed to be of the form (13). We will see that the Ansatz (14) provides soliton and kink type solutions of the model (10)-(11), in this way justifying *a posteriori* the assumption made for the relevant tau functions.

Assuming that the fields φ_a $(a = 1, 2)$ are real, from (14) one can write

$$
\begin{aligned}
\varphi_{1,2} &= \frac{4}{\beta_0}\arctan[F(\tau_{1,2}, \tau_0)] \qquad (15)\\
F &\equiv \frac{e_1\Big([Re(\tau_{1,2})]^2 + [Im(\tau_{1,2})]^2\Big) - \Big(Re(\tau_{1,2})Re(\tau_0) + Im(\tau_{1,2})Im(\tau_0)\Big)}{\Big[Im(\tau_{1,2}) * Re(\tau_0) - Re(\tau_{1,2}) * Im(\tau_0)\Big]},\\
e_1 &= \pm 1 \qquad (16)
\end{aligned}
$$

In terms of the tau functions the system of equations (10)-(11) becomes

$$
\begin{aligned}
&\frac{2i}{\beta_0^2}\Big[\frac{\partial^2\tau_1}{\tau_1} - \frac{(\partial\tau_1)^2}{\tau_1^2} - \frac{\partial^2\tau_0}{\tau_0} + \frac{(\partial\tau_0)^2}{\tau_0^2}\Big] + \frac{\beta_1\mu_1\Delta_{11}}{2i}\Big[\frac{(\tau_2\tau_0)^{4\nu_1} - \tau_1^{8\nu_1}}{(\tau_2\tau_0)^{2\nu_1}\tau_1^{4\nu_1}}\Big] + \\
&\frac{\beta_2\mu_2\Delta_{12}}{2i}\Big[\frac{(\tau_1\tau_0)^{4\nu_2} - \tau_2^{8\nu_2}}{(\tau_1\tau_0)^{2\nu_2}\tau_2^{4\nu_2}}\Big] - \frac{\beta_3\mu_3\Delta_{13}}{2i}\Big[\frac{(\tau_0)^{4\nu_3(r+s)} - \tau_1^{4r\nu_3}\tau_2^{4s\nu_3}}{(\tau_2)^{2s\nu_3}(\tau_1)^{2r\nu_3}\tau_0^{2\nu_3(r+s)}}\Big] = 0, \qquad (17)
\end{aligned}
$$

$$\frac{2i}{\beta_0^2}\left[\frac{\partial^2\tau_2}{\tau_2}-\frac{(\partial\tau_2)^2}{\tau_2^2}-\frac{\partial^2\tau_0}{\tau_0}+\frac{(\partial\tau_0)^2}{\tau_0^2}\right]+\frac{\beta_1\mu_1\Delta_{21}}{2i}\left[\frac{(\tau_2\tau_0)^{4\nu_1}-\tau_1^{8\nu_1}}{(\tau_2\tau_0)^{2\nu_1}\tau_1^{4\nu_1}}\right]+$$
$$\frac{\beta_2\mu_2\Delta_{22}}{2i}\left[\frac{(\tau_1\tau_0)^{4\nu_2}-\tau_2^{8\nu_2}}{(\tau_1\tau_0)^{2\nu_2}\tau_2^{4\nu_2}}\right]-\frac{\beta_3\mu_3\Delta_{23}}{2i}\left[\frac{(\tau_0)^{4\nu_3(r+s)}-\tau_1^{4r\nu_3}\tau_2^{4s\nu_3}}{(\tau_2)^{2s\nu_3}(\tau_1)^{2r\nu_3}\tau_0^{2\nu_3(r+s)}}\right]=0. \quad (18)$$

We will see that the 1-soliton and 1-kink type solutions are related to half-integer or integer values of the parameters ν_i and the values $r, s = 0, 1$. In the next subsections we write the 1-antisoliton, 1-antikink and bounce type solutions, and in order to perform the cumbersome computations we resort to the MAPLE program.

2.1. One Soliton Associated to φ_1

Consider the tau functions

$$\tau_0 = 1 + i\,d\exp[\gamma(x-vt)];\quad \tau_1 = 1 - i\,d\exp[\gamma(x-vt)];\quad \tau_2 = 1 + i\,d\exp[\gamma(x-vt)].$$

This choice satisfies the system of equations (17)-(18) for the set of parameters

$$\nu_1 = 1/2,\ \delta_1 = 2,\ \delta_2 = 1,\ \nu_2 = 1,\ \nu_3 = 1,\ r = 1. \quad (19)$$

provided that

$$13\mu_3 = 5\mu_2 - 4\mu_1,\ \ \gamma_1^2 = \frac{1}{13}(6\mu_2 + 3\mu_1). \quad (20)$$

Now, taking $e_1 = 1$ in Eq. (16) and the relation (15) one has

$$\varphi_1 = -\frac{4}{\beta_0}\arctan\{d\ \exp[\gamma_1(x-vt)]\},\ \ \varphi_2 = 0. \quad (21)$$

This solution is precisely the sine-Gordon 1-antisoliton associated to the field φ_1 with mass $M_1 = \frac{8\gamma_1}{\beta_0^2}$. We plot a soliton of this type in Fig. 3.

2.2. One Soliton Associated to φ_2

Next, let us consider the tau functions

$$\tau_0 = 1 + i\,d\exp[\gamma(x-vt)],\quad \tau_1 = 1 + i\,d\exp[\gamma(x-vt)],\quad \tau_2 = 1 - i\,d\exp[\gamma(x-vt)]$$

This set of tau functions solves the system (17)-(18) for the choice of parameters

$$\nu_1 = 1,\ \delta_1 = 1,\ \delta_2 = 2,\ \nu_2 = 1/2,\ \nu_3 = 1,\ s = 1 \quad (22)$$

provided that

$$13\mu_3 = 5\mu_1 - 4\mu_2,\ \ \gamma_2^2 = \frac{1}{13}(6\mu_1 + 3\mu_2) \quad (23)$$

Now, choose $e_1 = 1$ in (16) and through (15) one can get

$$\varphi_2 = -\frac{4}{\beta_0}\arctan\{d\exp[\gamma_2(x-vt)]\},\ \ \varphi_1 = 0 \quad (24)$$

Similarly, this is the sine-Gordon 1-antisoliton associated to the field φ_2 with mass $M_2 = \frac{8\gamma_2}{\beta_0^2}$ and its profile is of the type shown in Fig 3.

2.3. Two One-Solitons Associated to $\varphi \equiv \varphi_{1,2}$

Now, let us consider the tau functions

$\tau_0 = 1 + i\, d\, \exp[\gamma(x - vt)], \quad \tau_1 = 1 - i\, d\, \exp[\gamma(x - vt)], \quad \tau_2 = 1 - i\, d\, \exp[\gamma(x - vt)].$

This choice satisfies (17)-(18) for

$$\nu_1 = 1, \; \delta_1 = 1/2, \; \nu_2 = 1, \; \delta_2 = 1/2, \; \nu_3 = 1/2, \; r = s = 1, \tag{25}$$

provided that

$$d^2 = 1, \quad 38\gamma_3^2 = 25\mu_1 + 13\mu_2 + 19\mu_3 \tag{26}$$

Now, taking $e_1 = 1$ in (15) one has

$$\varphi_1 = \varphi_2 \equiv \hat{\varphi}_1, \tag{27}$$

$$\hat{\varphi}_1 = -\frac{4}{\beta_0}\arctan\{d\ \exp[\gamma_3(x - vt)]\}. \tag{28}$$

This is a sine-Gordon 1-antisoliton associated to both fields $\varphi_{1,2}$ in the particular case when they are equal to each other. It possesses a mass $M_3 = \frac{8\gamma_3}{\beta_0^2}$.

In view of the symmetry (12) we are able to write

$$d^2 = 1, \quad 38\gamma_4^2 = 25\mu_2 + 13\mu_1 + 19\mu_3, \tag{29}$$

and then on has another soliton of this type

$$\varphi_1 = \varphi_2 \equiv \hat{\varphi}_2, \tag{30}$$

$$\hat{\varphi}_2 = -\frac{4}{\beta_0}\arctan\{d\ \exp[\gamma_4(x - vt)]\}. \tag{31}$$

It possesses a mass $M_4 = \frac{8\gamma_4}{\beta_0^2}$. This 1-antisoliton is of the type shown in Fig. 3.

The GSG system (10)-(11) reduces to the usual SG equation for each choice of the parameters (19), (22) and (25), respectively. Then, the $n-$soliton solutions in each case can be constructed as in the ordinary sine-Gordon model by taking appropriate tau functions in (13)-(14).

The baryon number associated to each of the above 1-soliton solutions has been computed in connection to QCD$_2$, and it takes the same value $B = N_c$ (in this normalization the quark has baryon number $B_{quark} = 1$) [6].

A modified model with rich soliton dynamics is the so-called stepwise sine-Gordon model in which the system parameter depends on the sign of the SG field [36]. It would be interesting to consider the above GSG model along the lines of this reference.

2.4. Mass Splitting of Solitons

It is interesting to write some relations among the various soliton masses

$$M_3^2 = \frac{1}{76}(109M_2^2 + 5M_1^2); \quad M_4^2 = \frac{1}{76}(109M_1^2 + 5M_2^2); \tag{32}$$

If $\mu_1 = \mu_2$ then we have the degeneracy $M_1 = M_2$, and $M_3 = M_4 = \sqrt{3/2}M_1$. Notice that if $M_1 \neq M_2$ then $M_3 < M_1 + M_2$ and $M_4 < M_1 + M_2$, and the third and fourth solitons are stable in the sense that energy is required to dissociate them.

2.5. Kinks of the Reduced Two-Frequency Sine-Gordon Model

In the system (10)-(11) we perform the following reduction $\varphi \equiv \varphi_1 = \varphi_2$ such that

$$\Phi_1 = \Phi_2, \quad \Phi_3 = q\, \Phi_1, \tag{33}$$

with q being a real number. Therefore, using the constraint (4) one can deduce the relationships

$$\delta_1 = \frac{q}{2}, \quad \delta = 1. \tag{34}$$

Moreover, for consistency of the system of equations (10)-(11) we have to impose the relationships

$$\nu_1\mu_1\Delta_{11} + \nu_2\mu_2\Delta_{12} \quad = \quad \nu_1\mu_1\Delta_{21} + \nu_2\mu_2\Delta_{22}, \tag{35}$$

$$\Delta_{13} \quad = \quad \Delta_{23}. \tag{36}$$

These relations imply

$$\delta^2 = 1, \quad \mu_1 = \delta\ \mu_2 \tag{37}$$

.

Taking into account the relations (34) and (37) together with (5) we get

$$\mu_1 = \mu_2, \quad \delta = 1, \quad \nu_1 = \nu_2, \quad \nu_3 = \frac{q}{2}\nu_1, \quad r = s = 1. \tag{38}$$

Thus the system of Eqs.(10)-(11) reduce to

$$\partial^2\Phi \quad = \quad -\frac{\mu_1}{\nu_1}\sin(\nu_1\Phi) - \frac{\mu_3\delta_1}{\nu_1}\sin(q\,\nu_1\Phi), \quad \Phi \equiv \beta_0\varphi. \tag{39}$$

This is the so-called *two-frequency sine-Gordon* model (DSG) and it has been the subject of much interest in the last decades, from the mathematical and physical points of view. It encounters many interesting physical applications, see e.g. [11, 12, 30, 31].

If the parameter q satisfies

$$q = \frac{n}{m} \in \mathbb{Q} \tag{40}$$

with $m,\ n$ being two relative prime positive integers, then the potential $\frac{\mu_1}{\nu_1^2}(1-\cos(\nu_1\Phi)) + \frac{\mu_3}{2\nu_1^2}(1-\cos(q\nu_1\Phi))$ associated to the model (39) is periodic with period

$$\frac{2\pi}{\nu_1}m = \frac{2\pi}{q\,\nu_1}n. \tag{41}$$

As mentioned above the theory (39) possesses topological excitations. The fundamental topological excitations degenerates in the $\mu_1 = 0$ limit to an $n-$soliton state of the relevant sine-Gordon model and similarly in the limit $\mu_3 = 0$ it will be an m-soliton state. For general values of the parameters $\mu_1,\ \mu_3,\ \delta_1,\ \nu_1$ the solitons are in some sense "confined" inside the topological excitations which become in this form some composite objects. On the other hand, if $q \notin \mathbb{Q}$ then the potential is not periodic, so, there are no topologically charged excitations and the solitons are completely confined [9, 10].

The model (39) in the limit $\mu_1 = 0$ reduces to

$$\partial^2 \varphi = -\frac{\mu_3 q}{2\nu_1 \beta_0} \sin(q\nu_1 \beta_0 \varphi). \tag{42}$$

For later discussion we record here the mass of the soliton associated to this equation,

$$M_{\mu_3} = \frac{8}{(q\nu_1\beta_0)^2}\sqrt{q^2\mu_3/2}. \tag{43}$$

Correspondingly in the limit $\mu_3 = 0$ one has

$$\partial^2 \varphi = -\frac{mu_1}{\nu_1 \beta_0} \sin(\nu_1 \beta_0 \varphi) \tag{44}$$

with associated soliton mass

$$M_{\mu_1} = \frac{8}{(\nu_1\beta_0)^2}\sqrt{\mu_1} \tag{45}$$

Notice that other possibilities to perform the reduction of type (33) encounter some inconsistencies, e.g. the attempt to implement the reduction $\Phi_1 = \Phi_3,\ \Phi_2 = q'\,\Phi_1$ implies $\delta_{1,2}^2 < 0$ which is a contradiction since $\delta_{1,2}$ are real numbers by definition. The same inconsistency occurs when one tries to reduce the $sl(3)$ GSG model to a three-frequency SG model. We expect that the three and higher frequency models [37] will be related to $sl(N)$, $N \geq 4$, GSG models.

In the following we will provide some kink solutions for particular set of parameters. Consider

$$\nu_1 = 1/2,\ \delta_1 = \delta_2 = 1,\ \nu_2 = 1/2,\ \nu_3 = 1/2 \text{ and } q = 2,\ n = 2,\ m = 1 \tag{46}$$

which satisfy (38) and (40), respectively. This set of parameters provide the so-called *double sine-Gordon model* (DSG). Its potential $-[4\mu_1(\cos\frac{\Phi}{2} - 1) + 2\mu_3(\cos\Phi - 1)]$ has period 4π and has extrema at $\Phi = 2\pi p_1$, and $\Phi = 4\pi p_2 \pm 2\cos^{-1}[1 - |\mu_1/(2\mu_3)|]$ with $p_1, p_2 \in \mathbb{Z}$; the second extrema exists only if $|\mu_1/(2\mu_3)| < 1$. From the mathematical point of view the DSG model belongs to a class of theories with partial integrability [38]. Depending on the values of the parameters $\beta_0,\ \mu_1,\ \mu_3$ the quantum field theory version of the DSG model presents a variety of physical effects, such as the decay of the false vacuum, a phase transition, confinement of the kinks and the resonance phenomenon due to unstable bound states of excited kink-antikink states (see [12] and references therein). The semi-classical spectrum of neutral particles in the DSG theory is investigated in [39]. Let us mention that the DSG model has recently been in the center of some controversy regarding the computation of its semiclassical spectrum, see [12, 13].

Interestingly the functions[1]

$$\begin{aligned} \tau_0 &= 1 + i\,d\exp[\gamma(x - vt)] + h\exp[2\gamma(x - vt)], \\ \tau_1 &= 1 - i\,d\exp[\gamma(x - vt)] + h\exp[2\gamma(x - vt)], \end{aligned} \tag{47}$$

[1]These functions are obtained by adding the term $\exp[2\gamma(x - vt)]$ to the relevant tau functions for one solitons used above. This procedure adds a new method of solving DSG which deserve further study. The multi-frequency SG equations can be solved through the Jacobi elliptic function expansion method, see e.g. [40].

satisfy the equation (39) for the parameters (46) provided

$$e^{-i\Phi/2} = \tau_1/\tau_0 \quad (48)$$

$$\gamma^2 = \mu_1 + 2\mu_3, \quad h = -\frac{\mu_1}{4}, \quad e_1 = -1 \quad (49)$$

The general solution of this type can be written as

$$\Phi := 4\arctan\left[\frac{1}{d}\,\frac{1 + h\,\exp[2\gamma(x - vt)]}{\exp[\gamma(x - vt)]}\right] \quad (50)$$

2.5.1. DSG Kink ($h < 0,\ \mu_i > 0$)

For the choice of parameters $h < 0,\ \mu_i > 0$ in (49) the equation (50) provides

$$\varphi := \frac{4}{\beta_0}\arctan\left[\frac{-2|h|^{1/2}}{d}\sinh[\gamma_K(x - vt) + a_0]\right], \quad \gamma_K \equiv \pm\sqrt{\mu_1 + 2\mu_3}, \quad (51)$$

$$a_0 = \frac{1}{2}\ln|h|.$$

This is the DSG 1-kink solution with mass

$$M_K = \frac{16}{\beta_0^2}\gamma_K\left[1 + \frac{\mu_1}{\sqrt{2\mu_3(\mu_1 + 2\mu_3)}}\ln\left(\frac{\sqrt{\mu_1 + 2\mu_3} + \sqrt{2\mu_3}}{\sqrt{\mu_1}}\right)\right]. \quad (52)$$

Notice that in the limit $\mu_1 \to 0$ the kink mass becomes $M_K = \frac{16}{\beta_0^2}\sqrt{2\mu_3}$, which is twice the soliton mass (43) of the model (42) for the parameters $\nu_1 = 1/2,\ q = 2$. Similarly, in the limit $\mu_3 \to 0$ the kink mass becomes $\frac{8}{(\beta_0/2)^2}\sqrt{\mu_1}$, which is the soliton mass (45) of the model (44) for $\nu_1 = 1/2,\ q = 2$; thus in this case the coupling constant is $\beta_0/2$. As discussed above these solitons get in some sense "confined" inside the kink if the parameters satisfy $\mu_i \neq 0$. The 1-antikink is plotted in Fig. 4. Moreover, the relevant baryon number associated to this DSG kink becomes $B_{kink} = 4N_c$ [6].

2.5.2. Bounce-Like Solution ($h > 0,\ \mu_1 < 0$)

For the parameters $h > 0,\ \mu_1 < 0$ one gets from (50)

$$\varphi := \frac{4}{\beta_0}\arctan\left[\frac{2h^{1/2}}{d}\cosh[\gamma'(x - vt) + a_0']\right], \quad \gamma' = 2\mu_3 - |\mu_1|, \quad a_0' = \frac{1}{2}\ln h \quad (53)$$

This is the bounce-like solution and interpolates between the two vacuum values 2π and $4\pi - 2\arccos(1 - |\mu_1/2\mu_3|)$ and then it comes back. Since 2π is a false vacuum position this solution is not related to any stable particle in the quantum theory [12]. In Fig. 2 we plot this profile.

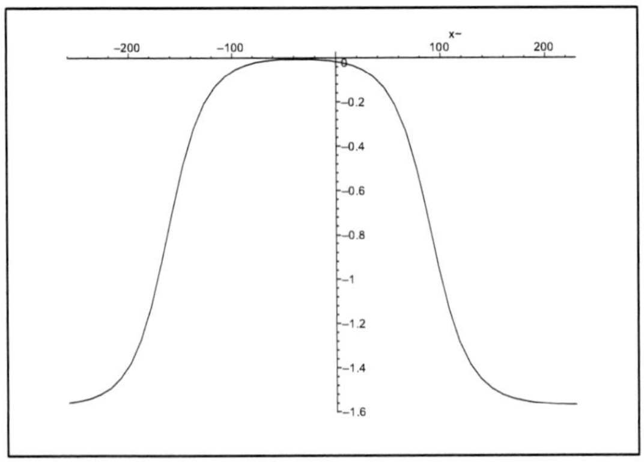

Figure 2. Bounce-like solution $(\frac{\beta_0}{4}\varphi)$ plotted for $\mu_2 = -0.0000001,\ \mu_3 = 0.001,\ d = -2.$

3. Classical GSG as a Reduced Toda Model Coupled to Matter

In this section we provide the algebraic construction of the $sl(3)$ affine Toda model coupled to matter fields (ATM) and closely follows refs. [15, 33, 41] but the reduction process to arrive at the classical GSG model is new. The previous treatments of the $sl(3)$ ATM model used the symplectic and on-shell decoupling methods to unravel the classical GSG and generalized massive Thirring (GMT) dual theories describing the strong/weak coupling sectors of the ATM model [14, 15, 42]. The ATM model describes some scalars coupled to spinor (Dirac) fields in which the system of equations of motion has a local gauge symmetry. In this way one includes the spinor sector in the discussion and conveniently gauge fixing the local symmetry by setting some spinor bilinears to constants we are able to decouple the scalar (Toda) fields from the spinors, the final result is a direct construction of the classical generalized sine-Gordon model (cGSG) involving only the scalar fields. In the spinor sector we are left with a system of equations in which the Dirac fields couple to the cGSG fields.

The zero curvature condition (119) gives the following equations of motion [41]

$$\frac{\partial^2\theta_a}{4i\,e^{\eta}} = m_\psi^1[e^{\eta-i\phi_a}\widetilde{\psi}_R^l\psi_L^l + e^{i\phi_a}\widetilde{\psi}_L^l\psi_R^l] + m_\psi^3[e^{-i\phi_3}\widetilde{\psi}_R^3\psi_L^3 + e^{\eta+i\phi_3}\widetilde{\psi}_L^3\psi_R^3];\quad a = 1,2 \tag{54}$$

$$-\frac{\partial^2\widetilde{\nu}}{4} = i m_\psi^1 e^{2\eta-\phi_1}\widetilde{\psi}_R^1\psi_L^1 + i m_\psi^2 e^{2\eta-\phi_2}\widetilde{\psi}_R^2\psi_L^2 + i m_\psi^3 e^{\eta-\phi_3}\widetilde{\psi}_R^3\psi_L^3 + \mathbf{m}^2 e^{3\eta} \tag{55}$$

$$-2\partial_+\psi_L^1 = m_\psi^1 e^{\eta+i\phi_1}\psi_R^1, \qquad -2\partial_+\psi_L^2 = m_\psi^2 e^{\eta+i\phi_2}\psi_R^2, \tag{56}$$

$$2\partial_-\psi_R^1 = m_\psi^1 e^{2\eta-i\phi_1}\psi_L^1 + 2i\Big(\frac{m_\psi^2 m_\psi^3}{i m_\psi^1}\Big)^{1/2} e^{\eta}(-\psi_R^3\widetilde{\psi}_L^2 e^{i\phi_2} - \widetilde{\psi}_R^2\psi_L^3 e^{-i\phi_3}), \tag{57}$$

$$2\partial_-\psi_R^2 = m_\psi^2 e^{2\eta-i\phi_2}\psi_L^2 + 2i\Big(\frac{m_\psi^1 m_\psi^3}{i m_\psi^2}\Big)^{1/2} e^{\eta}(\psi_R^3\widetilde{\psi}_L^1 e^{i\phi_1} + \widetilde{\psi}_R^1\psi_L^3 e^{-i\phi_3}), \tag{58}$$

$$-2\partial_+\psi_L^3 = m_\psi^3 e^{2\eta+i\phi_3}\psi_R^3 + 2i\Big(\frac{m_\psi^1 m_\psi^2}{i m_\psi^3}\Big)^{1/2} e^{\eta}(-\psi_L^1\psi_R^2 e^{i\phi_2} + \psi_L^2\psi_R^1 e^{i\phi_1}), \tag{59}$$

$$2\partial_-\psi_R^3 = m_\psi^3 e^{\eta-i\phi_3}\psi_L^3, \qquad 2\partial_-\widetilde{\psi}_R^1 = m_\psi^1 e^{\eta+i\phi_1}\widetilde{\psi}_L^1, \tag{60}$$

$$-2\partial_+\widetilde{\psi}_L^1 = m_\psi^1 e^{2\eta-i\phi_1}\widetilde{\psi}_R^1 + 2i\Big(\frac{m_\psi^2 m_\psi^3}{im_\psi^1}\Big)^{1/2} e^{\eta}(-\psi_L^2\widetilde{\psi}_R^3 e^{-i\phi_3} - \widetilde{\psi}_L^3\psi_R^2 e^{i\phi_2}), \quad (61)$$

$$-2\partial_+\widetilde{\psi}_L^2 = m_\psi^2 e^{2\eta-i\phi_2}\widetilde{\psi}_R^2 + 2i\Big(\frac{m_\psi^1 m_\psi^3}{im_\psi^2}\Big)^{1/2} e^{\eta}(\psi_L^1\widetilde{\psi}_R^3 e^{-i\phi_3} + \widetilde{\psi}_L^3\psi_R^1 e^{i\phi_1}), \quad (62)$$

$$2\partial_-\widetilde{\psi}_R^2 = m_\psi^2 e^{\eta+i\phi_2}\widetilde{\psi}_L^2, \qquad -2\partial_+\widetilde{\psi}_L^3 = m_\psi^3 e^{\eta-i\phi_3}\widetilde{\psi}_R^3, \quad (63)$$

$$2\partial_-\widetilde{\psi}_R^3 = m_\psi^3 e^{2\eta+i\phi_3}\widetilde{\psi}_L^3 + 2i\Big(\frac{m_\psi^1 m_\psi^2}{im_\psi^3}\Big)^{1/2} e^{\eta}(\widetilde{\psi}_R^1\widetilde{\psi}_L^2 e^{i\phi_2} - \widetilde{\psi}_R^2\widetilde{\psi}_L^1 e^{i\phi_1}), \quad (64)$$

$$\partial^2\eta = 0, \quad (65)$$

where $\phi_1 \equiv 2\theta_1 - \theta_2$, $\phi_2 \equiv 2\theta_2 - \theta_1$, $\phi_3 \equiv \theta_1 + \theta_2$. Therefore, one has

$$\phi_3 = \phi_1 + \phi_2 \quad (66)$$

The θ fields are considered to be in general complex fields. In order to define the classical generalized sine-Gordon model we will consider these fields to be real.

Apart from the *conformal invariance* the above equations exhibit the $\Big(U(1)_L\Big)^2 \otimes \Big(U(1)_R\Big)^2$ *left-right local gauge symmetry*

$$\theta_a \to \theta_a + \xi_+^a(x_+) + \xi_-^a(x_-), \quad a = 1,2 \quad (67)$$

$$\widetilde{\nu} \to \widetilde{\nu}\,; \qquad \eta \to \eta \quad (68)$$

$$\psi^i \to e^{i(1+\gamma_5)\Xi_+^i(x_+)+i(1-\gamma_5)\Xi_-^i(x_-)}\,\psi^i, \quad (69)$$

$$\widetilde{\psi}^i \to e^{-i(1+\gamma_5)(\Xi_+^i)(x_+)-i(1-\gamma_5)(\Xi_-^i)(x_-)}\,\widetilde{\psi}^i, \quad i = 1,2,3; \quad (70)$$

$$\Xi_\pm^1 \equiv \pm\xi_\pm^2 \mp 2\xi_\pm^1, \; \Xi_\pm^2 \equiv \pm\xi_\pm^1 \mp 2\xi_\pm^2, \; \Xi_\pm^3 \equiv \Xi_\pm^1 + \Xi_\pm^2.$$

One can get global symmetries for $\xi_\pm^a = \mp\xi_\mp^a$ = constants. For a model defined by a Lagrangian these would imply the presence of two vector and two chiral conserved currents. However, it was found only half of such currents [33]. This is a consequence of the lack of a Lagrangian description for the $sl(3)^{(1)}$ CATM in terms of the B and $F^\pm$ fields (see Appendix). So, the vector current

$$J^\mu = \sum_{j=1}^{3} m_\psi^j \bar{\psi}^j\gamma^\mu\psi^j \quad (71)$$

and the chiral current

$$J^{5\,\mu} = \sum_{j=1}^{3} m_\psi^j \bar{\psi}^j\gamma^\mu\gamma_5\psi^j + 2\partial_\mu(m_\psi^1\theta_1 + m_\psi^2\theta_2) \quad (72)$$

are conserved

$$\partial_\mu J^\mu = 0, \quad \partial_\mu J^{5\,\mu} = 0 \quad (73)$$

The conformal symmetry is gauge fixed by setting

$$\eta = \text{const.} \quad (74)$$

The off-critical model obtained in this way exhibits the vector and topological currents equivalence [41, 42]

$$\sum_{j=1}^{3} m_\psi^j \bar{\psi}^j \gamma^\mu \psi^j \equiv \epsilon^{\mu\nu} \partial_\nu (m_\psi^1 \theta_1 + m_\psi^2 \theta_2), \quad m_\psi^3 = m_\psi^1 + m_\psi^2, \quad m_\psi^i > 0. \tag{75}$$

Moreover, it has been shown that the soliton type solutions are in the orbit of the vacuum $\eta = 0$.

In the next steps we implement the reduction process to get the cGSG model through a gauge fixing of the ATM theory. The local symmetries (67)-(70) can be gauge fixed through

$$i\bar{\psi}^j \psi^j = iA_j = \text{const.}; \quad \bar{\psi}^j \gamma_5 \psi^j = 0. \tag{76}$$

From the gauge fixing (76) one can write the following bilinears

$$\widetilde{\psi}_R^j \psi_L^j + \widetilde{\psi}_L^j \psi_R^j = 0, \quad j = 1, 2, 3; \tag{77}$$

so, the eqs. (76) effectively comprises three gauge fixing conditions.

It can be directly verified that the gauge fixing (76) preserves the currents conservation laws (73), i.e. from the equations of motion (54)-(65) and the gauge fixing (76) together with (74) it is possible to obtain the currents conservation laws (73).

Taking into account the constraints (76) in the scalar sector, eqs. (54), we arrive at the following system of equations (set $\eta = 0$)

$$\partial^2 \theta_1 = M_\psi^1 \sin\phi_1 + M_\psi^3 \sin\phi_3, \tag{78}$$

$$\partial^2 \theta_2 = M_\psi^2 \sin\phi_2 + M_\psi^3 \sin\phi_3, \quad M_\psi^i \equiv 4A_i\, m_\psi^i, \quad i = 1, 2, 3. \tag{79}$$

Define the fields φ_1, φ_2 as

$$\varphi_1 \equiv a\theta_1 + b\theta_2, \quad a = \frac{4\nu_2 - \nu_1}{3\beta_0 \nu_1 \nu_2}, \quad d = \frac{4\nu_1 - \nu_2}{3\beta_0 \nu_1 \nu_2} \tag{80}$$

$$\varphi_2 \equiv c\theta_1 + d\theta_2, \quad b = -c = \frac{2(\nu_1 - \nu_2)}{3\beta_0 \nu_1 \nu_2}, \quad \nu_1, \nu_2 \in \mathbb{R} \tag{81}$$

Then, the system of equations (78)-(79) written in terms of the fields $\varphi_{1,2}$ becomes

$$\begin{aligned} \partial^2 \varphi_1 &= aM_\psi^1 \sin[\beta_0 \nu_1 (2\varphi_1 - \varphi_2)] + bM_\psi^2 \sin[\beta_0 \nu_2 (2\varphi_2 - \varphi_1)] + \\ &\quad (a+b) M_\psi^3 \sin\beta_0 [(2\nu_1 - \nu_2)\varphi_1 + (2\nu_2 - \nu_1)\varphi_2)], \end{aligned} \tag{82}$$

$$\begin{aligned} \partial^2 \varphi_2 &= cM_\psi^1 \sin[\beta_0 \nu_1 (2\varphi_1 - \varphi_2)] + dM_\psi^2 \sin[\beta_0 \nu_2 (2\varphi_2 - \varphi_1)] + \\ &\quad (c+d) M_\psi^3 \sin\beta_0 [(2\nu_1 - \nu_2)\varphi_1 + (2\nu_2 - \nu_1)\varphi_2)] \end{aligned} \tag{83}$$

The system of equations above considered for real fields $\varphi_{1,2}$ as well as for real parameters $M_\psi^i, a, b, c, d, \beta_0$ defines the *classical generalized sine-Gordon model* (cGSG). Notice that this classical version of the GSG model derived from the ATM theory is a submodel of the GSG model (10)-(11), defined in section 2., for the particular parameter values $r = \frac{2\nu_1 - \nu_2}{\nu_3}$, $s = \frac{2\nu_2 - \nu_1}{\nu_3}$ and the convenient identifications of the parameters in the coefficients of the sine functions of the both models.

The following reduced models can be obtained from the system (82)-(83):

i) *SG submodels*

i.1) For $\nu_2 = 2\nu_1$ one has $M^1_\psi = M^2_\psi$ and the system
$\varphi_2 = 0,\ \partial^2\varphi_1 = M^1_\psi \frac{3\nu_1}{\beta_0} \sin\beta_0 2\nu_1\varphi_1$.

i.2) For $\nu_1 = 2\nu_2$ one has $M^1_\psi = M^2_\psi$ and the system
$\varphi_1 = 0,\ \partial^2\varphi_2 = M^2_\psi \frac{3\nu_2}{\beta_0} \sin\beta_0 2\nu_2\varphi_2$.

i.3) For $\nu_2 = \nu_1 \equiv \nu$ and $\varphi_1 = \varphi_2 \equiv \hat{\varphi}_A$, $(A = 1, 2)$, one gets the sub-models

i.3a) $M^1_\psi = M^2_\psi,\ M^3_\psi = 0,\ \partial^2\hat{\varphi}_1 = aM^1_\psi \sin\beta_0\nu\hat{\varphi}_1$,

i.3b) $M^1_\psi = M^2_\psi = 0,\ \partial^2\hat{\varphi}_2 = aM^3_\psi \sin\beta_0\nu\hat{\varphi}_2$.

ii) *DSG sub-model*

For $\nu_1 = \nu_2$ and $M^1_\psi = M^2_\psi$ one gets the sub-model $\varphi_1 = \varphi_2 \equiv \varphi,\ \partial^2\varphi = aM^1_\psi \sin\beta_0\nu_1\varphi + aM^3_\psi \sin 2\beta_0\nu_1\varphi$.

The sub-models i.1)-i.2) each one contains the ordinary sine-Gordon model (SG) and they were considered in the subsections 2.1. and 2.2., respectively; the sub-model i.3) supports two SG models with different soliton masses which must correspond to the construction in subsection 2.3.; and the ii) case defines the double sine-Gordon model (DSG) studied in subsection 2.5.. Other meaningful reductions are possible arriving at either SG or DSG model. Notice that the reductions above are particular cases of the sub-models in subsections 2.1., 2.2., 2.3. and 2.5., respectively, for relevant parameter identifications.

The spinor sector in view of the gauge fixing (76) can be parameterized conveniently as

$$\begin{pmatrix} \psi^j_R \\ \psi^j_L \end{pmatrix} = \begin{pmatrix} \sqrt{A_j/2}\, u_j \\ i\sqrt{A_j/2}\, \frac{1}{v_j} \end{pmatrix};\quad \begin{pmatrix} \widetilde{\psi}^j_R \\ \widetilde{\psi}^j_L \end{pmatrix} = \begin{pmatrix} \sqrt{A_j/2}\, v_j \\ -i\sqrt{A_j/2}\, \frac{1}{u_j} \end{pmatrix}. \tag{84}$$

Therefore, in order to find the spinor field solutions one can solve the eqs. (56)-(64) for the fields u_j, v_j for each solution given for the cGSG fields $\varphi_{1,2}$ of the system (82)-(83).

3.1. Physical Solitons and Kinks of the ATM Model

The main feature of the one 'solitons' constructed in [33] is that for each positive root of $sl(3)$ there corresponds one soliton species associated to the fields ϕ_1, ϕ_2, ϕ_3, respectively. The relevant solutions for the spinor fields together with the 1-'solitons' satisfy the relationship (75). The class of 2-'soliton' solutions of $sl(3)$ ATM obtained in [33] behave as follows: i) they are given by 6 species associated to the pair (α_i, α_j), $i \leq j;\ i, j = 1, 2, 3$; where the α's are the positive roots of $sl(3)$ Lie algebra. Each species (α_i, α_i) solves the $sl(2)$ ATM submodel[2]. ii) they satisfy the $U(1)$ vector and topological currents equivalence (75). However, the possible kink type solutions associated in a *non-local* way to the spinor bilinears and the relevant gauge fixing of the local symmetry (67)-(70) have not been discussed in the literature. In order to consider the physical spectrum of solitons and study its properties, such as their masses and scattering time delays, it is mandatory to take into account these questions which are related to the counting of the true physical degrees of freedom of the theory. Therefore, one must consider the possible soliton type solutions

[2] $sl(2)$ ATM gauge unfixed $2-$'solitons' satisfy an analogous eq. to (75). Moreover, for φ real and $\widetilde{\psi} = \pm(\psi)^*$ one has, soliton-soliton SS, SS bounds and no $S\bar{S}$ (S =soliton, $\bar{S}$=anti-soliton) bounds [43] associated to the field φ.

associated to each spinor bilinear. The relation between this type of 'solitons', say $\hat{\phi}_j$, and their relevant fermion bilinears must be non-local as suggested by the equivalence equation (75). So, we may have soliton solutions of type

$$\hat{\phi}_j = \int^x dx' \, \bar{\psi}^j \gamma^0 \psi^j, \quad j = 1, 2, 3 \tag{85}$$

At this stage one is able to enumerate the physical 1-soliton (1-antisoliton) spectrum associated to the gauge fixed ATM model. In fact, we have three 'kinks' and their corresponding 'anti-kinks' associated to the fields ϕ_i (i=1,2,3), and three kink and antikink pairs of type $\hat{\phi}_j$, $j = 1, 2, 3$. Thus, we have six kink and their relevant antikink solutions, but in order to record the physical soliton and anti-soliton excitations one must take into account the four constraints (66) and (77). Therefore, we expect to find four pairs of soliton and anti-soliton physical excitations in the spectrum. This feature is nicely reproduced in the cGSG sector of the ATM model; in fact, in the last section we were able to write four usual sine-Gordon models as possible reductions of the cGSG model. Namely, one soliton associated to the fields φ_1, φ_2, respectively (subsections 2.1. and 2.2.) and 1-solitons associated to the field $\varphi_1 = \varphi_2 \equiv \varphi_A$, $A = 1, 2$, respectively (subsection 2.3.). In the 2-kink (2-antikink) sector a similar argument will provide us ten physical 2-solitons and their relevant 2-antisoliton excitations, i.e. six pairs of 2-kink and 2-antikink solutions of type ϕ and $\hat{\phi}$, respectively, which give twenty four excitations, and taking into account the constraints (66) and (77) we are left with ten pairs of 2-solitons and 2-antisolitons. In fact, these ten 2-solitons correspond to the pairs we can form with the four species of 1-solitons in all possible ways. The same argument holds for the corresponding ten 2-antisolitons.

In this way the system (82)-(83) gives rise to a richer (anti)soliton spectrum and dynamics than the θ_a field 'soliton' type solutions of the gauge unfixed model (54)-(64) found in [33]. Regarding this issue let us notice that in the procedure followed in ref. [33] the local symmetry (67)-(70) and the relevant gauge fixing has not been considered explicitly, therefore their 'solitons' do not correspond to the GSG solitons obtained above.

Notice that the tau functions in section 2. possess the function $\gamma(x - vt)$ in their exponents, whereas the corresponding ones in the ATM theory have two times this function [33, 43]. This fact is reflected in the GSG soliton solutions which are two times the relevant solutions of the ATM model. It has been observed already in the $sl(2)$ case that the θ 'soliton' of the gauge unfixed $sl(2)$ ATM model (see eq. (2.22) of [43]) is half the soliton of the usual SG model.

4. Topological Charges, Baryons as Solitons and Confinement

In this section we will examine the vacuum configuration of the cGSG model and the equivalence between the $U(1)$ spinor current and the topological current (75) in the gauge fixed model and verify that the charge associated to the $U(1)$ current gets confined inside the solitons and kinks of the GSG model obtained in section 2..

It is well known that in 1 + 1 dimensions the topological current is defined as $J^{\mu}_{\text{top}} \sim \epsilon^{\mu\nu}\partial_\nu \Phi$, where Φ is some scalar field. Therefore, the topological charge is $Q_{\text{top}} = \int J^0_{\text{top}} dx \sim \Phi(+\infty) - \Phi(-\infty)$. In order to introduce a topological current we follow the construction adopted in Abelian affine Toda models, so we define the field

$$\theta = \sum_{a=1}^{2} \frac{2\alpha_a}{\alpha_a^2} \theta_a \tag{86}$$

where α_a, $a = 1, 2$, are the simple roots of $sl(3)$. We then have that $\theta_a = (\theta|\lambda_a)$, where λ_a are the fundamental weights of $sl(3)$ defined by the relation [44]

$$2\frac{(\alpha_a|\lambda_b)}{(\alpha_a|\alpha_a)} = \delta_{ab}. \tag{87}$$

The fields ϕ_j in the equations (54)-(64) written as the combinations $(\theta|\alpha_j)$, $j = 1, 2, 3$, where the $\alpha_j^{\prime}s$ are the positive roots of $sl(3)$, are invariant under the transformation

$$\theta \rightarrow \theta + 2\pi\mu \qquad \text{or} \qquad \phi_j \rightarrow \phi_j + 2\pi(\mu|\alpha_j), \tag{88}$$

$$\mu \equiv \sum_{n_a \in \mathbb{Z}} n_a \frac{2\vec{\lambda}_a}{(\alpha_a|\alpha_a)}, \tag{89}$$

where μ is a weight vector of $sl(3)$, these vectors satisfy $(\mu|\alpha_j) \in \mathbb{Z}$ and form an infinite discrete lattice called the weight lattice [44]. However, this weight lattice does not constitute the vacuum configurations of the ATM model , since in the model described by (54)-(65) for any constants $\theta_a^{(0)}$ and $\eta^{(0)}$

$$\psi_j = \widetilde{\psi}_j = 0,\ \theta_a = \theta_a^{(0)},\ \eta = \eta^{(0)},\ \widetilde{\nu} = -\mathbf{m}^2 e^{\eta^{(0)}} x^+ x^- \tag{90}$$

is a vacuum configuration.

We will see that the topological charges of the physical one-soliton solutions of (54)-(65) which are associated to the new fields φ_a, $a = 1, 2$, of the cGSG model (82)-(83) lie on a modified lattice which is related to the weight lattice by re-scaling the weight vectors. In fact, the eqs. of motion (82)-(83) for the field defined by $\varphi \equiv \sum_{a=1}^{2} \frac{2\alpha_a}{\alpha_a^2} \varphi_a$, such that $\varphi_a = (\varphi|\lambda_a)$, are invariant under the transformation

$$\varphi \rightarrow \varphi + \frac{2\pi}{\beta_0} \sum_{a=1}^{2} \frac{q_a}{\nu_a} \frac{2\lambda_a}{(\alpha_a|\alpha_a)}, \quad q_a \in \mathbb{Z}. \tag{91}$$

So, the vacuum configuration is formed by an infinite discrete lattice related to the usual weight lattice by the relevant re-scaling of the fundamental weights $\lambda_a \rightarrow \frac{1}{\nu_a}\lambda_a$. The vacuum lattice can be given by the points in the plane φ_1 x φ_2

$$(\varphi_1\,,\,\varphi_2) = \frac{2\pi}{3\beta_0}\left(\frac{2q_1}{\nu_1} + \frac{q_2}{\nu_2}\,,\,\frac{q_1}{\nu_1} + \frac{2q_2}{\nu_2}\right), \quad q_a \in \mathbb{Z}. \tag{92}$$

In fact, this lattice is related to one in eq. (9) through appropriate parameter identifications. We shall define the topological current and charge, respectively, as

$$J^{\mu}_{\text{top}} = \frac{\beta_0}{2\pi}\epsilon^{\mu\nu}\partial_\nu\varphi, \quad Q_{\text{top}} = \int dx J^0_{\text{top}} = \frac{\beta_0}{2\pi}[\varphi(+\infty) - \varphi(\infty)]. \tag{93}$$

Taking into account the cGSG fields (82)-(83) and the spinor parameterizations (84) the currents equivalence (75) of the ATM model takes the form

$$\sum_{j=1}^{3} m_\psi^j \bar{\psi}^j \gamma^\mu \psi^j \equiv \epsilon^{\mu\nu}\partial_\nu(\zeta_\psi^1 \varphi_1 + \zeta_\psi^2 \varphi_2), \tag{94}$$

where $\zeta_\psi^1 \equiv \beta_0^2 \nu_1 \nu_2 (m_\psi^1 d + m_\psi^2 b)$, $\zeta_\psi^2 \equiv \beta_0^2 \nu_1 \nu_2 (m_\psi^2 a - m_\psi^1 b)$ and the spinors are understood to be written in terms of the fields u_j and v_j of (84).

Notice that the topological current in (94) is the projection of (93) onto the vector $\frac{2\pi}{\beta_0}\left(\zeta_\psi^1 \lambda_1 + \zeta_\psi^2 \lambda_2\right)$.

As mentioned in section 3. the gauge fixing (76) preserves the currents conservation laws (73). Moreover, the cGSG model was defined for the off critical ATM model obtained after setting $\eta =$ const. $= 0$. So, for the gauge fixed model it is expected to hold the currents equivalence relation (75) written for the spinor parameterizations u_j, v_j and the fields $\varphi_{1,2}$ as is presented in eq. (94). Therefore, in order to verify the $U(1)$ current confinement it is not necessary to find the explicit solutions for the spinor fields. In fact, one has that the current components are given by relevant partial derivatives of the linear combinations of the field solutions, $\varphi_{1,2}$, i.e. $J^0 = \sum_{j=1}^3 m_\psi^j \bar{\psi}^j \gamma^0 \psi^j = \partial_x(\zeta_\psi^1 \varphi_1 + \zeta_\psi^2 \varphi_2)$ and $J^1 = \sum_{j=1}^3 m_\psi^j \bar{\psi}^j \gamma^1 \psi^j = -\partial_t(\zeta_\psi^1 \varphi_1 + \zeta_\psi^2 \varphi_2)$. In particular the current components J^0, J^1 and their associated scalar field solutions are depicted in Figs. 3 and 4, respectively, for antisoliton and antikink solutions.

It is clear that the charge density related to this $U(1)$ current can only take significant values on those regions where the $x-$derivative of the fields $\varphi_{1,2}$ are non-vanishing. That is one expects to happen with the bag model like confinement mechanism in quantum chromodynamics (QCD). As we have seen the soliton and kink solutions of the GSG theory are localized in space, in the sense that the scalar fields interpolate between the relevant vacua in a limited region of space with a size determined by the soliton masses. The spinor $U(1)$ current gets the contributions from all the three spinor flavors. Moreover, from the equations of motion (56)-(64) one can obtain nontrivial spinor solutions different from vacuum (90) for each set of scalar field solutions φ_1, φ_2. For example, the solution $\varphi_1 =$soliton, $\varphi_2 = 0$ in section 2.1. implies $\phi_1 = \varphi_1$, $\phi_2 = -\varphi_1$, $\phi_3 = 0$ which substituting into the spinor equations of motion (56)-(64) will give nontrivial spinor field solutions. Therefore, the ATM model of section 3. can be considered as a multiflavor generalization of the two-dimensional hadron model proposed in [29, 30]. In the last reference a scalar field is coupled to a spinor such that the DSG kink arises as a model for hadron and the quark field is confined inside the bag.

5. Qualitons or Quark Solitons in Two-Dimensional QCD

Several properties of the ATM model deserve careful consideration in view of the relationships with two-dimensional QCD. In particular, it has been shown that the $sl(2)$ ATM model describes the low-energy spectrum of QCD_2 (1 flavor and N_c colors) [18]. In the context of bosonized QCD_2 the appearance of soliton solutions that have the quantum numbers of quarks as constituent of hadrons has been considered [8]. So, one can inquire about

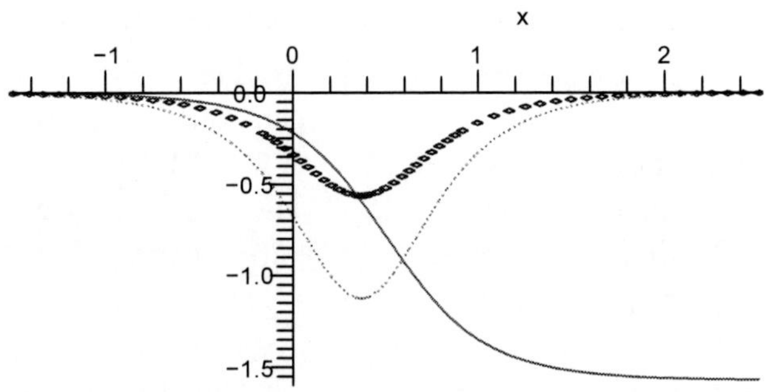

Figure 3. 1-antisoliton and confined current J^{μ}. The solid curve is the 1-antisoliton $(\frac{\beta_0}{4}\varphi)$, the dashdotted curve is J^0 and the curve with losangles is J^1. For $t = 1$, $\mu_1 = \mu_2 = 1, d = 1.5, v = 0.05, \beta_0 = 0.5$, $m_{\psi}^1 = m_{\psi}^2 = 1, \nu_1 = 1$, $\delta_1 = 1$, $\delta_2 = 2$.

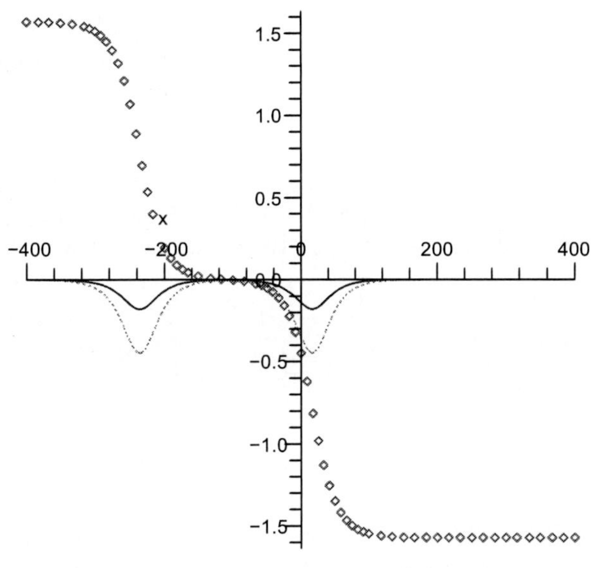

Figure 4. DSG kink solution and confined current J^{μ}. The curve with losangles is the antikink $(\frac{\beta_0}{4}\varphi)$, the dashdotted line is J^0, the solid curve is J^1. For $t = 1$, $\beta_0 = 10^8$, $m_{\psi}^{1,2} = \mu_1 = -0.0000001$, $\mu_3 := 0.001$, $d = 2$, $\delta_1 = \delta_2 = 1$, $\nu_1 = 1/2$.

these type of quark solitons in the context of the ATM model description of QCD$_2$. Since the ATM model describes the low-energy effective action in the strong coupling limit of QCQ$_2$, in order to disentangle the quark solitons one needs to restore, in some way, the heavy fields, i.e. the fields associated to the color degrees of freedom. For simplicity we choose the $sl(2)$ case in the following developments.

The Lagrangian of the $sl(2)$ ATM model is defined by [42, 43, 45]

$$\frac{1}{k}\mathcal{L} = \frac{1}{4}\partial_{\mu}\varphi\,\partial^{\mu}\varphi + i\bar{\psi}\gamma^{\mu}\partial_{\mu}\psi - m_{\psi}\,\bar{\psi}\,e^{2i\varphi\,\gamma_5}\,\psi, \tag{95}$$

where $k = \frac{\kappa}{2\pi}$, ($\kappa \in$ Z), φ is a real field, m_ψ is a mass parameter, and ψ is a Dirac spinor. Notice that $\bar{\psi} \equiv \widetilde{\psi}^T \gamma_0$. We shall take $\widetilde{\psi} = e_\psi \psi^*$ [43], where e_ψ is a real dimensionless constant. The conformal version (CATM) of (95) has been constructed in [41]. The integrability properties and the reduction processes: WZNW$\rightarrow$ CATM $\rightarrow$ ATM $\rightarrow$ sine-Gordon(SG) + free field, have been considered [42, 43, 45] . The $sl(n)$ ATM exhibits a generalized sine-Gordon/massive Thirring correspondence [14]. Moreover, (95) exhibits mass generation despite chiral symmetry [46] and confinement of fermions in a self-generated potential [43, 29].

The Lagrangian is invariant under $\varphi \rightarrow \varphi + n\pi$, thus the topological charge, $Q_{\text{topol.}} \equiv \int dx\, j^0$, $j^\mu = \frac{1}{\pi}\epsilon^{\mu\nu}\partial_\nu\, \varphi$, can assume nontrivial values. A reduction is performed imposing the constraint

$$\frac{1}{2\pi}\epsilon^{\mu\nu}\partial_\nu\, \varphi = \frac{1}{\pi}\bar{\psi}\gamma^\mu\psi, \tag{96}$$

where $J_\mu = \bar{\psi}\gamma^\mu\psi$ is the $U(1)$ Noether current. In fact, the soliton type solutions satisfy this relationship [43].

The Eq. (96) implies $\psi^\dagger\psi \sim \partial_x\varphi$, thus the Dirac field is confined to live in regions where the field φ is not constant. The $1(2)-$soliton(s) solution(s) for φ and ψ are of the sine-Gordon (SG) and massive Thirring (MT) types, respectively; they satisfy (96) for $|e_\psi| = 1$, and so are solutions of the reduced model [43]. Similar results hold in $sl(n)$ ATM [33, 14].

The equivalence (96) for multisolitons describes, $\varphi = \varphi_N$ [$Q_{\text{topol}} = N$ sign(e_ψ)] and Ψ $N-$solitons of the SG and MT type, respectively. Asymptotically one can write

$$\frac{1}{2\pi}\epsilon^{\mu\nu}\partial_\nu\, \varphi_N \approx \sum_{a=1}^{N}\frac{1}{\pi}\bar{\psi}_a\gamma^\mu\psi_a, \tag{97}$$

where the ψ_a's are the solutions for the individual localized lowest energy fermion states. In fact, (97) encodes the classical SG/MT correspondence [47]. Thus, the ATM model can accomodate $N_c = N-$fermion confined states with internal 'color' index a [29].

In order to gain insight into the QCD$_2$ origin of the ψ_a fields let us write the 'mass term' in the multifermion sector of ATM theory as [18]

$$\bar{\psi}_a\, e^{2i\varphi\,\gamma_5}\, \psi^a = \psi_L^{\dagger\, a}\psi_{R\,a}\, e^{2i\varphi} + \psi_R^{\dagger\, a}\psi_{L\,a}\, e^{-2i\varphi}. \tag{98}$$

The ATM mass term in the multifermion sector, Eq. (98), must be compared to the corresponding term in the bosonized QCD$_2$ in order to identify the fields related to the flavor and color degrees of freedom, respectively.

Therefore the total chiral invariant Lagragian including the kinetic terms for the quark fields becomes

$$\frac{1}{k}\mathcal{L} = \frac{1}{4}\partial_\mu\varphi\, \partial^\mu\varphi + ie_\psi\sum_a(\bar{\psi}_a\gamma^\mu\partial_\mu\psi_a - m_\psi\, \bar{\psi}_a\, e^{2i\varphi\,\gamma_5}\, \psi_a). \tag{99}$$

Although the QCD color degrees of freedom have a non-abelian symmetry we use abelian bosonization techniques in order to bosonize the fermions. This will be sufficient in order to reproduce various properties of the effective QCD$_2$ Lagrangian in this regime as presented in ref. [8]. So, let us introduce new boson field representations of the fermion bilinears as [48]

$$
\begin{aligned}
i : \bar{\psi}_a \gamma^\mu \partial_\mu \psi_a : &= -\frac{\alpha^2}{2\pi}(\partial_\nu \phi_a)^2 && (100)\\
: \bar{\psi}_a (1 \pm \gamma_5) \psi_a : &= -\frac{c\mu}{\pi} : e^{(\pm i \alpha \phi_a)} :, \quad : \bar{\psi}_a \gamma^\mu \psi_a := -\frac{\alpha}{\pi} \epsilon^{\mu\nu} \partial_\nu \phi_a, && (101)
\end{aligned}
$$

where $c = \frac{1}{2} \exp{(\gamma)}$, μ is an infrared regulator and α a real parameter.

In order to compare to the related QCD Lagrangian describing the regime $m_q >> e_c$ [8], which does not possess an exact chiral symmetry, we must introduce some chiral symmetry breaking terms in the Lagrangian (99). The most direct program for accomplishing this is simply to include certain chiral breaking terms in the bosonized version of the ATM+color model given in (99) in the form of

$$
\begin{aligned}
\mathcal{L}_{bos} &= \frac{k}{4} \partial_\mu \varphi \, \partial^\mu \varphi + \sum_a^{N_c} \{ \frac{k \alpha^2 e_\psi}{2\pi} \partial_\mu \phi_a \, \partial^\mu \phi_a + \frac{k m_\psi c \mu e_\psi}{\pi} \cos(2\varphi + 2\alpha\phi_a) - m_a \phi_a^2 \} - \\
& m_0 \varphi^2 - \sum_{a<b} m_{ab} \phi_a \phi_b - \sum_a m_{0a} \varphi \phi_a \qquad (102)
\end{aligned}
$$

Notice that we have included certain bilinear terms in the scalar fields as the symmetry breaking terms. Define the fields χ_a and Φ as

$$
\chi_a \equiv \frac{2}{\beta}(\alpha \phi_a + \varphi); \quad \Phi \equiv \frac{1}{\sqrt{2d}} (\varphi - \frac{k \beta e_\psi}{4\pi d} \sum_{a=1}^{N_c} \chi_a), \quad d \equiv \frac{k}{4} + \frac{k \beta^2 N_c e_\psi}{2\pi}. \qquad (103)
$$

So, providing the relationships

$$
\begin{aligned}
m_a &= const, \; m_{ab} = m_{ba} = const. \, (a < b), \; m_{0a} = const, \; \forall a, && (104)\\
m_{ab} &= \frac{m_{01}^2}{2m_0} - \frac{2\delta_c^2}{N_c}, \quad m_a = \frac{m_{01}^2}{4m_0} + \frac{\delta_c^2 (N_c - 1)}{N_c}, \quad \delta_c \equiv \frac{8 e_c^2 \alpha^2}{\beta^2}, && (105)\\
e_\psi &= \frac{-\pi m_{01}}{4 \alpha m_0}. && (106)
\end{aligned}
$$

the Lagrangian (102) becomes

$$
\begin{aligned}
\mathcal{L}_{bos} &= \frac{1}{2}(\partial_\mu \Phi)^2 - \frac{2m_0}{k}(1 + \frac{2N_c e_\psi}{\pi}) \Phi^2 + \sum_a \{ \frac{1}{2} (\partial_\mu \chi_a)^2 + 2M^2 (\cos \beta \, \chi_a) \} - \\
& \frac{k^2 e_\psi^2 \beta^2}{8\pi^2 d} \sum_{a<b} \partial_\mu \chi_a \partial^\mu \chi_b - 2e_c^2 (\frac{N_c - 1}{N_c}) \sum_a \chi_a^2 + \frac{4e_c^2}{N_c} \sum_{a<b} \chi_a \chi_b \qquad (107)
\end{aligned}
$$

where

$$
\beta^2 = 4\pi, \quad M^2 = \frac{c m_\psi \mu k e_\psi}{2\pi}, \quad e_\psi = \frac{N_c - \frac{1}{2}k\pi \pm \sqrt{N_c^2 + k\pi N_c - 2\pi k + \frac{1}{4}\pi^2 k^2}}{2k(N_c - 1)}. \qquad (108)
$$

The model (107) except the $a < b$ kinetic (the first term of the second line in (107)) and the Φ terms reproduces the QCD_2 bosonized Lagrangian (in the regime $m_q >> e_c$) presented in [8]. Notice that the Φ field completely decouples from the rest of the fields. Moreover, in the opposite limit, i.e. the strong coupling regime and large N limit we can verify that this field becomes a free massless field [18].

Besides, the low-energy spectrum of QCD_2 has ben studied by means of abelian [49] and non-abelian bosonizations [50, 2]. In this limit the baryons of QCD_2 are sine-Gordon solitons [2]. In the large N limit approach (weak e and small m_q) the SG theory also emerges [51].

The question of confinement of the "color" degrees of freedom associated to the field ψ in the ATM model by computing the string tension has been presented in [18]. In the *fundamental* representation of the quarks it has been taken $\frac{m_\psi}{4\pi} = m_q$ and $k = 2N/\pi$. Then from (108) one has $|e_\psi| = \frac{\pi}{2\sqrt{N^2-N}}$.

Following [8] we define the baryon number as

$$B = \frac{1}{\sqrt{\pi}} \sum_{k=1}^{N_c} \Big[\chi_k(+\infty) - \chi_k(-\infty)\Big] \tag{109}$$

We seek for solutions of the field equations of motion in the static case

$$\chi_a^{''} - 4M^2\sqrt{\pi}\,\mathrm{sen}\sqrt{4\pi}\chi_a + \rho \sum_{b>a}^{N_c} \chi_b^{''} - 4e_c^2\Big(\chi_a - \frac{1}{N_c}\sum_{b=1}^{N_c}\chi_b\Big) = 0, \tag{110}$$

$$\rho \equiv [N - 1 + 4\pi\sqrt{N(N-1)}]^{-1}. \tag{111}$$

Depending on the boundary conditions for the fields $\chi_k(\pm\infty)$ we may have certain nucleon states with $B = kN_c$, $k \in \mathbb{Z}$ (the baryon number is normalized to be N_c for the nucleon) or some quark solitons ($B = n$, $n =$integer non-multiple of N_c). These type of solutions can be discussed by analyzing the field equations for the static case [8]. In the low energy and strong coupling limit ($e_c >> m_q$) the nucleon states (baryons and multibaryon) are described by the generalized sine-Gordon solitons (see [6] and references therein), as can be inferred from the form of the eqs. of motion (110) in this limit. Whereas, the quark solitons exist for a sufficiently heavy quark m_q, but have infinite energy, corresponding to a string carrying the non-singlet color flux off to spatial infinity, i.e. they exist in the opposite limit $m_q >> e_c$. These quark soliton solutions disappear when the meson mass parameter M is reduced to become comparable to the gauge coupling strength e_c (it has the dimension of mass in QCD2). Let us search for solutions such that $\chi_a(-\infty) = 0$ for all a. Then, at $x = +\infty$ one has

$$4M^2\sqrt{\pi}\,\mathrm{sen}\sqrt{4\pi}\chi_a(\infty) + 4e_c^2\Big(\chi_a(\infty) - \frac{1}{N_c}\sum_{b=1}^{N_c}\chi_b(\infty)\Big) = 0. \tag{112}$$

The eq. (112) becomes the same as the one presented in [8] describing the boundary condition at $x = +\infty$. If we assume $\chi_a(\infty) = \chi$ for all a, one has that $\chi(\infty) = \frac{1}{2}\sqrt{\pi}n$, and $B = \frac{1}{2}nN_c$. But, in order to have positive eigenvalues of the squared mass matrix

$\frac{\partial^2 V}{\partial \chi_a \partial \chi_b}$, we must have even n, and thus integer baryon number $B = kN_c$ (baryons and multibaryons).

Following [8], in the search for quark solitons let us first concentrate on the case $N_c = 2$. Then, eq. (112) can be written as

$$\sin\sqrt{4\pi}\chi_1(\infty) = -\epsilon\sqrt{\pi}[\chi_1(\infty) - \chi_2(\infty)], \tag{113}$$

$$\sin\sqrt{4\pi}\chi_2(\infty) = -\epsilon\sqrt{\pi}[\chi_2(\infty) - \chi_1(\infty)], \tag{114}$$

where $\epsilon = \frac{e_c^2}{2\pi M^2}$. We may have non-baryonic solitons with $B = n$ for odd values of n (the quarks correspond to $n = 1$). For $\epsilon << 1$ we can find a series of solutions with positive second derivative matrix. This solution satisfies

$$\chi_2(\infty) = -\chi_2(\infty) + n\sqrt{\pi}, \tag{115}$$

which together with (113)-(114) provides

$$\sin\sqrt{4\pi}\chi_1 = -\epsilon(\sqrt{4\pi}\chi_1 - n\pi). \tag{116}$$

Let us define $\xi = \sqrt{4\pi}[\chi(\infty) - \frac{1}{2}n\sqrt{\pi}]$, then one has that the solutions are

$$\xi_l = \begin{cases} (\pi - \epsilon)(2l) & \text{for } n \text{ even} \\ (\pi - \epsilon)(2l+1) & \text{for } n \text{ odd} \end{cases}, \tag{117}$$

in the limit where ($l\epsilon << \frac{1}{4}$). The solutions (117) correspond to excitations of "colored" states and have infinite energy, with classical string tension

$$T \approx \begin{cases} \pi\, e_c^2\, (2l)^2 & \text{for } n \text{ even} \\ \pi\, e_c^2\, (2l+1)^2 & \text{for } n \text{ odd} \end{cases}. \tag{118}$$

The single constituent quark soliton corresponds to $n = 1, (2l + 1) = 1$. Thus, we have shown that QCD$_2$ has quark soliton solutions if the quark mass is sufficiently large. These quark solitons disappear when the quark mass m_q is reduced until the meson mass M becomes comparable to the dimensional gauge coupling strength e_c. The above picture can be directly generalized for any N_c, see more details in [8].

6. Discussion

The generalized sine-Gordon model GSG (10)-(11) provides a variety of solitons, kinks and bounce type solutions. The appearance of the non-integrable double sine-Gordon model as a sub-model of the GSG model suggests that this model is a non-integrable theory for the arbitrary set of values of the parameter space. However, a subset of values in parameter space determine some reduced sub-models which are integrable, e.g. the sine-Gordon submodels of subsections 2.1., 2.2. and 2.3..

In connection to the ATM spinors it was suggested that they are confined inside the GSG solitons and kinks since the gauge fixing procedure does not alter the $U(1)$ and topological currents equivalence (75). Then, in order to observe the bag model confinement mechanism it is not necessary to solve for the spinor fields since it naturally arises from the

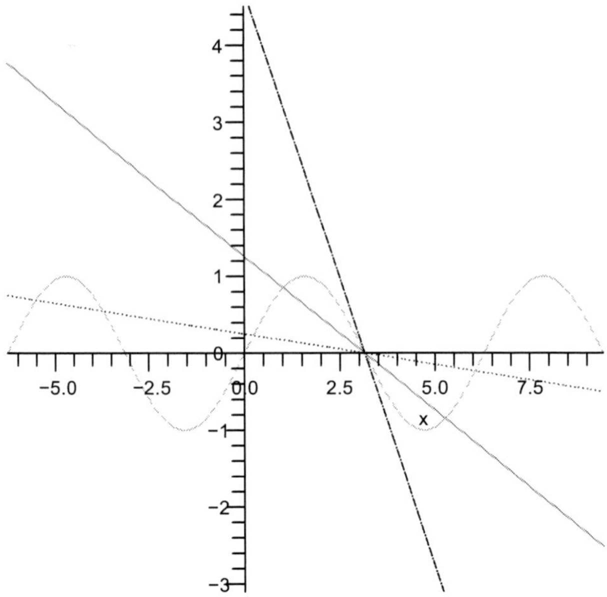

Figure 5. The axis is defined as $x \equiv \sqrt{4\pi}\chi_1(\infty)$. Comparison of the left-and right-hand sides of the soliton eq. (116), corresponding to a quark soliton with $n = 1$, for $\epsilon = 0.08$ (dotted line) and $\epsilon = 0.4$ (solid line). There are no solutions for $\epsilon > 1$ (dot-dashed line, drawn for $\epsilon = 1.5$).

currents equivalence relation. In this way our model presents a bag model like confinement mechanism as is expected in QCD.

The (generalized) massive Thirring model (GMT) is bosonized to the GSG model [16], therefore, in view of the solitons and kinks found above as solutions of the GSG model we expect that the spectrum of the GMT model will contain 4 solitons and their relevant anti-solitons, as well as the kink and antikink excitations. The GMT Lagrangian describes three flavor massive spinors with current-current interactions among themselves. So, the total number of solitons which appear in the bosonized sector suggests that the additional soliton (fermion) is formed due to the interactions between the currents in the GMT sector. However, in subsection 2.3. the soliton masses M_3 and M_4 become the same for the case $\mu_1 = \mu_2$, consequently, for this case we have just three solitons in the GSG spectrum, i.e., the ones with masses M_1, M_2 (subsections 2.1.-2.2.) and $M_3 = M_4$ (subsection 2.3.), which will correspond in this case to each fermion flavor of the GMT model. Moreover, the $sl(3)$ GSG model potential (6) has the same structure as the effective Lagrangian of the massive Schwinger model with $N_f = 3$ fermions, for a convenient value of the vacuum angle θ. The multiflavor Schwinger model resembles with four-dimensional QCD in many respects (see e.g. [52] and references therein).

The $sl(n)$ ATM models may be relevant in the construction of the low-energy effective theories of multiflavor QCD_2 with the dynamical fermions in the fundamental and adjoint representations. Notice that in these models the Noether and topological currents and the generalized sine-Gordon/massive Thirring models equivalences take place at the classical [15, 42] and quantum mechanical level [16, 43].

The interest in baryons with exotic quantum numbers has recently been stimulated by various reports of baryons composed by four quarks and an antiquark. The existence of these baryons cannot yet be regarded as confirmed, however, reports of their existence have stimulated new investigations about baryon structure (see e.g. [53] and references therein). Recently, the spectrum of exotic baryons in QCD_2, with $SU(N_f)$ flavor symmetry, has been discussed providing strong support to the chiral-soliton picture for the structure of normal and exotic baryons in four dimensions [6, 54]. The new puzzles in non-perturbative QCD are related to systems with unequal quark masses, so the QCD_2 calculation must take into account the $SU(N_f)$-breaking mass effects, i.e. for $N_f = 3$ it must be $m_s \neq m_{u,d}$. So, in view of our results above, the properties of the GSG and the ATM theories may find some applications in the study of mass splitting of baryons in QCD_2 and the understanding of the internal structure of baryons. Regarding this line of research, it has been shown that the GSG model describes the low-energy spectrum of normal and exotic baryons in QCD2 with unequal quark mass parameters [6].

Finally, we have considered the quark soliton (qualiton) solutions of QCD_2 in the regime $e_c << m_q$. In this context the role played by the sl(2) ATM model is clarified. In fact, the qualitons arise if the color degrees of freedom are restored by coupling them to the Toda field and convenient boundary conditions are imposed on the fields. So, we have shown that the sl(2) ATM model becomes a low-energy effective lagrangian describing the quark confinement mechanism in QCD_2. The equivalence between the Noether and topological currents (96) is a crucial property of the ATM model in order to provide the confinement mechanism. This picture can be directly generalized to any number of flavors N_f since a relationship analog to (96) holds in that case, e.g. the $N_f = 3$ case is presented in (75).

Acknowledgements

HB thanks IMPA (Rio de Janeiro) for hospitality and CNPq for partial support. HLC thanks FAPESP for support.

A. The Zero-Curvature Formulation of the ATM Model

We summarize the zero-curvature formulation of the $sl(3)$ ATM model [14, 15, 33]. Consider the zero curvature condition

$$\partial_+ A_- - \partial_- A_+ + [A_+, A_-] = 0. \tag{119}$$

The potentials take the form

$$A_+ = -BF^+B^{-1}, \quad A_- = -\partial_- BB^{-1} + F^-, \tag{120}$$

with

$$F^+ = F_1^+ + F_2^+, \quad F^- = F_1^- + F_2^-, \tag{121}$$

where B and $F_i^{\pm}$ contain the fields of the model

$$F_1^+ = \sqrt{im_\psi^1}\psi_R^1 E_{\alpha_1}^0 + \sqrt{im_\psi^2}\psi_R^2 E_{\alpha_2}^0 + \sqrt{im_\psi^3}\widetilde{\psi}_R^3 E_{-\alpha_3}^1, \tag{122}$$

$$F_2^+ = \sqrt{im_\psi^3}\psi_R^3 E_{\alpha_3}^0 + \sqrt{im_\psi^1}\widetilde{\psi}_R^1 E_{-\alpha_1}^1 + \sqrt{im_\psi^2}\widetilde{\psi}_R^2 E_{-\alpha_2}^1, \tag{123}$$

$$F_1^- = \sqrt{im_\psi^3}\psi_L^3 E_{\alpha_3}^{-1} + \sqrt{im_\psi^1}\widetilde{\psi}_L^1 E_{-\alpha_1}^0 + \sqrt{im_\psi^2}\widetilde{\psi}_L^2 E_{-\alpha_2}^0, \tag{124}$$

$$F_2^- = \sqrt{im_\psi^1}\psi_L^1 E_{\alpha_1}^{-1} + \sqrt{im_\psi^2}\psi_L^2 E_{\alpha_2}^{-1} + \sqrt{im_\psi^3}\widetilde{\psi}_L^3 E_{-\alpha_3}^0, \tag{125}$$

$$B = e^{i\theta_1 H_1^0 + i\theta_2 H_2^0}\, e^{\widetilde{\nu} C}\, e^{\eta Q_{ppal}} \equiv b\, e^{\widetilde{\nu} C}\, e^{\eta Q_{ppal}}. \tag{126}$$

$E_{\alpha_i}^n, H_1^n, H_2^n$ and C ($i = 1,2,3$; $n = 0, \pm 1$) are some generators of $sl(3)^{(1)}$; Q_{ppal} being the principal gradation operator. The commutation relations for an affine Lie algebra in the Chevalley basis are

$$[H_a^m, H_b^n] = mC\frac{2}{\alpha_a^2}K_{ab}\delta_{m+n,0} \tag{127}$$

$$[H_a^m, E_{\pm\alpha}^n] = \pm K_{\alpha a} E_{\pm\alpha}^{m+n} \tag{128}$$

$$[E_\alpha^m, E_{-\alpha}^n] = \sum_{a=1}^{r} l_a^\alpha H_a^{m+n} + \frac{2}{\alpha^2} mC\delta_{m+n,0} \tag{129}$$

$$[E_\alpha^m, E_\beta^n] = \varepsilon(\alpha,\beta)E_{\alpha+\beta}^{m+n}; \qquad \text{if } \alpha+\beta \text{ is a root} \tag{130}$$

$$[D, E_\alpha^n] = nE_\alpha^n, \qquad [D, H_a^n] = nH_a^n. \tag{131}$$

where $K_{\alpha a} = 2\alpha.\alpha_a/\alpha_a^2 = n_b^\alpha K_{ba}$, with n_a^α and l_a^α being the integers in the expansions $\alpha = n_a^\alpha \alpha_a$ and $\alpha/\alpha^2 = l_a^\alpha \alpha_a/\alpha_a^2$, and $\varepsilon(\alpha,\beta)$ the relevant structure constants.

Take $K_{11} = K_{22} = 2$ and $K_{12} = K_{21} = -1$ as the Cartan matrix elements of the simple Lie algebra $sl(3)$. Denoting by α_1 and α_2 the simple roots and the highest one by $\psi(= \alpha_1 + \alpha_2)$, one has $l_a^\psi = 1 (a = 1,2)$, and $K_{\psi 1} = K_{\psi 2} = 1$. Take $\varepsilon(\alpha,\beta) = -\varepsilon(-\alpha,-\beta)$, $\varepsilon_{1,2} \equiv \varepsilon(\alpha_1,\alpha_2) = 1$, $\varepsilon_{-1,3} \equiv \varepsilon(-\alpha_1,\psi) = 1$ and $\varepsilon_{-2,3} \equiv \varepsilon(-\alpha_2,\psi) = -1$.

One has $Q_{ppal} \equiv \sum_{a=1}^2 \mathrm{s}_a \lambda_a^v.H + 3D$, where λ_a^v are the fundamental co-weights of $sl(3)$, and the principal gradation vector is $\mathrm{s} = (1,1,1)$ [55].

References

[1] E. Abdalla, M.C.B. Abdalla, and K.D. Rothe, *Non-perturbative Methods in Two-dimensional Quantum Field Theory*, 2nd ed. World Scientific, Singapore, 2001.

[2] Y. Frishman and J. Sonnenschein, *Phys. Reports* **223** (1993) 309.

[3] E. Witten, *Commun. Math. Phys.* **92** (1984) 455.

[4] G.D. Date, Y. Frishman and J. Sonnenschein, *Nucl. Phys.* **B283** (1987) 365.

[5] E. Witten, *Nucl. Phys.* **B223** (1983) 422;
G. Adkins, C. Nappi and E. Witten, *Nucl. Phys.* **B228** (1983) 552.

[6] H. Blas, *JHEP* **0703** (2007) 055; [arXiv:hep-th/0702197].
H. Blas, *"Generalized sine-Gordon model and baryons in two-dimensional QCD"*, pp. 39-79, Editors: T. P. Harrison and R. N. Gonzales, Nova Science Pub. 2008.[arXiv:hep-th/0803.0536]

[7] D.B. Kaplan, *Phys. Lett.* **235B** (1990) 163; *Nucl. Phys.* **B231** (1991) 137; S. Weinberg, *Phys. Rev. Lett.* **65** (1990) 1181.

[8] J.R. Ellis, Y. Frishman, A. Hanany and M. Karliner, *Nucl. Phys.* **B382** (1992) 189. J.R. Ellis, Y. Frishman and M. Karliner, *Phys. Lett.* **272B** (1991) 333.

[9] G.Delfino and G. Mussardo, *Nucl. Phys.* **B516** (1998) 6675.

[10] Z. Bajnok, L.Palla, G. Takács and F. Wágner, *Nucl. Phys.* **B601** (2001) 503.

[11] D. K. Campbell, M. Peyrard and P. Sodano, *Physica* **D19** (1986) 165.

[12] G. Mussardo, V. Riva and G. Sotkov, *Nucl. Phys.* **B687** (2004) 189. D. Controzzi and G. Mussardo, *Phys. Rev. Lett.* **92** (2004) 021601.

[13] G. Takács and F. Wagner, *Nucl. Phys.* **B741** (2006) 353.

[14] J. Acosta, H. Blas, *J. Math. Phys.* **43** (2002) 1916, H.Blas, "Generalized sine-Gordon and massive Thirring models," in *New Developments in Soliton Research*, pp.123-147. Editor: L.V. Chen, Nova Science Pub. 2006; [arXiv:hep-th/0407020].

[15] H. Blas, *JHEP* **0311** (2003) 054, see also hep-th/0407020.

[16] H. Blas, *Eur. Phys. J.* **C37** (2004) 251; "Bosonization, soliton particle duality and Mandelstam-Halpern operators," in *Trends in Boson Research*, pp. 79-108. Editor: A.V. Ling ; Nova Science Pub. 2005;[arxiv:hep-th/0409269].

[17] H. Blas, H. L. Carrion and M. Rojas, *JHEP* **0503** (2005) 037; H. Blas, *JHEP* **0506** (2005) 022; H.Blas and H. L. Carrion, *Non-commutative (generalized) sine-Gordon/massive Thirring duality and soliton solutions*, to appear.

[18] H. Blas, *Phys. Rev.* **D66** (2002) 127701.

[19] D. Diakonov and V. Y. Petrov, *Nucl. Phys.* **B272** (1986) 457; D. Diakonov, "From pions to pentaquarks," arXiv:hep-ph/0406043.

[20] R. Rajaraman, *Phys. Rev. Lett.* **42** (1979) 200.

[21] N. Riazi, A. Azizi and S. M. Zebarjad, *Phys. Rev.* **D66** (2002) 065003.

[22] L. Pogosian and T. Vachaspati, *Phys. Rev.* **D64** (2001) 105023; L. Pogosian, *Phys. Rev.* **D65** (2002) 065023.

[23] A. A. Izquierdo, M. A. G. Leon and J. M. Guilarte, *Phys. Rev.* **D65** (2002) 085012.

[24] D. Bazeia, H. Boschi-Filho and F. A. Brito, *JHEP* **9904** (1999) 028.

[25] D. Bazeia, L. Losano and R. Menezes, *Physica* **D208** (2005) 236.

[26] K Pohlmeyer, *Commun. Math. Phys.* **46** (1976) 207,
A. Mikhailov,"A nonlocal Poisson bracket of the sine-Gordon model", hep-th/0511069

[27] K. Okamuraa and R. Suzukib, *A Perspective on Classical Strings from Complex Sine-Gordon Solitons*, [arXiv:hep-th/0609026],
H.-Yu. Chen, N. Dorey and K. Okamura, *JHEP* **09** (2006) 024.

[28] A. Mikhailov, Bihamiltonian structure of the classical superstring in $AdS5 \times S^5$, [arXiv:hep-th/0609108].
M. Spradlin and A. Volovich, Dressing the Giant Magnon, [arxiv:hep-th/0607009].

[29] S.-J. Chang, S.D. Ellis and B.W. Lee, *Phys. Rev.* **D11** (1975) 3572.

[30] T. Uchiyama, *Phys. Rev.* **D14** (1976) 3520.

[31] Y.S. Kivshar and B.A. Malomed, *Rev. Mod. Phys.* **61** (1989) 763.

[32] C.T. Zhang, *Phys. Rev.* **A35** (1987) 886.
B.A. Malomed, *Phys. Lett.* **123A** (1987) 459.

[33] A.G. Bueno, L.A. Ferreira and A.V. Razumov, *Nucl. Phys.* **B626** (2002) 463.

[34] L. A. Ferreira, J. L. Miramontes and J. Sanchez Guillen, *J. Math. Phys.* **38** (1997) 882.

[35] H. Blas, H. L. Carrion, *JHEP* **01** (2007) 027. [arxiv:hep-th/0610107].

[36] N. Riazi and A. Sheykhi, *Phys. Rev.* **D74** (2006) 025003.

[37] G. Z. Tóth, *J. Physics* **A37** (2004) 9631.

[38] J. Weiss, *J. Math. Phys.* **25** (1984) 2226.

[39] G. Mussardo, *Neutral Bound States in Kink-like Theories,* [arxiv:hep-th/0607025].

[40] S. Liu, Z. Fu and S. Liu, *Phys. Lett.* **351A** (2006) 59.
M.Wang and X. Li, *Chaos, Solitons and Fractals* **C27** (2006) 477.

[41] L.A. Ferreira, J-L. Gervais, J. Sánchez Guillen and M.V. Saveliev, *Nucl. Phys.* **B470** (1996) 236.

[42] H. Blas and B.M. Pimentel, *Annals Phys.* **282** (2000) 67.

[43] H. Blas and L.A. Ferreira, *Nucl. Phys.* **B571** (2000) 607.

[44] J.M. Humphreys, Introduction to Lie algebras and representation theory, *Graduate Texts in Mathematics,* Vol. 9, Springer-Verlag, 1972.

[45] H. Blas, *Nucl. Phys.* **B596** (2001) 471; see also [arXiv:hep-th/0005037].

[46] E. Witten, *Nucl. Phys.* **B145** (1978) 110.

[47] S.J. Orfanidis, *Phys. Rev.* **D14** (1976) 472.

[48] M. Stone, *Bosonization*, 1st edition (World Scientific, Singapore, 1994).

[49] V. Baluni, *Phys. Lett.* **90B** (1980) 407; P. J. Steinhardt, *Nucl. Phys.* **B176** (1980) 100.

[50] D. Gonzales and A.N. Redlich, *Nucl. Phys.* **B256** (1985) 621; I. Affleck, *Nucl. Phys.* **B265** (1986) 448.

[51] L.L. Salcedo, S. Levit and J.W. Negele, *Nucl. Phys.* **B361** (1991) 585.

[52] Y. Hosotani and R. Rodriguez, *J. Physics* **A31** (1998) 9925.
J. E. Hetrick, Y. Hosotani and S. Iso, *Phys. Lett.* **350B** (1995) 92.

[53] S. Kabana, *J. Phys.* **G31** (2005) S1155.

[54] J. R. Ellis and Y. Frishman, *JHEP* **0508** (2005) 081.

[55] V.G. Kac, *Infinite-Dimensional Lie Algebras*, 3rd edn. Cambridge Univ. Press, Cambridge, 1990.

INDEX

A

B

C

D

E

F

G

H

I

J

K

L

M

N

O

P

T

U

V

W

Y